Essentials of Padé Approximants

Essentials of Padé Approximants

GEORGE A. BAKER, JR.

Applied Mathematics Department
Brookhaven National Laboratory
Upton, New York

and

Baker Laboratory
Cornell University
Ithaca, New York

ACADEMIC PRESS   New York   San Francisco   London   1975

A Subsidiary of Harcourt Brace Jovanovich, Publishers

ACADEMIC PRESS, INC.
111 Fifth Avenue, New York, New York 10003

United Kingdom Edition published by
ACADEMIC PRESS, INC. (LONDON) LTD.
24/28 Oval Road, London NW1

Library of Congress Cataloging in Publication Data

Baker, George A   Date
    Essentials of Pade approximants.

    Bibliography: p.
    1.   Padé approximants.   I.   Title.
QC20.7.P3B34   515'.235   74-1632
ISBN 0-12-074855-X

Contents

Preface

The main purpose of this book is to present a unified account of the essential parts of our present knowledge of Padé approximants. Insofar as I have succeeded, this book should serve both as a text for students and as a reference for the more advanced research worker. A considerable number (but by no means a complete list) of references has been included.

I have tried to keep the required background needed to read this book to a minimum. In the main, it should be readable by an advanced undergraduate in mathematics or a graduate student in theoretical phsyics, chemistry, or engineering. The main prerequisite is a course in functions of a complex variable.

A Padé approximant is the ratio of two polynomials constructed from the coefficients of the Taylor series expansion of a function. Since it provides an approximation to the function throughout the whole complex plane, the study of Padé approximants is simultaneously a topic in mathematical approximation theory and analytic function theory. It has wide applicability to those areas of knowledge that involve analytic techniques.

The subject matter is divided into four general areas. In the first area, algebraic properties, I have tried to summarize and unify most of the widely scattered work, both classical and modern. Here is found the work on recursion relations that is fundamental to the construction of computer algorithms for Padé approximants and for many special functions. Throughout its history, the study of Padé approximants has been found to be closely related to a surprising number of other subjects. The connection with continued fractions and orthogonal polynomials is described in detail. Since explicit formulas for the Padé approximants can be obtained, and since it encompasses a large number of useful special functions, Gauss's hypergeometric function has been treated extensively. The generalization to the case where there is Taylor series information at more than one point is also treated. Extremely basic to the properties of the Padé approximants are their invariance properties. Some of these properties are closely related to rotations of the Riemann sphere.

The second main area is that of general convergence theory, which begins with a number of numerical examples to illustrate the possibilities and problems. Then results are given for point-by-point convergence. Results are first given for the special cases of horizontal and vertical sequences and then for the convergence of general sequences. The main problem in the convergence theory of Padé approximants is controlling the location of poles and zeros. Their distribution is discussed. Without direct consideration of the distribution of poles and zeros, a reasonably thorough discussion is given of convergence in (Hausdorff) measure and convergence in the mean on the Riemann sphere.

The third main area treated concerns two special widely occurring classes of functions. These classes are the series of Stieltjes and of Pólya. For these classes, an extremely detailed theory of the Padé approximants is available. Series of Stieltjes are directly related to the moment problem of probability theory. Here, converging upper and lower bounds for the function value can be obtained for real argument, or a converging lens-shaped inclusion region for complex argument. These results extend to asymptotic as well as convergent series. For Pólya frequency series, not only do the Padé approximants converge, but so do the numerators and denominators separately.

The final main area concerns a number of generalizations and applications of Padé approximants. These topics are not treated in detail and in the main represent fairly recent developments that are still subjects of current research interest. We discuss generalizations to series of Stieltjes whose form is more complex but whose theory is directly analogous. Also treated are Padé methods of summing series whose coefficients are formal expressions that are infinite and Padé methods of summing series whose coefficients are matrices. Two areas in which the Padé approximant has been most widely and successfully applied are the statistical mechanics of critical phenomena and the field of scattering physics. Some of the highlights of these applications are presented. In conclusion, electrical circuit synthesis and a few more topics are treated in the last chapter.

The bulk of the work for this book was done during my sabbatical visit in 1971–1972 to Cornell University on leave from Brookhaven National Laboratory. I am grateful to the University for the hospitality I received. I am also grateful to my father, Professor G. A. Baker, University of California at Davis, who has read the manuscript and has suggested many improvements in it. Finally, I am grateful to numerous colleagues for many suggestions and discussions that have both broadened the scope and improved the presentation.

PART I

ALGEBRAIC PROPERTIES

INTRODUCTION

A. The Taylor Series Problem

The relation between the coefficients of the Taylor series expansion of a function and the values of the function is both a profound mathematical question and an important practical one. It is basic to the study of mathematical analysis, and to the practical calculation of mathematical models of nature throughout much of physical and biological science. Much has been learned about the answer to this question, and much yet remains to be learned. The classical answer is that if the Taylor series expansion converges absolutely, then it uniquely defines the value of a function which is differentiable an arbitrary number of times. Conversely, if a function is differentiable an arbitrary number of times, it uniquely defines the Taylor series expansion. Practically, we are approximating the function by longer and longer polynomials. This approach, however, has undesirable limitations for practical calculations. Consider the example,

$$f(x) = \left(\frac{1 + 2x}{1 + x}\right)^{1/2} = 1 + \frac{1}{2}x - \frac{5}{8}x^2 + \frac{13}{16}x^3 - \frac{141}{128}x^4 + \cdots \quad (1.1)$$

It is easy to show that the Taylor series representation fails to converge for any value of $x > \frac{1}{2}$. Yet $f(x)$ is a perfectly smooth and mild function for $0 \leqslant x \leqslant +\infty$ ranging from 1 to $\sqrt{2}$. The classical answer is to develop new Taylor series representations from the old one by computing $f(x)$ and its derivatives at a new point $0 < x_0 < \frac{1}{2}$. This new representation will carry us to larger x, but we will not get to $x = \infty$. In fact, by this method we can never reach $x = \infty$, and to make any progress at all in that direction is extremely tedious. For this example we can employ a special trick to transform the series into one which can be approximated by longer and longer polynomials. Suppose we make the change of variable

$$x = w/(1 - 2w) \quad \text{or} \quad w = x/(1 + 2x) \tag{1.2}$$

then

$$f(x(w)) = (1 - w)^{-1/2} = 1 + \frac{1}{2}w + \frac{3}{8}w^2 + \frac{5}{16}w^3 + \frac{35}{128}w^4 + \cdots$$

$$\tag{1.3}$$

The point $x = \infty$ goes into $w = \frac{1}{2}$ under this change of variables. It can easily be shown that the Taylor series representation (1.3) now converges at $w = \frac{1}{2}$ ($x = \infty$). The first few successive approximations to $f(\infty)$ are

$$1, \quad 1.125, \quad 1.34375, \quad 1.38281, \quad 1.39990, \ldots \tag{1.4}$$

which are converging to $\sqrt{2} = 1.414 \ldots$. In terms of the original variable x the successive approximations provided by (1.3) are

$$1, \quad \frac{1 + (5/2)x}{1 + 2x}, \quad \frac{1 + (9/2)x + (43/8)x^2}{(1 + 2x)^2}, \ldots \tag{1.5}$$

which are rational fractions in x.

B. Padé Approximants

The Padé approximants are a particular type of rational fraction approximation to the value of a function. The idea is to match the Taylor series expansion as far as possible. For example, for the example of (1.1) we would like to pick an approximation of the form

$$(a + bx)/(c + dx) \tag{1.6}$$

so that it would tend to a finite limit as x tends to infinity. If we use the first three coefficients, we produce the approximation

$$\frac{1 + (7/4)x}{1 + (5/4)x} = 1 + \frac{1}{2}x - \frac{5}{8}x^2 + \frac{25}{32}x^3 - \frac{125}{128}x^4 + \cdots \tag{1.7}$$

which has the value 1.4 at $x = \infty$, which is better than any of the approximations (1.4). The next such approximation is

$$\frac{1 + (13/4)x + (41/16)x^2}{1 + (11/4)x + (29/16)x^2} \rightarrow \frac{41}{29} = 1.413793103 \qquad (1.8)$$

which is to be compared with $\sqrt{2} = 1.414213562$. Further such approximations converge quite well and are

$$1.414201183, \quad 1.414213198, \quad 1.414213552, \ldots \qquad (1.9)$$

The last approximation quoted uses the first 11 Taylor series coefficients and is off by only 10^{-8}.

We can also form the same type of approximation to the expansion (1.3) in powers of w. Then we obtain

$$1, \quad \frac{1 - \frac{1}{4}w}{1 - \frac{3}{4}w}, \quad \frac{1 - \frac{3}{4}w + \frac{1}{16}w^2}{1 - \frac{5}{4}w + \frac{5}{16}w^2}, \ldots \qquad (1.10)$$

which agree through the first, third, and fifth coefficients of (1.3), respectively. It is no accident that when we evaluate (1.10) at $w = \frac{1}{2}$ (the value corresponding to $x = \infty$) we get $1, 1.4, 41/29, \ldots$. These values are identical with those obtained by analyzing the series expansion in x. This *invariance property* is a general and important property of Padé approximants and is the basis of their ability to sum the x series in our example and give excellent results, even at $x = \infty$.

We also note that the successive approximations (1.8) and (1.9) increase monotonically. Although this property is not a general one, it can be proven to hold in a wide variety of cases. It is an extremely important property, practically, as one can frequently prove that certain Padé approximants form converging upper and lower bounds.

We will now define the Padé approximants, and the Padé table.

Definition. We denote the L, M Padé approximant to $A(x)$ by

$$[L/M] = P_L(x)/Q_M(x) \qquad (1.11)$$

where $P_L(x)$ is a polynomial of degree at most L and $Q_M(x)$ is a polynomial of degree at most M. The formal power series

$$A(x) = \sum_{j=0}^{\infty} a_j x^j \qquad (1.12)$$

determines the coefficients of $P_L(x)$ and $Q_M(x)$ by the equation

$$A(x) - P_L(x)/Q_M(x) = O(x^{L+M+1}) \tag{1.13}$$

Since we can obviously multiply the numerator and denominator by any constant and leave $[L/M]$ unchanged, we impose the normalization condition

$$Q_M(0) = 1.0 \tag{1.14}$$

Finally we require that P_L and Q_M have no common factors.

If we write the coefficients of $P_L(x)$ and $Q_M(x)$ as

$$P_L(x) = p_0 + p_1 x + \cdots + p_L x^L,$$
$$Q_M(x) = 1 + q_1 x + \cdots + q_M x^M \tag{1.15}$$

then by (1.14) we may multiply (1.13) by $Q_M(x)$, which linearizes the coefficient equations. We can write out (1.13) in more detail as

$$
\begin{aligned}
a_0 &= p_0 \\
a_1 + a_0 q_1 &= p_1 \\
a_2 + a_1 q_1 + a_0 q_2 &= p_2 \\
&\ \ \vdots \\
a_L + a_{L-1} q_1 + \cdots + a_0 q_L &= p_L \\
a_{L+1} + a_L q_1 + \cdots + a_{L-M+1} q_M &= 0 \\
&\ \ \vdots \\
a_{L+M} + a_{L+M-1} q_1 + \cdots + a_L q_M &= 0
\end{aligned}
\tag{1.16}
$$

where we define

$$a_n \equiv 0 \quad \text{if} \quad n < 0 \quad \text{and} \quad q_j \equiv 0 \quad \text{if} \quad j > M \tag{1.17}$$

The foundation for the development of Padé approximants was laid by Cauchy in his famous "Cours d'Analyse" (1821).* In it he studied "recursion series," which are those series whose coefficients, except for the first few, satisfy a linear recursion relation. (Cauchy credits Daniel Bernoulli for his work on finding the modulus of the smallest root of a polynomial in this regard.) Cauchy also gives a formula generalizing the Lagrange polynomial interpolation formula which fits the values of a

---

*I am indebted to Professor H. Wallin for these early references.

function at n points in rational fraction form. It is this formula of Cauchy's, in fact, which was Jacobi's (1846) original starting point and led to the original discovery of Padé approximants. He developed a variety of determinantal formulations for the fitting problem. He further considered the limiting case when all the points at which the function values were fit were coincident. In this case the equations are equivalent to (1.16). Jacobi was the first to give the Padé approximant in the modern sense. Later Frobenius (1881) gave a comprehensive investigation of the algebraic properties of the Padé approximants and gave identities relating Padé approximants whose numerators and denominators have degrees differing by at most unity. Padé (1892) arranged the approximants in a semi-infinite array or table and investigated the structure of this table, as well as special properties of the approximants to e^x. We note in passing that nowhere in his thesis does Padé refer to any of this earlier literature and was presumably unaware of it.

The definition we have given differs in several respects from the classical definitions. First, as regards notation, classically the approximants were denoted as

$$[M, L] = [L/M] \tag{1.18}$$

Unfortunately, some authors choose to use instead the notation

$$[L, M] = [L/M] \tag{1.19}$$

We introduced (1.11) in order to have a notation which would avoid this source of confusion. As a matter of convention we reserve the letter L for the degree of the numerator and M for the degree of the denominator. We further use conventionally

$$L + M = N, \quad L - M = J \tag{1.20}$$

for the sum and difference of these degrees. A further, mathematically important way in which our definition differs from the classical one is in the normalization condition (1.14). Frobenius (1881) and Padé (1892) simply required that $Q_M(x) \not\equiv 0$. That these definitions can be different can be seen in the example

$$A(x) = 1 + x^2 + \cdots \tag{1.21}$$

For $L = M = 1$ one can easily verify that

$$P_1(x) = Q_1(x) = x, \quad P_1(x)/Q_1(x) = 1 \tag{1.22}$$

satisfies

$$Q_M(x)A(x) - P_L(x) = O(x^{N+1}) \qquad (1.23)$$

but not (1.13). In fact by our definition [1/1] does not exist for this series.

C. Solution for the Padé Approximants

The following theorem, due to Frobenius (1881) and Padé (1892), holds under both definitions (we prove it here for our definition).

Theorem 1.1 (uniqueness).   When it exists, the $[L/M]$ Padé approximant to any formal power series $A(x)$ is unique.

Proof: Assume that there are two such Padé approximants $X(x)/Y(x)$ and $U(x)/V(x)$, where the degree of X and U is less than or equal to L and that of Y and U is less than or equal to M. Then we must have, by (1.13),

$$X(x)/Y(x) - U(x)/V(x) = O(x^{L+M+1}) \qquad (1.24)$$

since both approximate the same series. If we multiply (1.24) by $Y(x)V(x)$, we obtain

$$X(x)V(x) - U(x)Y(x) = O(x^{L+M+1}) \qquad (1.25)$$

but the left-hand side of (1.25) is a polynomial of degree at most $L + M$, and thus is identically zero. Since neither Y nor V is identically zero, we conclude

$$X/Y = U/V \qquad (1.26)$$

Since, by definition, both X and Y, and U and V are relatively prime and $Y(0) = V(0) = 1.0$, we have shown that the two, supposedly different Padé approximants, are the same. ∎

The above theorem holds whether or not the defining equations are nonsingular. If they are nonsingular, then we can solve them directly and obtain (Jacobi, 1846)

$$[L/M] = \frac{\det \begin{vmatrix} a_{L-M+1} & a_{L-M+2} & \cdots & a_{L+1} \\ \vdots & \vdots & \ddots & \vdots \\ a_L & a_{L+1} & \cdots & a_{L+M} \\ \sum\limits_{j=M}^{L} a_{j-M} x^j & \sum\limits_{j=M-1}^{L} a_{j-M+1} x^j & \cdots & \sum\limits_{j=0}^{L} a_j x^j \end{vmatrix}}{\det \begin{vmatrix} a_{L-M+1} & a_{L-M+2} & \cdots & a_{L+1} \\ \vdots & \vdots & \ddots & \vdots \\ a_L & a_{L+1} & \cdots & a_{L+M} \\ x^M & x^{M-1} & \cdots & 1 \end{vmatrix}}$$

$$(1.27)$$

where (1.17) holds and, if the lower index on a sum exceeds the upper, the sum is replaced by zero.

D. The Padé Table

Although Frobenius (1881) organized the Padé approximants in a doubly indexed array, Padé (1892) was the first to emphasize the importance of displaying them in tabular form and to study the structure of such a table.

By the Padé table we mean the array

$$
\begin{array}{cccccc}
[0/0] & [0/1] & [0/2] & [0/3] & [0/4] & \cdots \\
[1/0] & [1/1] & [1/2] & [1/3] & [1/4] & \cdots \\
[2/0] & [2/1] & [2/2] & [2/3] & [2/4] & \cdots \\
[3/0] & [3/1] & [3/2] & [3/3] & [3/4] & \cdots \\
[4/0] & [4/1] & [4/2] & [4/3] & [4/4] & \cdots \\
\vdots & \vdots & \vdots & \vdots & \vdots & \cdots
\end{array}
$$

$$(1.28)$$

The partial sums of the Taylor series occupy the first column of the table.

It should be noted that our table is the transpose of the original presentation of Padé (1892) and many subsequent workers.

By way of an example of the Padé table, we give the upper left-hand corner for the function e^x in Table 1.1. We observe that evaluating the [1/1], [2/2], [3/3], and [4/4] at $x = 1$ yields the approximations

$$e \approx 3, \quad 19/7, \quad 193/71, \quad 2721/1001 \tag{1.29}$$

The last quoted approximation is only off 1 in the eighth figure.

E. An Application of Padé Approximants to Physics

We conclude this chapter with an example of how the Padé approximant method can be used to solve a physical problem. In the theory of critical phenomena one expects on general physical grounds that the magnetic susceptibility χ will diverge as the temperature is reduced to the Curie temperature. The Curie temperature is the lowest temperature for which the system does not spontaneously magnetize itself. The simplest hypothesis is that χ is proportional to $(1 - T_c/T)^{-1-g}$. We wish then to investigate this hypothesis for some model system. Domb and Sykes (1961) give high-temperature series expansions to considerable length for the Ising model of ferromagnetism. For illustration we choose a calculation (Baker, 1961) on a two-dimensional square lattice, because when it was done only part of the answer was known exactly. Subsequent work (Fisher, 1962, 1967) has verified the results completely. The expansion is given in terms of

$$w = \tanh(J/kT), \quad J = \text{exchange integral} \tag{1.30}$$

and it is known (Onsager, 1944) that

$$w_c = \sqrt{2} - 1 = 0.414213562 \ldots \tag{1.31}$$

but the exact value of g is known with less full rigor. In order to test the hypothesis, we note that if it is true, then $d[\ln \chi(w)]/dw$ will have a simple pole at w_c with residue $-(1 + g)$. From the expansion

$$d \ln \chi / dw = 4 + 8w + 28w^2 + 48w^3 + 164w^4 + 296w^5 + 956w^6$$

$$+ 1760w^7 + 5428w^8 + 10568w^9 + 31{,}068w^{10} + 62{,}640w^{11}$$

$$+ 179{,}092w^{12} + 369{,}160w^{13} + 1{,}034{,}828w^{14} + \cdots \tag{1.32}$$

we can compute the [1/1] to [7/7] diagonal Padé approximants (Baker, 1961). The results are given in Table 1.2. We see rather rapid convergence

TABLE 1.1 Padé Approximants to e^x

L\M	0	1	2	3	4
0	$\dfrac{1}{1}$	$\dfrac{1}{1-x}$	$\dfrac{2}{2-2x+x^2}$	$\dfrac{6}{6-6x+3x^2-x^3}$	$\dfrac{24}{24-24x+12x^2-4x^3+x^4}$
1	$\dfrac{1+x}{1}$	$\dfrac{2+x}{2-x}$	$\dfrac{6+2x}{6-4x+x^2}$	$\dfrac{24+6x}{24-18x+6x^2-x^3}$	$\dfrac{120+24x}{120-96x+36x^2-8x^3+x^4}$
2	$\dfrac{2+2x+x^2}{2}$	$\dfrac{6+4x+x^2}{6-2x}$	$\dfrac{12+6x+x^2}{12-6x+x^2}$	$\dfrac{60+24x+3x^2}{60-36x+9x^2-x^3}$	$\dfrac{360+120x+12x^2}{360-240x+72x^2-12x^3+x^4}$
3	$\dfrac{6+6x+3x^2+x^3}{6}$	$\dfrac{24+18x+16x^2+x^3}{24-6x}$	$\dfrac{60+36x+9x^2+x^3}{60-24x+3x^2}$	$\dfrac{120+60x+12x^2+x^3}{120-60x+12x^2-x^3}$	$\dfrac{840+360x+60x^2+4x^3}{840-480x+120x^2-16x^3+x^4}$
4	$\dfrac{24+24x+12x^2+4x^3+x^4}{24}$	$\dfrac{120+96x+36x^2+8x^3+x^4}{120-24}$	$\dfrac{360+240x+72x^2+12x^3+x^4}{360-120x+12x^2}$	$\dfrac{840+480x+120x^2+16x^3+x^4}{840-360x+60x^2-4x^3}$	$\dfrac{1680+840x+180x^2+20x^3+x^4}{1680-840x+180x^2-20x^3+x^4}$

TABLE 1.2 Results Derived from $d \ln \chi / dw$

L	Location	Residue
1	0.28571428	−0.6530612
2	0.41118648	−1.6545587
3	0.40926772	−1.6257290
4	0.41644866	−1.7973526
5	0.41216606	−1.6823402
6	0.41412464	−1.7458396
7	0.41421058	−1.7496448
∞	0.414213562	

of the location of the Curie point to the known value and an estimate of $g \approx 3/4$. Other Padé procedures which assume the exact value of w_c confirm $g = 3/4$ more closely.

The range of problems to which Padé approximants can be applied is very large. Some of these applications have been discussed in the recent book "The Padé Approximant in Theoretical Physics" by Baker and Gammel (1970), and we will discuss some of them later.

2

THE STRUCTURE OF THE PADÉ TABLE

A. The C Table

By the structure of the Padé table, introduced in Chapter 1, we will deduce from general arguments which Padé approximants are equal to each other and which do not exist. The original work in this direction is due to Padé (1892). Our results are slightly different, due to a slightly different definition. From Eq. (1.27) it is evident that $C(L/M) \neq 0$ (the coefficient of 1 in the denominator) is a sufficient condition for the existence of the $[L/M]$ Padé approximant, where we define [with Frobenius (1881)]

$$C(r/s) = \det \begin{vmatrix} a_{r-s+1} & a_{r-s+2} & \cdots & a_r \\ \vdots & \vdots & \ddots & \vdots \\ a_r & a_{r+1} & \cdots & a_{r+s-1} \end{vmatrix} \qquad (2.1)$$

With this remark in mind, we will first study the structure of the corresponding table of the C's. As an example of what to expect, let us examine a portion of the C table for Gragg's example (1972)

$$A(x) = (1 - z + z^3)/(1 - 2z + z^2)$$

$$= 1 + z + z^2 + 2z^3 + 3z^4 + 4z^5 + 5z^6 + \cdots \qquad (2.2)$$

We have displayed it in Table 2.1, where we see that the zeros appear in square blocks, entirely surrounded by nonzero entries. This feature is a general one and we will now prove it. Preliminary to proving this assertion, we will prove Sylvester's determinant identity, which is a special case of a more general identity [see Muir (1960, Sec. 148)]. First, if we define a determinant of the $n \times n$ matrix A in the usual way as

$$\det|A| = \sum_{\sigma} \operatorname{sgn} \sigma \prod_{i=1}^{n} a_{i,\,\sigma(i)} \tag{2.3}$$

where the sum is over all permutations σ, and sgn σ is $+1$ or -1 as the permutation σ is even or odd, then we obtain in the usual way the Laplace expansion of the determinant

$$\det|A| = \sum_{w \in Q_{sn}} (-1)^{\sigma(u)+\sigma(w)} \det|A_{uw}| \det|A_{u'w'}| \tag{2.4}$$

The notation of (2.4) is as follows. For $w = (i_1, \ldots, i_s) \in Q_{sn}$ let w' be the increasing sequence of integers remaining in $1, \ldots, n$ after the integers of w are deleted. Set

$$\sigma(w) = i_1 + \cdots + i_s \tag{2.5}$$

By u we mean any fixed member of Q_{sn}. The A_{uw} denotes the $s \times s$ submatrix of A formed from intersection of the subset w of the columns and of the subset u of the rows. Put otherwise, a determinant can be expanded by multiplying all the s-order determinants formed from any preselected s rows by their appropriately signed complimentary minors.

Suppose now we consider an $(n + 2) \times (n + 2)$ matrix of the form

$$A = \begin{pmatrix} M & h & g \\ f & e & d \\ c & b & a \end{pmatrix} \tag{2.6}$$

TABLE 2.1 C Table for Eq. (2.2)

L\M	0	1	2	3	4	5	6	7	8
0	1	1	-1	-1	1	1	-1	-1	1 $\cdots$
1	1	1	0	-1	0	0	-1	1	1 $\cdots$
2	1	1	1	-1	0	0	-1	0	1 $\cdots$
3	1	2	-1	-1	1	1	-1	-1	1 $\cdots$
4	1	3	-1	0	0	0	0	0	0 $\cdots$
5	1	4	-1	0	0	0	0	0	0 $\cdots$
6	1	5	-1	0	0	0	0	0	0 $\cdots$
7	1	6	-1	0	0	0	0	0	0 $\cdots$
8	1	7	-1	0	0	0	0	0	0 $\cdots$

where M is an $n \times n$ matrix, c and f are $(1 \times n)$, and g and h are $(n \times 1)$. We seek to re-express $\det|A| \det|M|$. To this end, we write, using the usual rules of determinants,

$$\det|A| \det|M| = \det \begin{vmatrix} M & h & g & 0 \\ f & e & d & 0 \\ c & b & a & c \\ 0 & 0 & 0 & M \end{vmatrix} = \det \begin{vmatrix} M & h & g & 0 \\ f & e & d & 0 \\ c & b & a & c \\ M & h & g & M \end{vmatrix}$$

$$= \det \begin{vmatrix} M & h & g & 0 \\ f & e & d & 0 \\ 0 & b & a & c \\ 0 & h & g & M \end{vmatrix} \tag{2.7}$$

If we expand the last determinant in (2.7) by the Laplace expansion using the first $n + 1$ rows, then the only terms that survive without a column of zeros in one or the other factor are

$$\det \begin{vmatrix} M & h \\ f & e \end{vmatrix} \det \begin{vmatrix} a & c \\ g & M \end{vmatrix} - \det \begin{vmatrix} M & g \\ f & d \end{vmatrix} \det \begin{vmatrix} b & c \\ h & M \end{vmatrix} \tag{2.8}$$

which can easily be transformed into

$$\det|A| \det|A_{n+1, n+2; n+1, n+2}|$$

$$= \det|A_{n+2; n+2}| \det|A_{n+1; n+1}| - \det|A_{n+2; n+1}| \det|A_{n+1; n+2}| \tag{2.9}$$

where the subscripts now denote which rows and columns are deleted. When we recognize that any two elements not in the same row or column can be moved by elementary row and column operations to the $(n + 1, n + 1)$ and $(n + 2, n + 2)$ positions with at most a sign change, we can then re-express (2.9) to give Sylvester's determinant identity,

$$\det|A| \det|A_{rs; pq}| = \det|A_{rp}| \det|A_{sq}| - \det|A_{rq}| \det|A_{sp}| \tag{2.10}$$

where $r < s$, and $p < q$ defines the order.

We may now apply this identity to derive a recursion relation among the

$C(L/M)$. We obtain, by using the first and last rows and columns of $C(L/M + 1)$,

$$C(L/M + 1)C(L/M - 1) = C(L + 1/M)C(L - 1/M) - [C(L/M)]^2$$

(2.11)

Referring to Fig. 2.1, we see that identity (2.11) relates the determinants in a plus- or star-shaped pattern in the C table, the difference of the products of the opposite ends is equal to the square of the central value. We can now prove the theorem on the structure of the C table.

	L-I/M		
L/M-I	L/M	L/M+I	
	L+I/M		

Fig. 2.1 The relative position of the elements in the C table that enter into identity (2.11).

Theorem 2.1.   Every zero entry in the $C(L/M)$ table for a formal power series

$$A(x) = 1 + \sum_{k=1}^{\infty} a_k x^k$$

occurs in a square block of zero entries and is completely bordered by nonzero entries.

Proof: By definition, the entire first column is $C(L/0) = 1.0$ since these are zero by zero matrices. The first horizontal row is composed of lower right triangular matrices and is easily computed to be

$$C(0/M) = (-1)^{M(M-1)/2}$$

(2.12)

Thus the left and top borders of the C table are all nonzero. This result is indicated in Fig. 2.2 by the shaded borders. We now proceed to search the table in the pattern indicated in Fig. 2.2. If and when we first encounter a zero entry, call it $C(L/M)$, then we will know that $C(L - 1/M - 1)$, $C(L/M - 1)$, and $C(L - 1/M)$ are all nonzero. Now we proceed to consider successively $C(L + 1/M)$, $C(L + 2/M)$, ..., until a nonzero entry is found, if such exists. We will assume for the time being that

$C(L + S/M)$ is the first such nonzero entry. We will discuss the case $S = \infty$ later. We now show that all the $C(L + s/\ M - 1), s = 0, \dots, S$ are different from zero. For, suppose $C(L + r/\ M - 1) = 0, 0 \leqslant r \leqslant S$. Then by applying (2.11) for $L \to L + r - 1$ and $M \to M - 1$, since we also have $C(L + r - 1/M) = 0$, we infer that, since the first two terms vanish, $C(L + r - 1/\ M - 1) = 0$. We can repeat this argument until we show that $C(L/M - 1) = 0$, but this would be a contradiction, since we already know that it is nonzero. Therefore it must be that

$$C(L + s/\ M - 1) \neq 0, \qquad s = -1, 0, \dots, S \qquad (2.13)$$

which takes care of the left border.

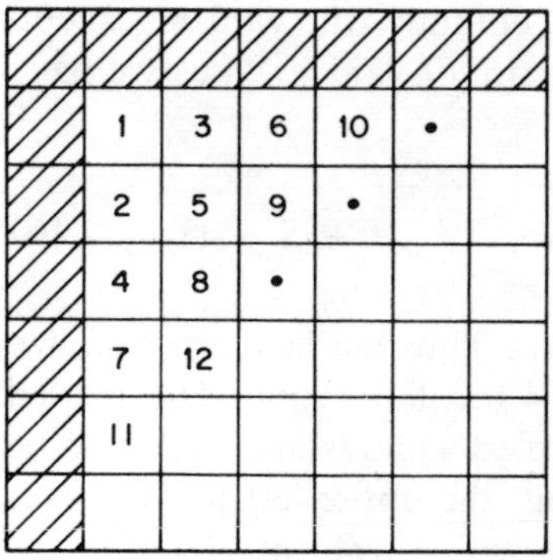

Fig. 2.2 The pattern (numerical order) of search for a zero entry in the C table. The shaded areas, $C(L/0)$ and $C(0/M)$, are always nonzero.

Next we show that the entries $C(L/M + r) = 0$ for $r = 0, 1, \dots, S - 1$. To this end, it is helpful to write out $C(L/M + r)$ explicitly as

$$\det \begin{vmatrix} a_{L-M-r+1} & \cdots & a_{L-M} & a_{L-M+1} & \cdots & a_{L-1} & a_L \\ a_{L-M-r+2} & \cdots & a_{L-M+1} & a_{L-M+2} & \cdots & a_L & a_{L+1} \\ \vdots & \ddots & \vdots & \vdots & \ddots & \vdots & \vdots \\ a_L & \cdots & a_{L+r-1} & a_{L+r} & \cdots & a_{L+M+r-2} & a_{L+M+r-1} \end{vmatrix}$$

$$(2.14)$$

If we fix our attention on the last M columns, we see the successive submatrices for $C(L/M), C(L + 1/M), \dots, C(L + r/M)$ all have zero determinant. Since the last but one $(M - 1)$ columns give the submatrices for $C(L - 1/M - 1), \dots, C(L + r/M - 1)$, all of which are nonzero, by the theory of algebraic linear equations, there exists a unique set of multiples of columns $r, \dots, M + r - 1$ which makes elements one to $M - 1$ all zero. Since $C(L/M) = 0$, this same set of multiples must also make element M of the last column vanish as well. However, determinants $C(L + 1/M)$, $C(L + 2/M)$, ... , $C(L + r/M)$ are also zero, so therefore all the elements in the last column vanish by this set of elementary column operations. Hence (2.14) must vanish also for all $r \leqslant S - 1$. When we recognize that relation (2.11) (see Fig. 2.1) has symmetry between horizontal and vertical directions except for minus signs, we can apply the same arguments we used on the left-hand border to the top border and show that all the determinants

$$C(L - 1/\ M + r) \neq 0, \qquad r = 1, 0, \ldots, S \qquad (2.15)$$

Now consider $C(L/M + S)$. If we apply the same reduction as we did to (2.14), then again every element in the last column will vanish, except, since $C(L + S/M) \neq 0$, the last one, which cannot vanish. Thus we have reduced

$$C(L/M + S) = (\text{nonzero}) \times C(L - 1/\ M + S - 1) \qquad (2.16)$$

which we have just shown in (2.15) is nonzero. Thus we have now shown that the situation is as in Fig. 2.3, where $S = 4$ for illustration. The shaded squares are known to be nonzero and the labeled squares are known to be zero. If we now apply (2.11) successively, from the upper left to the lower right we can fill with zeros an $S \times S$ square with corner at (L/M). The remaining border on the right and bottom may not contain any zero entry or we would obtain a contradiction as we did on the left border. Thus our block of zeros is completely surrounded by nonzero entries. In the case $S = \infty$ all our arguments still hold and the lower right corner is a square block of zeros, still bordered by nonzero entries.

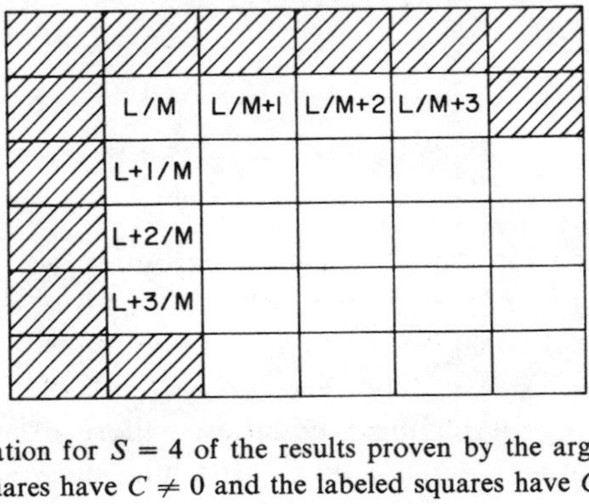

Fig. 2.3 An illustration for $S = 4$ of the results proven by the arguments leading to Eq. (2.16). The shaded squares have $C \neq 0$ and the labeled squares have $C = 0$.

In either case the uninvestigated portion of the C table is left a simply connected region bordered by nonzero entries. We may therefore continue our search (Fig. 2.2) for zeros as before, skipping over, of course, all those squares already checked. Again it must be true of any new zero square encountered, call it $C(L_1/M_1)$, that $C(L_1 - 1/ M_1 - 1)$, $C(L_1/M_1 - 1)$, and $C(L_1 - 1/M_1)$ are all nonzero. Thus we can prove again the existence of a bordered square block of zeros (it might, of course, be only 1×1; it might also happen, of course, that there are no elements at all.) Since we can ultimately scan the C table to any degree required, we have thus established Theorem 2.1. ∎

B. Block Structure of the Padé Table

The following theorem is due to Padé (1892), but was proved using the modern definition of Padé approximants by Chisholm (1966).

Theorem 2.2. The function $f_0(x)$ is of the form

$$f_0(x) = \frac{\sum_0^l c_t x^t}{1 + \sum_1^m e_u x^u} \tag{2.17}$$

if and only if the Padé approximants are given by

$$[L/M] = f_0(x) \tag{2.18}$$

for all $L \geqslant l$ and $M \geqslant m$.

Proof: If the expansion of (2.17) is

$$f_0(x) = \sum_{t=0}^{\infty} d_t x^t \tag{2.19}$$

then

$$\left(\sum_{u=0}^{m} e_u x^u \right) \sum_{t=0}^{\infty} d_t x^t = \sum_{t=0}^{l} c_t x^t \tag{2.20}$$

From Eq. (2.20) it is clear that Eq. (1.16), which defines the $[L/M]$ Padé approximants, is satisfied if we choose any polynomial $\pi_h(x)$ with $\pi_h(0)$ = 1.0 of degree h and pick

$$P_L(x) = \sum_{r=0}^{L} p_r x^r \equiv \pi_h(x) \sum_{t=0}^{l} c_t x^t, \qquad L \geqslant l + h$$

$$Q_n(x) = \sum_{s=0}^{M} q_s x^s \equiv \pi_h(x) \sum_{u=0}^{m} e_u x^u, \qquad M \geqslant m + h \qquad (2.21)$$

Thus there are solutions to the Padé equations of the form $P_L(x)/Q_M(x)$ for $f_0(x)$ of form (2.17). However by the uniqueness theorem (1.1), these solutions are the only ones. Thus Eq. (2.17) implies (2.18).

If (2.18) holds for all $L \geqslant l$ and $M \geqslant m$, then the Padé equations imply

$$\left(\sum_{u=0}^{m} e_u x^u \right) \sum_{t=0}^{\infty} g_t x^t - \sum_{t=0}^{l} c_t x^t = O(x^{L+M+1}) \qquad (2.22)$$

If we now examine Eq. (1.16), we find that in each equation, since $e_0 = 1.0$, the coefficient of the g_t with highest subscript is exactly 1.0. By considering L and M large enough, we can reach the coefficient of any power of x by treating the right-hand side of (2.22) as zero. Thus (2.22) provides for the unique recursive solution for the g_t, given by Eq. (2.18). Since these g_t are exactly the d_t of (2.19), we have shown also that (2.18) implies (2.17). ∎

We now come to Padé's principal theorem (1892, 1900). Our version is slightly different from his due to our use of the modern definition.

Theorem 2.3 (Padé). The Padé table can be completely dissected into $r \times r$ blocks with horizontal and vertical sides, $r \geqslant 1$. Let $[\lambda/\mu]$ denote the unique minimal ($\lambda + \mu$ = minimum) member of a particular $r \times r$ block. Then:

(1) The $[\lambda/\mu]$ exists and the numerator and denominator are of full nominal degree.

(2) $[\lambda + p/\mu + q] = [\lambda/\mu]$ for $p + q \leqslant r - 1, p \geqslant 0, q \geqslant 0$.

(3) $[\lambda + p/\mu + q]$ do not exist for $p + q \geqslant r$, $r - 1 \geqslant p \geqslant 1$, $r - 1 \geqslant q \geqslant 1$.

(4) The equations for the $[\lambda + p/\mu]$, $0 \leqslant p \leqslant r - 1$, and $[\lambda/\mu + q]$, $0 \leqslant q \leqslant r - 1$, are nonsingular, and those for the other block members are singular.

(5) $C(\lambda + p/\mu + q) = 0$ for $1 \leqslant p \leqslant r - 1, 1 \leqslant q \leqslant r - 1$; and $C \neq 0$ otherwise.

Proof: The proof of this theorem rests on Theorem 2.1 for the ·$C(L/M)$ table. Consider any $(r - 1) \times (r - 1)$ square block of zeros in the C table. Adjoin to this block the left and top border, including the corner. We can make this addition without overlap since each square of zeros is completely bordered by nonzero entries. This construction gives property (5). For the block dissection so defined, property (4) follows by representation (1.27) for the Padé approximant, because the $C(L/M)$ are the coefficients of unity in the denominator of (1.27), thus yielding a unique solution of the (therefore) nonsingular Padé equations (1.16). Since the $C(\lambda/\mu)$ is explicitly nonzero by Theorem 2.1, $[\lambda/\mu]$ exists, thereby establishing the first half of property (1). Now the coefficient of x^μ in the denominator of $[\lambda/\mu]$ is, by Eq. (1.27), equal to $C(\lambda + 1/\mu)$, which is nonzero, by Theorem 2.1. The coefficient of x^λ in the numerator of $[\lambda/\mu]$ is, by Eq. (1.27), $(-1)^\mu C(\lambda/\mu + 1)$, which is also nonzero, by Theorem 2.1. Thus we have established property (1).

Now we must necessarily have

$$Q_\mu(x) A(x) - P_\lambda(x) = Kx^{\lambda+\mu+s} + \cdots \qquad (2.23)$$

where $s \geqslant 1$ by the Padé equations, and $K \neq 0$. It may be that $s = \infty$. In this case Theorem 2.2 implies property (2) and that $r = \infty$. There are then no cases included in property (3), so the proof of the theorem is complete for that special case. Let us assume therefore that $\infty > s \geqslant 1$. In the case $r = 1$ there are no cases to consider under properties (2) and (3) besides the triviality $[\lambda/\mu] = [\lambda/\mu]$. Therefore we assume that $r > 1$. From inspection of (1.27) we see that the coefficient of $x^{\lambda+1}$ in $[\lambda + 1/\mu]$ is $C(\lambda + 1/\mu + 1)$, which vanishes by our definition of $[\lambda/\mu]$ as $r > 1$. Thus the $[\lambda + 1/\mu]$ is of the form of the $[\lambda/\mu]$ and hence is equal to it by the uniqueness theorem 1.1. Hence $s \geqslant 2$. Explicitly for $Q_\mu(x)$ we have

$$\sum_{j=0}^{\mu} q_j a_{J-j} = 0, \qquad J = \lambda + 1, \ldots, \lambda + \mu + 1 \qquad (2.24)$$

Now since $C(\lambda + t/\ \mu + 1) = 0, t = 1, \ldots, r$, if we apply the elementary column operations described by (2.24) to $C(\lambda + 2/\ \mu + 1)$, we make the last row all zero because of (2.24) for the $A_{\mu+1, j}, j = 1, \ldots, \mu$, and the $A_{\mu+1, \mu+1}$ must vanish because $C(\lambda + 2/\ \mu + 1) = 0$ and $C(\lambda + 1/\mu)$ $\neq 0$ by hypothesis. We can, however, continue to iterate this argument and thus prove by induction that (2.24) holds for $J = \lambda + 1, \ldots, \lambda + \mu + r$. It cannot hold for $J = \lambda + \mu + r + 1$ since we have, by hypothesis on the block size and Theorem 2.1, that $C(\lambda + r + 1/\ \mu + 1) \neq 0$, which would be a contradiction. Thus by (2.24) we conclude that $s = r$ in (2.23). Hence

$[\lambda/\mu]$ is a solution to the Padé equations for $L + M \leqslant \lambda + \mu + r - 1$, and property (2) follows by the uniqueness theorem (1.1).

By hypothesis and Theorem 2.1 the determinant of the Padé equations vanishes under the conditions described in property (3), yet by (2.23) ($s = r$) the Padé equations are not satisfied. Therefore they are inconsistent and no such Padé approximants exist, which is property (3).

In order to complete the proof of this theorem, we finally have to show that the determinantal solution (1.27) for $[\lambda/\mu]$ has no common factors between numerator and denominator when $C(\lambda/\mu) \neq 0$. Suppose it did; then by canceling we could obtain the result that $[\lambda/\mu] = [\lambda - 1/\mu - 1]$ by the uniqueness theorem. However, then the Padé equations imply

$$\sum_{j=0}^{\mu-1} q_j a_{J-j} = 0, \qquad J = \lambda, \dots, \lambda + \mu \qquad (2.25)$$

which in turn implies $C(\lambda/\mu) = 0$, contrary to assumption. Therefore $C(\lambda/\mu) \neq 0$ implies that $[\lambda/\mu]$ has no common factor between numerator and denominator. ■

We have illustrated the contents of Padé's block theorem in Fig. 2.4 for $r = 5$. The labeled entries are in the block; the shaded squares have singular, but consistent Padé equations and the Padé approximant for them is the $[\lambda/\mu]$. The cross-hatched squares have singular, inconsistent Padé equations. The others in the block have nonsingular equations with $[\lambda/\mu]$ as the solution. In fact $[\lambda/\mu]$ is the only distinct approximant in the block.

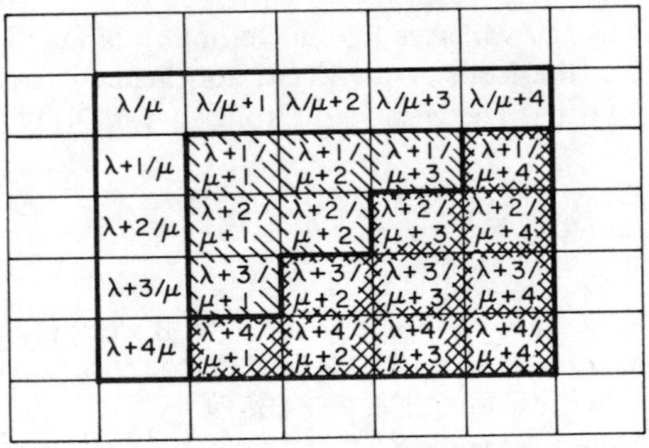

Fig. 2.4 An illustration of Padé's block theorem for a 5×5 block in the Padé table. All the labeled entries are in the block. The shaded squares have singular, but consistent equations. The cross-hatched squares have inconsistent equations. The plain squares in the block have nonsingular equations. Padé approximants exist for every square in the block except the cross-hatched ones, and are all equal to $[\lambda/\mu]$.

In Gragg's (1972) example the Padé table block dissection can be deduced from Table 2.1. It has, besides many 1×1 blocks, a 2×2 block with minimal element [0/1] and another at [1/6]. It has a 2×2 block at [0/3] and an $\infty \times \infty$ block at [3/2]. The [1/2], [2/7], [2/4], [2/5], and [1/5] consequently do not exist. Furthermore, [0/1] = [1/1] = [0/2], [1/6] = [2/6] = [1/7], and [0/3] = [0/4] = [0/5] = [1/3] = [1/4] = [2/3]. Finally, [3/2] = [3 + r/2 + s] for all $r, s \geqslant 0$. Figure 2.5 illustrates this example. The unique Padé approximants which appear in the blocks of the Padé table shown in Fig. 2.5 are (Gragg, 1972)

$$[0/0] = 1, \quad [1/0] = 1 + x, \quad [2/0] = 1 + x + x^2$$

$$[3/0] = 1 + x + x^2 + 2x^3, \quad [4/0] = 1 + x + x^2 + 2x^3 + 3x^4$$

$$[5/0] = 1 + x + x^2 + 2x^3 + 3x^4 + 4x^5, \quad [0/1] = 1/(1 - x)$$

$$[2/1] = (1 - x - x^2)/(1 - 2x),$$

$$[3/1] = (1 - \tfrac{1}{2}x - \tfrac{1}{2}x^2 + \tfrac{1}{3}x^3)/(1 - \tfrac{3}{2}x)$$

$$[4/1] = (1 - \tfrac{1}{3}x - \tfrac{1}{3}x^2 + \tfrac{2}{3}x^3 + \tfrac{1}{3}x^4)/(1 - \tfrac{4}{3}x)$$

$$[5/1] = (1 - \tfrac{1}{4}x - \tfrac{1}{4}x^2 + \tfrac{3}{4}x^3 + \tfrac{1}{2}x^4 + \tfrac{1}{4}x^5)/(1 - \tfrac{5}{4}x)$$

$$[2/2] = (1 - x^2)/(1 - x - x^2), \quad [3/2] = (1 - x + x^3)/(1 - 2x + x^2)$$

$$[0/3] = 1/(1 - x - x^3), \quad [0/6] = 1/(1 - x - x^3 + x^6)$$

$$[1/6] = (1 - x)/(1 - 2x + x^2 - x^3 + x^4 + x^6)$$

$$[0/7] = 1/(1 - x - x^3 + x^6 + x^7), \; \ldots,$$

Following Padé (1892), we introduce the following definition.

0/0	0/1	0/1	0/3	0/3	0/3	0/6	0/7	0/8
1/0	0/1	✕	0/3	0/3	✕	1/6	1/6	1/8
2/0	2/1	2/2	0/3	✕	✕	1/6	✕	2/8
3/0	3/1	3/2	3/2	3/2	3/2	3/2	3/2	3/2
4/0	4/1	3/2	3/2	3/2	3/2	3/2	3/2	3/2
5/0	5/1	3/2	3/2	3/2	3/2	3/2	3/2	3/2

Fig. 2.5 The block structure of the Padé table for Gragg's example.

Definition. The power series $A(x)$ and its associated Padé table are called normal if $C(\lambda/\mu) \neq 0$ for all $\lambda, \mu \geqslant 0$.

It follows at once by Theorem 2.3 that for a normal series the Padé table has no blocks of order $r > 1$. Furthermore, in a normal Padé table the approximant occupying the square $[L/M]$ has, in its simplest terms, a numerator and denominator whose degrees are exactly L and M, respectively.

Normal series will be much more convenient to work with than those without this property. In particular, every Padé approximant exists.

It is evident from the Padé equations that for purely even or odd series the Padé table cannot be normal [see example (1.21)], but the Padé table must be filled with (at least) 2×2 blocks.

C. Existence of Infinitely Many Padé Approximants

We are now in a position to prove an existence theorem for some especially interesting subsequences of Padé approximants. As we have seen from Padé's theorem (2.3), not all Padé approximants necessarily exist. The following theorem is due to Baker (1973).

Theorem 2.4 (existence).   Given any formal power series with $a_0 \neq 0$, then: for every fixed M there exists an infinite sequence of L_j for which the $[L_j/M]$ exists; for every fixed L there exists an infinite sequence of M_j for which the $[L/M_j]$ exists; and for every fixed J there exists an infinite sequence of M_j for which the $[M_j + J/M_j]$ exists

Proof: Consider first the case with M fixed. We divide the class of all formal power series into two parts. In the first part are all rational fractions with the degree of the denominator less than or equal to M. For these power series the conclusion follows by Theorem 2.2. For the other part there can be no blocks of infinite order r which overlap the Mth vertical column, since that would imply that the power series belonged to the first part, contrary to assumption. Therefore there must exist an infinite number of different blocks which intersect the Mth column. Referring to Fig. 2.4, we see that there always exists at least one Padé approximant along any vertical column in each block. Hence the conclusion follows. The argument for fixed L is the same as that just given, except that the roles of numerator and denominator are reversed.

For the case of J fixed we again divide the class of all formal power series ($a_0 \neq 0$) into two parts. In the first part we put all series equivalent to rational fractions. Since there are an infinite number of Padé approximants of the form $[M + J/M]$ where the degree of the numerator and that of the denominator both exceed those of any given rational fraction, Theorem 2.2 supplies the conclusion. Let us now consider the remainder of the series. The only way the $[M + J/M]$ can fail to exist is for the Padé equations (1.16) to be inconsistent, which in turn implies $C(M + J/M) = 0$. If there do not exist an infinite number of $[M + J/M]$, then there exists an M_0 such that $C(M + J/M) = 0$ for $M \geqslant M_0$. However, if that is so, then by repeatedly using the determinant identity (2.11) we can show that

$$C(M + J + r/ M + s) = 0, \qquad r \geqslant 0, \quad s > 0 \qquad (2.26)$$

which in turn implies (Theorem 2.3) that the series belongs to the first part, contrary to assumption. Thus there exist an infinite number of M_j such that $[M_j + J/M_j]$ exist. ∎

3

IDENTITIES

A. Two-Term Identities

Frobenius (1881) supplied most of the identities known until the introduction of the high-speed digital computer renewed interest in the use of Padé approximants. The first group of identities we will discuss are the two-term identities which hold between adjacent entries of the Padé table. It is understood in this chapter that the identities are only implied when all the quantities actually exist.

By the Padé equations (1.13) we can write for any formal power series $A(x)$

$$A(x) - P_L^{(J)}(x)/Q_M^{(J)}(x) = O(x^{L+M+1})$$

$$A(x) - P_{L+1}^{(J)}(x)/Q_{M+1}^{(J)}(x) = O(x^{L+M+3}) \tag{3.1}$$

where $L - M = J$ by (1.20), and thus

$$P_{L+1}^{(J)}(x)/Q_{M+1}^{(J)}(x) - P_L^{(J)}(x)/Q_M^{(J)}(x) = O(x^{L+M+1}) \tag{3.2}$$

or

$$P_{L+1}^{(J)}(x)Q_M^{(J)}(x) - P_L^{(J)}(x)Q_{M+1}^{(J)}(x) = O(x^{L+M+1}) \tag{3.3}$$

However, the left-hand side of (3.3) is a polynomial of degree at most $(L + M + 1)$ and so the right-hand side of (3.3) is at most the single term x^{L+M+1}. From the determinantal solution Eq. (1.27) [we will here adopt the normalization of Eq. (1.27) for $P_L^{(J)}(x)$ and $Q_M^{(J)}(x)$, i.e., $Q_M^{(J)}(O)$

$= C(L/M)]$ we can compute the coefficient as

$$- C(L + 1/\quad M + 2)C(L + 1/M) + C(L/M + 1)C(L + 2/\quad M + 1)$$

$$(3.4)$$

since the coefficient of x^L in $P_L^{(J)}(x)$ is $(-1)^M C(L/M + 1)$ and the coefficient of x^M in $Q_M^{(J)}(x)$ is $(-1)^M C(L + 1/M)$. By Sylvester's determinant identity, we can re-express (3.4) and thus obtain

$$\frac{P_{L+1}^{(J)}(x)}{Q_{M+1}^{(J)}(x)} - \frac{P_L^{(J)}(x)}{Q_M^{(J)}(x)} = \frac{[C(L + 1/\quad M + 1)]^2 x^{L+M+1}}{Q_{M+1}^{(J)}(x)Q_M^{(J)}(x)} \qquad (3.5)$$

By way of illustration, suppose we use the series Eq. (1.32). Then we can easily compute from Eqs. (1.27) and (2.1) that

$$C(2/2) = -400$$

$$[2/2]-[1/1]= \frac{-1600 - 3072x - 1792x^2}{-400 + 32x + 2288x^2} - \frac{32 - 48x}{8 - 28x}$$

$$= \frac{\begin{array}{c} -12{,}800 + 20{,}224 + 71{,}680x^2 + 50{,}176x^3 \\ +12{,}800 - 20{,}224x - 71{,}680x^2 + 109{,}824x^3 \end{array}}{(-400 + 32x + 2288x^2)(8 - 28x)}$$

$$= \frac{(-400)^2 x^3}{(-400 + 32x + 2288x^2)(8 - 28x)} \qquad (3.6)$$

in accord with identity (3.5).

By working in exactly the same way, we can give the following additional two-term identities:

$$\frac{P_{L+1}^{(J+1)}(x)}{Q_M^{(J+1)}(x)} - \frac{P_L^{(J)}(x)}{Q_M^{(J)}(x)} = \frac{C(L + 1/M)C(L + 1/\quad M + 1)x^{L+M+1}}{Q_M^{(J+1)}(x)Q_M^{(J)}(x)} \qquad (3.7)$$

$$\frac{P_L^{(J-1)}(x)}{Q_{M+1}^{(J-1)}(x)} - \frac{P_L^{(J)}(x)}{Q_M^{(J)}(x)} = \frac{C(L/M + 1)C(L + 1/\quad M + 1)x^{L+M+1}}{Q_{M+1}^{(J-1)}(x)Q_M^{(J)}(x)} \qquad (3.8)$$

$$\frac{P_L^{(J-1)}(x)}{Q_{M+1}^{(J-1)}(x)} - \frac{P_{L+1}^{(J+1)}(x)}{Q_M^{(J+1)}(x)} = \frac{[C(L + 1/\quad M + 1)]^2 x^{L+M+2}}{Q_{M+1}^{(J-1)}(x)Q_M^{(J+1)}(x)} \qquad (3.9)$$

If, instead of using nearest-neighbor entries in the Padé table, we use next-nearest neighbors, we can obtain more complex identities. For example,

$$\frac{P_{L+1}^{(J+1)}(x)}{Q_M^{(J+1)}(x)} - \frac{P_{L-1}^{(J-1)}(x)}{Q_M^{(J-1)}(x)}$$

$$= \frac{\begin{array}{c} C(L/M+1)C(L+1/M)x^{L+M} \\ + C(L/M)C(L+1/\ M+1)x^{L+M+1} \end{array}}{Q_M^{(J+1)}(x)Q_M^{(J-1)}(x)} \tag{3.10}$$

and

$$\frac{P_L^{(J-1)}(x)}{Q_{M+1}^{(J-1)}(x)} - \frac{P_L^{(J+1)}(x)}{Q_{M-1}^{(J+1)}(x)}$$

$$= \frac{\begin{array}{c} C(L/M+1)C(L+1/M)x^{L+M} \\ - C(L/M)C(L+1/\ M+1)x^{L+M+1} \end{array}}{Q_{M+1}^{(J-1)}(x)Q_{M-1}^{(J+1)}(x)} \tag{3.11}$$

In the last two identities we must use the determinantal identity (2.10) instead of the special case (2.11).

B. Cross Ratios

From the foregoing two-term identities we can derive the cross-ratio identities (Baker, 1970). A cross ratio is a ratio of the form

$$\frac{(z_1 - z_2)(z_3 - z_4)}{(z_1 - z_3)(z_2 - z_4)} = R \tag{3.12}$$

Each of the four z_i appears in both the numerator and the denominator. If we compute (3.12) when the $\bar{z}_i$ are four adjacent entries in the Padé table, as, for example, in Fig. 3.1, then by identities (3.7) and (3.8) we can write

$$\frac{\{[L/M] - [L/M + 1]\}\{[L + 1/M] - [L + 1/\quad M + 1]\}}{\{[L/M] - [L + 1/M]\}\{[L/M + 1] - [L + 1/\quad M + 1]\}}$$

$$= \frac{\begin{array}{c}\{-C(L/M + 1)C(L + 1/\quad M + 1)x^{L+M+1}\}\\ \times\{-C(L + 1/\quad M + 1)C(L + 2/\quad M + 1)x^{L+M+2}\}\end{array}}{\begin{array}{c}\{-C(L + 1/M)C(L + 1/\quad M + 1)x^{L+M+1}\}\\ \times\{-C(L + 1/\quad M + 1)C(L + 1/\quad M + 2)x^{L+M+2}\}\end{array}}$$

$$= \frac{C(L/M + 1)C(L + 2/\quad M + 1)}{C(L + 1/M)C(L + 1/\quad M + 2)} = \text{const} \tag{3.13}$$

where the denominator polynomials canceled and so did the remaining power of x. The variant patterns shown in Fig. 3.2 lead to, by identities (3.7) and (3.9),

$$\frac{\{[L/M] - [L + 1/\quad M + 1]\}\{[L + 1/M] - [L/M + 1]\}}{\{[L/M] - [L/M + 1]\}\{[L + 1/M] - [L + 1/\quad M + 1]\}}$$

$$= \frac{[C(L + 1/\quad M + 1)]^2 x}{C(L/M + 1)C(L + 2/\quad M + 1)} \propto x \tag{3.14}$$

and by using (3.13) and (3.14)

$$\frac{\{[L/M] - [L + 1/\quad M + 1]\}\{[L + 1/M] - [L/M + 1]\}}{\{[L/M] - [L + 1/M]\}\{[L/M + 1] - [L + 1/\quad M + 1]\}}$$

$$= \frac{[C(L + 1/\quad M + 1)]^2 x}{C(L + 1/M)C(L + 1/\quad M + 2)} \propto x \tag{3.15}$$

Some other patterns of interest are shown in Fig. 3.3:

$$\frac{\{[L/M] - [L + 1/\quad M - 1]\}\{[L/M + 1] - [L + 1/M]\}}{\{[L/M] - [L/M + 1]\}\{[L + 1/\quad M - 1] - [L + 1/M]\}}$$

$$= \frac{C(L + 1/M)C(L + 1/\quad M + 1)x}{C(L/M + 1)C(L + 2/M)} \propto x \tag{3.16}$$

and

$$\frac{\{[L/M] - [L-1/ \quad M+1]\}\{[L+1/M] - [L/M+1]\}}{\{[L/M] - [L+1/M]\}\{[L-1/ \quad M+1] - [L/M+1]\}}$$

$$= \frac{C(L/M+1)C(L+1/ \quad M+1)x}{C(L+1/M)C(L/M+2)} \propto x \qquad (3.17)$$

Fig. 3.1 The entries in the Padé table involved in the cross-ratio identity (3.13). The dashed lines indicate the difference factors involved.

Fig. 3.2 The entries in the Padé table involved in the cross-ratio identities (3.14) and (3.15). The dashed lines indicate the difference factors involved.

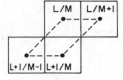
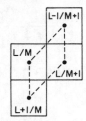

Fig. 3.3 The entries in the Padé table involved in the cross-ratio identities (3.16) and (3.17). The dashed lines indicate the difference factors involved.

Many more identities of this type can easily be derived from the two-term relations given. It will be noted that in every case the cross ratio is a much simpler function of x than is any of the component parts. The power of x is easily deduced from the error made in the Padé approximation, and only the constant remains to be worked out in any case.

C. Three-Term Identities

We can derive the three-term identities directly from the cross ratios and the two-term identities. Suppose we solve Eq. (3.12) for one of the z's, say z_1, in terms of the other three z's and R. We easily get

$$z_1 = \frac{z_2(z_3 - z_4) - z_3 R(z_2 - z_4)}{(z_3 - z_4) - R(z_2 - z_4)} \qquad (3.18)$$

If we now substitute Padé approximants for the z_i as

$$z_i = P_i/Q_i \tag{3.19}$$

then Eq. (3.18) becomes

$$\frac{P_1}{Q_1} = \frac{P_2(P_3Q_4 - Q_3P_4) - P_3R(P_2Q_4 - P_4Q_2)}{Q_2(P_3Q_4 - Q_3P_4) - Q_3R(P_2Q_4 - P_4Q_2)} \tag{3.20}$$

Now by the two-term identities the quantities in parentheses are simple powers for adjacent entries in the Padé table. Likewise R is only a simple power for an appropriately chosen cross ratio. Thus, aside from selecting a constant by which to multiply the numerator and denominator of (3.20) to preserve the normalization of Q_1, Eq. (3.20) provides a linear relation between the numerators of three nearby Padé approximants. Indeed, Eq. (3.20) shows that we obtain the very same relation between the corresponding denominators.

If we apply (3.20) to the pattern of Fig. 3.1 [Eq. (3.13)] and then solve successively for each corner, we get the four Frobenius *triangle identities*, since P_4 is always the opposite corner. Since the relations are linear, and are satisfied by both numerator and denominator, we can state the resultant relations for

$$S(L/M) = G(x)P_L^{(J)}(x) + H(x)Q_M^{(J)}(x) \tag{3.21}$$

where G and H are arbitrary, as

$$C(L + 1/M)S(L - 1/M) - C(L/M + 1)S(L/M - 1)$$

$$= C(L/M)S(L/M) \tag{3.22}$$ *****

$$C(L/M + 1)S(L + 1/M) - C(L + 1/M)S(L/M + 1)$$

$$= C(L + 1/ \quad M + 1)xS(L/M) \tag{3.23}$$ *****

$$C(L + 1/M)S(L/M) - C(L/M)S(L + 1/M)$$

$$= C(L + 1/ \quad M + 1)xS(L/M - 1) \tag{3.24}$$ *****

$$C(L/M + 1)S(L/M) - C(L/M)S(L/M + 1)$$

$$= C(L + 1/ \quad M + 1)xS(L - 1/M) \tag{3.25}$$ *****

The normalizing constant can be verified in every case by taking $G = 0$, $H = 1$, and equating the coefficients of the highest power of x. In Eq. (3.24), Sylvester's identity (2.11) was used. The pattern of asterisks in the margin shows the relative position in the Padé table of the terms in the identity.

One can derive, either by combining the triangle identities appropriately, or by considering the cross ratios for the patterns of Fig. 3.4 directly, relations between the $S(L/M)$ for three entries in a row (horizontal, vertical, or diagonally directed). They are

$$C(L/M)C(L/M + 1)S(L + 1/M)$$

$$- [C(L/M + 1)C(L + 1/M) + C(L/M)C(L + 1/\quad M + 1)x]$$

$$\times S(L/M) + C(L + 1/M)C(L + 1/\quad M + 1)xS(L - 1/M) = 0$$

$$(3.26)$$

$$C(L/M)C(L + 1/M)S(L/M + 1)$$

$$- [C(L/M + 1)C(L + 1/M) - C(L/M)C(L + 1/\quad M + 1)x]$$

$$\times S(L/M) + C(L/M + 1)C(L + 1/\quad M + 1)xS(L/M - 1) = 0$$

$$(3.27)$$

$$C(L/M)^2 S(L + 1/\quad M + 1) - \{C(L/M)C(L + 1/\quad M + 1)$$

$$+ [C(L + 1/\quad M + 2)C(L/M - 1)$$

$$- C(L + 2/\quad M + 1)C(L - 1/M)]x\}\ S(L/M)$$

$$+ C(L + 1/\quad M + 1)^2 x^2 S(L - 1/\quad M - 1) = 0 \qquad (3.28)$$

$$C(L + 1/M)^2 S(L - 1/\quad M + 1)$$

$$- [C(L + 1/M - 1)C(L/M + 2) - C(L - 1/\quad M + 1)C(L + 2/M)$$

$$+ C(L + 1/M)C(L/M + 1)x]S(L/M)$$

$$+ C(L/M + 1)^2 S(L + 1/\quad M - 1) = 0 \qquad (3.29)$$

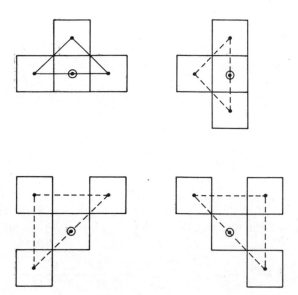

Fig. 3.4 The entries in the Padé table involved in the polynomial identities (3.26)–(3.29).

D. Five-Term Identities

From the determinantal solution Eq. (1.27) we see that (2.11) is an equation between $Q_M^{(J)}(0)$ for Padé table entries in a pattern like Fig. 2.1. The Frobenius *star* or quadratic identity shows that this result holds, not just for $x = 0$, but in fact for any $S(L/M)$ given by Eq. (3.21). If we multiply Eq. (3.24) by $S(L - 1/M)$ and Eq. (3.25) by $S(L/M - 1)$ and subtract, we get by using Eq. (3.22)

$$S(L + 1/M)S(L - 1/M) - S(L/M + 1)S(L/M - 1) = S(L/M)^2$$

$$(3.30)$$

This identity involves five nearby Padé table entries. We observe that the coefficients in this identity do not depend on the power series coefficients, unlike the other identities derived.

There is one further known identity, due to Wynn (1966), which can be written without a knowledge of the series coefficients. It also relates five Padé table entries arranged in the pattern of Fig. 2.1. We can derive it from our previous identities as follows. Multiply Eq. (3.22) by $[C(L + 1/M + 1)x]$ and Eq. (3.23) by $C(L/M)$ and subtract them. This operation yields

$$C(L/M + 1)C(L/M)S(L + 1/M)$$

$$- xC(L + 1/\quad M + 1)C(L + 1/M)S(L - 1/M)$$

$$= C(L/M)C(L + 1/M)S(L/M + 1)$$

$$- xC(L/M + 1)C(L + 1/\quad M + 1)S(L/M - 1) \quad (3.31)$$

If we now specialize $S(L/M) = Q_M^{(J)}(x)$ and multiply Eq. (3.31) by $Q_M^{(J)}(x)/[x^{L+M+1}C(L/M)C(L/M + 1)C(L + 1/M)C(L + 1/M + 1)]$ and use the two term identities (3.7) and (3.8) we obtain Wynn's identity,

$$\frac{1}{[L + 1/M] - [L/M]} + \frac{1}{[L - 1/M] - [L/M]}$$

$$= \frac{1}{[L/M + 1] - [L/M]} + \frac{1}{[L/M - 1] - [L/M]} \quad (3.32)$$

which directly relates the values of the Padé approximants without reference to the series coefficients from which they were formed. This feature is a substantial practical advantage, since the determinants are tedious to evaluate.

E. Three-Term Identity Coefficients from Padé Coefficients

We can make most of the identities more convenient to use if we identify the coefficients in the identities with the coefficients of the Padé approximants themselves. If we let

$$Q_M^{(J)}(x) = \sum_{j=0}^{M} q_j(L/M)x^j, \qquad P_L^{(J)}(x) = \sum_{j=0}^{L} p_j(L/M)x^j \quad (3.33)$$

then by Eq. (1.27)

$$q_0 = C(L/M), \qquad q_M(L/M) = (-1)^M C(L + 1/M)$$

$$p_L(L/M) = (-1)^M C(L/M + 1) = -q_{M+1}(L - 1/M + 1) \quad (3.34)$$

For example, we can rewrite the triangle identity (3.22) as

$$q_M(L/M)S(L - 1/M)$$

$$= p_L(L/M)S(L/M - 1) - p_L(L/M - 1)S(L/M)$$

$$= p_L(L/M)S(L/M - 1) + q_M(L - 1/M)S(L/M) \quad (3.35) \quad {}^{*}_{*}$$

The coefficients on the first right-hand side conform to expectations since for $S = P$ the coefficient x^L must cancel. The second right-hand side makes clear the results for $S = Q$. Similarly, we can rewrite the other triangle inequalities [Eqs. (3.23)–(3.25)]

$$q_M(L/M)S(L/M + 1)$$

$$= p_L(L/M)S(L + 1/M) - p_{L+1}(L + 1/M)xS(L/M)$$

$$= p_L(L/M)S(L + 1/M) + q_{M+1}(L/M + 1)xS(L/M) \quad (3.36) \quad {}^{**}_{*}$$

$$q_M(L/M)S(L/M) = -p_L(L/M - 1)S(L + 1/M)$$

$$+p_{L+1}(L + 1/M)xS(L/M - 1) \qquad (3.37) \quad {}^{**}_{*}$$

$$p_L(L/M)S(L/M) = q_M(L - 1/M)S(L/M + 1)$$

$$-q_{M+1}(L/M + 1)xS(L - 1/M) \qquad (3.38) \quad {}^{*}_{**}$$

In this reduction we have always used self-coefficients, that is, coefficients from one or another of the approximants in the identities. The same re-expression can also be done for the other three-term identities Eq. (3.26)–(3.29). Thus,

$$q_M(L - 1/M)p_L(L/M)S(L + 1/M) - [p_L(L/M)q_M(L/M)$$

$$+ q_M(L - 1/M)p_{L+1}(L + 1/M)x]S(L/M)$$

$$+ q_M(L/M)p_{L+1}(L + 1/M)S(L - 1/M) = 0 \qquad (3.39) \quad {}^{*}_{**}$$

$$p_L(L/M - 1)q_M(L/M)S(L/M + 1) + [p_L(L/M)q_M(L/M)$$

$$-p_L(L/M - 1)q_{M+1}(L/M + 1)x]S(L/M)$$

$$+p_L(L/M)q_{M+1}(L/M + 1)S(L/M - 1) = 0 \qquad (3.40) \quad {}^{**}_{*|*}$$

$$q_0(L/M)^2 S(L+1/ \quad M+1) - \{ q_0(L/M)q_0(L+1/ \quad M+1)$$

$$+ [p_{L+1}(L+1/ \quad M+1)q_{M-1}(L-1/ \quad M-1)$$

$$- q_{M+1}(L+1/ \quad M+1)p_{L-1}(L-1/ \quad M-1)]x\}S(L/M)$$

$$+ x^2 q_0(L+1/ \quad M+1)^2 S(L-1/ \quad M-1) = 0 \qquad (3.41) \quad *_{*_*}$$

$$q_0(L/M)q_M(L/M)p_{L+1}(L+1/ \quad M-1)S(L-1/ \quad M+1)$$

$$- [q_0(L+1/ \quad M-1)p_L(L/M)q_{M+1}(L-1/ \quad M+1)$$

$$- q_0(L-1/ \quad M+1)q_M(L/M)p_{L+1}(L+1/ \quad M-1)$$

$$+ q_0(L/M)q_M(L/M)q_{M+1}(L-1/ \quad M+1)x]S(L/M)$$

$$+ q_0(L/M)p_L(L/M)q_{M+1}(L-1/ \quad M+1)S(L+1/M-1) = 0$$

$$(3.42) \quad *^{*^*}$$

In Eq. (3.42) we have used Eq. (2.11) to eliminate the seeming dependence of Eq. (3.29) on one more power series coefficient than was needed to compute any of the Padé approximants directly. We have made the coefficient identification in such a way as to make it transparent that Eq. (3.42) is independent of the normalization of P and Q. If we divide Eq. (3.42) by $q_0(L/M)^2 q_0(L+1/ \quad M-1)q_0(L-1/ \quad M+1)$, then only the ratios q_j/q_0 and p_j/q_0 ever appear.

F. Compact Expressions for Padé Approximants

A more compact way to write Eq. (1.27) for the $[M-1/M]$ has been given by Nuttall (1967) and generalized to the $[L/M]$ by Baker (1970). To derive these expressions, we start by using Jacobi's (1846) reduction of the determinant for $Q_M(x)$ in Eq. (1.27) from $(M+1) \times (M+1)$ to $M \times M$. To do this, we multiply the second row by x and subtract from the first row. Then we multiply the third by x and subtract from the second, and so on, so that Eq. (1.27) becomes

$$[L/M] = \frac{\det\begin{vmatrix} a_{L-M+1} - xa_{L-M+2} & \cdots & a_L - xa_{L+1} & a_{L+1} \\ \vdots & \ddots & \vdots & \vdots \\ a_L - xa_{L+1} & \cdots & a_{L+M-1} - xa_{L+M} & a_{L+M} \\ -a_{L-M+1}x^{L+1} & \cdots & -a_L x^{L+1} & \sum_{j=0}^{L} a_j x^j \end{vmatrix}}{\det\begin{vmatrix} a_{L-M+1} - xa_{L-M+2} & \cdots & a_L - xa_{L+1} & a_{L+1} \\ \vdots & \ddots & \vdots & \vdots \\ a_L - xa_{L+1} & \cdots & a_{L+M-1} - xa_{L+M} & a_{L+M} \\ 0 & \cdots & 0 & 1 \end{vmatrix}}$$

(3.43)

By expanding the demonitor along the last row, we get only a contribution from the coefficient of 1, which is a determinant of order $M \times M$. If we now add x^{j-M-1} times the jth column in the numerator to the last column in the numerator, we obtain

$$[L/M] = \frac{\det\begin{vmatrix} a_{L-M+1} - xa_{L-M+2} & \cdots & a_L - xa_{L+1} & a_{L-M+1}x^{-M} \\ \vdots & \ddots & \vdots & \vdots \\ a_L - xa_{L+1} & \cdots & a_{L+M-1} - xa_{L+M} & a_L x^{-M} \\ -a_{L-M+1}x^{L+1} & \cdots & -a_L x^{L+1} & \sum_{j=0}^{L-M} a_j x^j \end{vmatrix}}{\det\begin{vmatrix} a_{L-M+1} - xa_{L-M+2} & \cdots & a_L - xa_{L+1} \\ \vdots & \ddots & \vdots \\ a_L - xa_{L+1} & \cdots & a_{L+M-1} - xa_{L+M} \end{vmatrix}}$$

(3.44)

which, by expansion of the numerator along the last row and last column and using the standard formula for the matrix inverse, is

$$[L/M] = \sum_{j=0}^{L-M} a_j x^j + x^{L-M+1}[\mathbf{w}^T(L/M)\mathbf{W}^{-1}(L/M)\mathbf{w}(L/M)] \quad (3.45)$$

where $\mathbf{W}^{-1}$ is the inverse of the matrix

$$\mathbf{W}(L/M) = \begin{pmatrix} a_{L-M+1} - xa_{L-M+2} & \cdots & a_L - xa_{L+1} \\ \vdots & \ddots & \vdots \\ a_L - xa_{L+1} & \cdots & a_{L+M-1} - xa_{L+M} \end{pmatrix}$$

$$(3.46)$$

and we define the vector

$$\mathbf{w}(L/M) = (a_{L-M+1}, a_{L-M+2}, \ldots, a_L) \quad (3.47)$$

If $j < 0$, then we take $a_j \equiv 0$. The sum in Eq. (3.45) is omitted if $L < M$. Equation (3.45) holds even if $M > L + 1$, in spite of the appearance of negative powers of x. A sufficient number of the low-order coefficients of x in the matrix element vanish to compensate.

An additional compact expression involving an off-diagonal matrix element of $\mathbf{W}^{-1}$ can be easily derived in the same way. It is

$$[L/M] = \sum_{j=0}^{L+n} a_j x^j$$

$$+ x^{L+n+1}[\mathbf{w}^T(L+M/M)\mathbf{W}^{-1}(L/M)\mathbf{w}(L+n/M)] \quad (3.48)$$

where $0 \leqslant n \leqslant M$.

G. Connection between Padé Tables of $f(x)$ and $(1 + \alpha x)f(x)$

The identities given in this section between the numerator polynomials of Padé approximants to $f(x)$ and those to $(1 + \alpha x)f(x)$ are due to Arms and Edrei (1970). The ones for the denominator are new in this context, but in the context of Toeplitz forms [see Grenander and Szegö (1958)] are the expansion of the "kernel polynomial."

We introduce the notation

$$f(x) = a_0 + a_1 z + a_2 z^2 + \cdots$$

$$g(x) = (1 + \alpha x)f(x) = g_0 + g_1 x + g_2 x^2 + \cdots \quad (3.49)$$

where $g_j = a_j + \alpha a_{j-1}$. We append a subscript f or g to the notation of Eq. (3.33) to distinguish which function we are approximating. Let us consider

$$\Lambda = \det \begin{vmatrix} 1 & -\alpha & \cdots & (-\alpha)^M \\ a_{L-M} & a_{L-M+1} & \cdots & a_{L+1} \\ \vdots & \vdots & \ddots & \vdots \\ a_{L-1} & a_L & \cdots & a_{L+M} \\ a_{j-M-1} & a_{j-M} & \cdots & a_j \end{vmatrix} \qquad (3.50)$$

By elementary column operations we can convert Λ into

$$\Lambda = \det \begin{vmatrix} 1 & 0 & \cdots & 0 \\ a_{L-M} & g_{L-M+1} & \cdots & g_{L+1} \\ \vdots & \vdots & \ddots & \vdots \\ a_{L-1} & g_L & \cdots & g_{L+M} \\ a_{j-M-1} & g_{j-M} & \cdots & g_j \end{vmatrix} = p_{j,g}(L/M) \quad (3.51)$$

by expansion of the determinant along the first row and by Eq. (1.27). By using Sylvester's determinant identity (2.10) on form (3.50) we obtain

$$\Lambda\Lambda_{1,M+2;1,M+2} = \Lambda_{1,1}\Lambda_{M+2,M+2} - \Lambda_{1,M+2}\Lambda_{M+2,1} \qquad (3.52)$$

and by reducing the results as we did (3.50) to (3.51), we can use Eq. (1.27) to identify the results as

$$C_f(L/M)p_{j,g}(L/M) = p_{j,f}(L/M)C_g(L/M)$$

$$+ p_{j-1,f}(L-1/M)\alpha C_g(L+1/M) \quad (3.53)$$

Multiplying (3.53) by x^j and adding, we get the numerator polynomial identity

$$\frac{P_{L,g}^{(J)}(x)}{C_g(L/M)} = \frac{P_{L,f}^{(J)}(x)}{C_f(L/M)} + \alpha x \frac{P_{L-1,f}^{(J-1)}(x)}{C_f(L-1/M)}$$

$$\times \frac{Q_{M,f}^{(J)}(-1/\alpha)C_f(L-1/M)}{Q_{M,f}^{(J-1)}(-1/\alpha)C_f(L/M)} \qquad (3.54)$$

In expressing the right-hand side of Eq. (3.54) exclusively in terms of α and the Padé and C tables for $f(x)$, we have used the identity

$$\alpha^M Q_{M,f}^{(J-1)}(-1/\alpha) = C_g(L/M) \qquad (3.55)$$

which follows immediately from the work of the previous section, Eq. (3.43).

In order to derive an identity for the denominator polynomials, we consider

$$\Lambda = \det \begin{vmatrix} 1 & a_{L-M} & a_{L-M+1} & \cdots & a_L \\ -\alpha & a_{L-M+1} & a_{L-M+2} & \cdots & a_{L+1} \\ \vdots & \vdots & \vdots & \ddots & \vdots \\ (-\alpha)^M & a_L & a_{L+1} & \cdots & a_{L+M} \\ 0 & x^M & x^{M-1} & \cdots & 1 \end{vmatrix}$$

$$= \det \begin{vmatrix} 1 & a_{L-M} & a_{L-M+1} & \cdots & a_L \\ 0 & g_{L-M+1} & g_{L-M+2} & \cdots & g_{L+1} \\ \vdots & \vdots & \vdots & \ddots & \vdots \\ 0 & g_L & g_{L+1} & \cdots & g_{L+M} \\ 0 & x^M & x^{M-1} & \cdots & 1 \end{vmatrix}$$

$$= Q_{M,g}^{(J)}(x) \qquad (3.56)$$

by elementary row operations and by Eq. (1.27). If we apply Sylvester's determinant identity to the first line of Eq. (3.56) in the form of Eq. (3.52), we obtain directly

$$Q_{M,g}^{(J)}(x)C_f(L/M) = Q_{M,f}^{(J)}(x)C_g(L/M) + \alpha x C_f(L/M+1)Q_{M-1,g}^{(J+1)}(x) \qquad (3.57)$$

If we use Eq. (3.57) repeatedly to re-express the $Q_{M,g}^{(J)}$ in terms of $Q_{M,f}^{(J)}$, then we can derive the identity for the denominator polynomials,

$$\mathscr{Q}_{M,g}^{(J)}(x) = \frac{C_f(L/M + 1)}{C_f(L - 1/M)}$$

$$\times \sum_{j=1}^{M} \frac{x^j C_f(L - 1/\quad M - j)\mathscr{Q}_{M-j,f}^{(J-1+j)}(-1/\alpha)\mathscr{Q}_{M-j,f}^{(J+j)}(x)}{C_f(L/M + 1 - j)\mathscr{Q}_{M,f}^{(J-1)}(-1/\alpha)} \qquad (3.58)$$

where we have represented the normalized polynomial, i.e., $\mathscr{Q}(0) = 1$, by

$$\mathscr{Q}_{M,h}^{(J)}(x) = Q_{M,h}^{(J)}(x)/C_h(L/M) \qquad (3.59)$$

We can derive an alternate identity for the denominator polynomials by using the Padé equations to replace P by Qf in Eq. (3.54). This operation yields

$$\mathscr{Q}_{M,g}^{(J)}(z) = \frac{\mathscr{Q}_{M,f}^{(J-1)}(-1/\alpha)\mathscr{Q}_{M,f}^{(J)}(z) + \alpha z \mathscr{Q}_{M,f}^{(J-1)}(z)\mathscr{Q}_{M,f}^{(J)}(-1/\alpha)}{\mathscr{Q}_{M,f}^{(J-1)}(-1/\alpha)(1 + \alpha z)} \qquad (3.60)$$

a simpler result. It is to be noted that the numerator of Eq. (3.60) vanishes for $z = -1/\alpha$, so that the expression is a polynomial, since $1 + \alpha z$ exactly divides the numerator.

These identities will be the key to the discussion of Pólya frequency series, which we will consider in Chapter 18.

RELATION BETWEEN
PADÉ APPROXIMANTS AND
CONTINUED FRACTIONS

A. Fundamental Recursion Formulas

Those continued fractions that are related to Taylor series are special cases of the Padé table. In our presentation of the theory of continued fractions we have selected only some of the more fundamental material since it is our purpose to show the relation to the Padé table, and not to develop the theory of continued fractions. We will usually assume in this chapter that nothing by which we need to divide vanishes, and defer to the very end a discussion of other cases.

By a continued fraction we mean the limit of successive truncations (that is, treating $a_j, j > n$, as zero) of

$$G = b_0 + \cfrac{a_1}{b_1 + \cfrac{a_2}{b_2 + \cfrac{a_3}{b_3 + \cdots}}} \qquad (4.1)$$

The first few truncations are

$$\frac{A_0}{B_0} = \frac{b_0}{1}, \qquad \frac{A_1}{B_1} = \frac{a_1 + b_1 b_0}{b_1}, \qquad \frac{A_2}{B_2} = \frac{b_0 a_2 + a_1 b_2 + b_0 b_1 b_2}{a_2 + b_1 b_2} \qquad (4.2)$$

and

$$\frac{A_n}{B_n} = b_0 + \cfrac{a_1}{b_1 + \cfrac{a_2}{b_2 + \cdots \cfrac{}{+ \cfrac{a_n}{b_n}}}}$$

(4.3)

Equation (4.3) defines the nth convergent to G. Thus we define

$$G = \lim_{n \to \infty} (A_n / B_n)$$

(4.4)

Just as we saw in Chapter 3 for Padé approximants, there are also simple recursion relations for successive convergents for continued fractions. In fact we have

$$A_{n+1} = b_{n+1} A_n + a_{n+1} A_{n-1}, \qquad B_{n+1} = b_{n+1} B_n + a_{n+1} B_{n-1} \quad (4.5)$$

A glance at Eq. (4.2) will verify Eq. (4.5) for the case $n = 1$. If we make the induction hyphothesis and suppose (4.5) is true for all $n \leqslant k$, then we observe from (4.3) that we can obtain $b_{k+2}^{-1} A_{k+2}$ and $b_{k+2}^{-1} B_{k+2}$ from A_{k+1} and B_{k+1} by replacing b_{k+1} by $b_{k+1} + (a_{k+2}/b_{:+2})$. Thus from (4.5) for $n = k$ we have, with the foregoing replacement

$$b_{k+2}^{-1} A_{k+2} = [b_{k+1} + (a_{k+2}/b_{k+2})] A_k + a_{k+1} A_{k-1}$$

$$b_{k+2}^{-1} B_{k+2} = [b_{k+1} + (a_{k+2}/b_{k+2})] B_k + a_{k+1} B_{k-1}$$

(4.6)

On multiplication of (4.6) by b_{k+2} and using (4.5) for $n = k$, we have

$$A_{k+2} = b_{k+2} A_{k+1} + a_{k+2} A_k, \qquad B_{k+2} = b_{k+2} B_{k+1} + a_{k+2} B_k \quad (4.7)$$

Thus by induction (4.5) holds for all n, since it holds for $n = 1$. These relations are called the *fundamental recurrence formulas*, and were first established by Wallis (1655) and studied by Euler (1737). From them we can derive the determinant formula which will be of great use later in relating the coefficients of the continued fraction to those of the corresponding Taylor series. Consider

$$\det \begin{vmatrix} A_{n-1} & A_n \\ B_{b-1} & B_n \end{vmatrix} = \det \begin{vmatrix} A_{n-1} & b_n A_{n-1} + a_n A_{n-2} \\ B_{n-1} & b_n B_{n-1} + a_n B_{n-2} \end{vmatrix}$$

$$= -a_n \det \begin{vmatrix} A_{n-2} & A_{n-1} \\ B_{n-2} & B_{n-1} \end{vmatrix} \tag{4.8}$$

By successive applications we have

$$A_{n-1}B_n - B_{n-1}A_n = (-1)^n a_1 a_2 \cdots a_n, \qquad n = 1, 2, \ldots \tag{4.9}$$

where the case $n = 1$ can be verified directly from Eq. (4.2).

B. Equivalence Transformations

In the absence of the anomalous vanishing of various of the parameters, we can transform (4.1) by means of *equivalence transformations*. Such a transformation consists in multiplying successive numerators and denominators by nonzero numbers. Thus

$$G = b_0 + \cfrac{c_1 a_1}{c_1 b_1 + \cfrac{c_1 c_2 a_2}{c_2 b_2 + \cfrac{c_2 c_3 a_3}{c_3 b_3 + \cdots}}} \tag{4.10}$$

Although G is unchanged, the normalization of A_n and B_n will in general be changed by the factor $(c_1 c_2 \cdots c_n)$. One important case is if $b_n \neq 0$ and $c_n = b_n^{-1}$. Then (4.10) becomes

$$G = b_0 + \cfrac{a_1'}{1 + \cfrac{a_2'}{1 + \cfrac{a_3'}{1 + \cdots}}} \tag{4.11}$$

where

$$a_1' = a_1/b_1, \qquad a_j' = a_j/b_j b_{j-1}, \qquad j \geq 2 \tag{4.12}$$

Consequently, we may usually content ourselves with the form (4.11) rather than the more general form (4.1). Another type of transformation that can be made on a continued fraction is a contraction. Then we consider only the sequence (A_{2n}/B_{2n}) or the sequence (A_{2n+1}/B_{2n+1}). We obtain from (4.11) ($b_0 = 1$)

$$\frac{A_2}{B_2} = 1 + \frac{a_1'}{1 + a_2'} \tag{4.13}$$

Plainly, if we substitute for a_2',

$$\frac{a_2'}{1 + \dfrac{a_3'}{1 + a_4'}} = \frac{a_2'(1 + a_4')}{1 + a_4' + a_3'} = a_2' - \frac{a_2'a_3'}{1 + a_3' + a_4'} \tag{4.14}$$

then we get

$$\frac{A_4}{B_4} = 1 + \cfrac{a_1'}{1 + a_2' - \cfrac{a_2'a_3'}{1 + a_3' + a_4'}} \tag{4.15}$$

but this process may be repeated successively on $a_4', a_6', \ldots$. Thus the *even part of G* is

$$G_e = 1 + \cfrac{a_1'}{1 + a_2' - \cfrac{a_2'a_3'}{1 + a_3' + a_4' - \cfrac{a_4'a_5'}{1 + a_5' + a_6' - \cdots}}} \tag{4.16}$$

In a similar fashion the *odd part* is

$$G_o = 1 + a_1' - \cfrac{a_1'a_2'}{1 + a_2' + a_3' - \cfrac{a_3'a_4'}{1 + a_4' + a_5' - \cfrac{a_5'a_6'}{1 + a_6' + a_7' - \cdots}}} \tag{4.17}$$

As a simple example of these various forms, suppose

$$G = 1 + \cfrac{1}{2 + \cfrac{1}{2 + \cfrac{1}{2 + \cdots}}} \tag{4.18}$$

then we get

$$\frac{A_0}{B_0} = \frac{1}{1}, \quad \frac{A_1}{B_1} = \frac{3}{2}, \quad \frac{A_2}{B_2} = \frac{7}{5}, \quad \frac{A_3}{B_3} = \frac{17}{12}, \quad \frac{A_4}{B_4} = \frac{41}{29} \quad (4.19)$$

Since $b_0 = 1$, $a_i = 1$, $b_i = 2$, $i \geqslant 1$, the recursion relations become

$$A_{n+1} = 2A_n + A_{n-1}, \qquad B_{n+1} = 2B_n + B_{n-1} \qquad (4.20)$$

The general term generated by these recursion relations can be easily found. One simply assumes that A_n or B_n is a sum of $\alpha x^n + \beta y^n$ and solves for α, β, x, and y. The result is

$$A_n = \tfrac{1}{2}(1 + \sqrt{2})^{n+1} + \tfrac{1}{2}(1 - \sqrt{2})^{n+1},$$

$$B_n = \left[\tfrac{1}{4}(1 + \sqrt{2})^{n+1} - \tfrac{1}{4}(1 - \sqrt{2})^{n+1} \right]\sqrt{2} \qquad (4.21)$$

From (4.21) one easily computes the limit as n goes to infinity of the convergents to be

$$G = \sqrt{2} \qquad (4.22)$$

It is not a coincidence that the even convergents (4.19) are the same as the [0/0], [1/1], [2/2] Padé approximants to (1.1) as derived in (1.7) and (1.8). One can verify by direct calculation from (4.21) that Eq. (4.9) yields

$$A_{n-1}B_n - B_{n-1}A_n = (-1)^n \qquad (4.23)$$

The transformed version of (4.18) into form (4.11) is

$$G = 1 + \cfrac{\tfrac{1}{2}}{1 + \cfrac{\tfrac{1}{4}}{1 + \cfrac{\tfrac{1}{4}}{1 + \cdots}}} \qquad (4.24)$$

The even and odd parts of G are now easily worked out as

$$G_e = 1 + \cfrac{\tfrac{1}{2}}{\tfrac{5}{4} - \cfrac{\tfrac{1}{16}}{\tfrac{3}{2} - \cfrac{\tfrac{1}{16}}{\tfrac{3}{2} - \cdots}}} \qquad G_o = \tfrac{3}{2} - \cfrac{\tfrac{1}{8}}{\tfrac{3}{2} - \cfrac{\tfrac{1}{16}}{\tfrac{3}{2} - \cfrac{\tfrac{1}{16}}{\tfrac{3}{2} - \cdots}}} \qquad (4.25)$$

C. Convergence Theorems

Successive truncations of (4.25) are easily seen to yield the appropriate result from (4.19), as far as it goes. By working in a more formalized but similar way to this example we can give a convergence theorem for continued fractions due to Van Vleck (Van Vleck, 1904; Pringsheim, 1910).

Theorem 4.1. If the a_n of the continued fraction

$$G = \cfrac{1}{1 + \cfrac{a_2}{1 + \cfrac{a_3}{1 + \cdots}}}$$

(4.26)

have the property

$$\lim_{n \to \infty} a_n = a \tag{4.27}$$

and a does not lie in the region $-\infty \leqslant a \leqslant -\frac{1}{4}$; then there exists an n such that

$$G_\nu = \cfrac{a_\nu}{1 + \cfrac{a_{\nu+1}}{1 + \cfrac{a_{\nu+2}}{1 + \cdots}}}$$

(4.28)

converges for all $\nu > n$.

Proof: The plan of the proof is to introduce a set of extra variables in terms of which we can easily express the solution of the recursion relation (4.5) for the numerator and denominator of the nth convergent. We show that these auxiliary variables tend to a limit and use this property to show that the convergents tend to a limit and, hence, by (4.4) define G_ν.

First we consider the roots of the equation

$$y^2 - y - a = 0 \tag{4.29}$$

by analogy with our solution procedure of (4.20). The roots u, u' of (4.28) satisfy

$$u + u' = 1 \qquad uu' = -a \tag{4.30}$$

and we choose them such that $|u'| \leqslant |u|$. The discriminant of (4.29) is $D = 1 + 4a$. By the properties of the solution of a quadratic equation, if D is real and $-\infty \leqslant D \leqslant 0$, then $|u'| = |u|$ since either $u' = u$ or they are complex conjugates. Otherwise we may select $|u'| < |u|$. We now suppose that we are treating the case where

$$0 < |u'/u| = M < 1 \tag{4.31}$$

Let us next introduce the sequence of numbers, for $\nu \geqslant n_0 > 0$,

$$u_\nu + u'_\nu = u_{\nu+1} + u'_{\nu+1} = 1, \qquad u_\nu u'_{\nu+1} = -a_{\nu+1} \tag{4.32}$$

where we require $|u'_\nu| \leqslant |u_\nu|$ asymptotically. We need first to show that these procedures are always possible. First we select a sequence of integers n_λ such that for all $\nu \geqslant n_\lambda$

$$|1 - (a_{\nu+1}/a)| \leqslant N^{\lambda-1}(1 - N)^2 \tag{4.33}$$

where $N = M^{1/3}$. This selection is possible by (4.27). Define

$$u_{n_0} = Nu \tag{4.34}$$

We can now compute all the u_ν, u'_ν by (4.32). Equation (4.32) leads directly to

$$u_\nu(1 - u_{\nu+1}) = -a_{\nu+1}, \qquad u_{\nu+1} = (u_\nu + a_{\nu+1})/u_\nu \tag{4.35}$$

By combining this result with (4.30), we can write

$$\begin{aligned}
u - u_{\nu+1} &= \frac{u_\nu(u - 1) - a_{\nu+1}}{u_\nu} = \frac{-u_\nu u' + uu' + a - a_{\nu+1}}{u_\nu} \\
&= \frac{u'(u - u_\nu) - [(a - a_{\nu+1})/a]uu'}{u - (u - u_\nu)} \\
&= u'\frac{[1 - (u_\nu/u)] - [1 - (a_{\nu+1}/a)]}{1 - [1 - (u_\nu/u)]}
\end{aligned} \tag{4.36}$$

or dividing by u and taking absolute values, we have

$$\left|1 - \frac{u_{\nu+1}}{u}\right| \leqslant M\frac{|1 - (u_\nu/u)| + |1 - (a_{\nu+1}/a)|}{|1 - |1 - (u_\nu/u)||} \tag{4.37}$$

For $\nu = n_0$ we have, by (4.34)

$$|1 - (u_\nu/u)| \leqslant 1 - N, \qquad 1 - |1 - (u_\nu/u)| \geqslant N \tag{4.38}$$

Thus by (4.37)

$$\left| 1 - \frac{u_{\nu+1}}{u} \right| \leqslant N^3 \frac{(1 - N) + N^{-1}(1 - N)^2}{N}$$

$$= N^2(1 - N) + N(1 - N)^2$$

$$= N(1 - N) < 1 - N \tag{4.39}$$

since $N < 1$. We can now, however, apply (4.38) and (4.39) successively and obtain

$$|1 - (u_{\nu+1}/u)| \leqslant N(1 - N), \qquad \nu = n_0, \dots, n_1 - 1 \tag{4.40}$$

On the next application of (4.37), taking account of (4.33), we have for $\nu = n_1$

$$\left| 1 - \frac{u_\nu}{u} \right| \leqslant N^3 \frac{N(1 - N) + (1 - N)^2}{N}$$

$$= N^3(1 - N) + N^2(1 - N)^2$$

$$= N^2(1 - N) < N(1 - N) \tag{4.41}$$

By proceeding indefinitely we have

$$\left| 1 - \frac{u_{n_\lambda + \mu}}{u} \right| < N^{\lambda+1}(1 - N), \qquad \mu = 1, 2, \dots, n_{\lambda+1} - n_\lambda \tag{4.42}$$

Consequently u_ν and u'_ν converge, that is,

$$\lim_{\nu \to \infty} u_\nu = u, \qquad \lim_{\nu \to \infty} u'_\nu = \lim_{\nu \to \infty} (1 - u_\nu) = 1 - u = u' \tag{4.43}$$

For the ratio u'_ν / u_ν we can write

$$\frac{u'}{u} - \frac{u'_\nu}{u_\nu} = \frac{1 - u}{u} - \frac{1 - u_\nu}{u_\nu} = \frac{1}{u} - \frac{1}{u_\nu} = \frac{u_\nu - u}{u u_\nu}$$

or

$$\left| \frac{u'}{u} - \frac{u'_\nu}{u_\nu} \right| \leqslant \frac{|u_\nu - u|}{|u u_\nu|} = \frac{|1 - (u_\nu/u)|}{|u| \, |1 - [(u_\nu/u) - 1]|} \leqslant \frac{1}{N|u|} \left| 1 - \frac{u_\nu}{u} \right| \tag{4.44}$$

by the second inequality of (4.38). Thus by (4.42)

$$\lim_{\nu \to \infty} (u'_\nu / u_\nu) = u' / u \tag{4.45}$$

and hence is less than unity for ν large enough.

We now turn to the computation of the convergents. Let n be selected so that for all $\nu \geqslant n$, $|u'_\nu / u_\nu| < M + \delta$ for a given δ, $0 < \delta < 1 - M$. Such a selection is possible by (4.45). If C_j stands for either A_j or B_j, then by (4.5)

$$C_{j+1} = C_j + a_{j+1} C_{j-1}$$

or if we use (4.30), we have

$$C_{j+1} - u_{j+1} C_j = u'_{j+1}(C_j - u_j C_{j-1}) \tag{4.46}$$

From (4.46) we can write a sequence of equations

$$C_{n+1} - u_{n+1} C_n = u_{n+1}(C_n - u_n C_{n-1})\frac{u'_{n+1}}{u_{n+1}}$$

$$C_{n+2} - u_{n+2} C_{n+1} = u_{n+1} u_{n+2}(C_n - u_n C_{n-1})\frac{u'_{n+1} u'_{n+2}}{u_{n+1} u_{n+2}}$$

$$\vdots \tag{4.47}$$

$$C_{m+1} - u_{m+1} C_m = u_{n+1} \cdots u_{m+1}(C_n - u_n C_{n-1})\frac{u'_{n+1} \cdots u'_{m+1}}{u_{n+1} \cdots u_{m+1}}$$

If we multiply by $u_{n+2} \cdots u_{m+1}$, $u_{n+3} \cdots u_{m+1}$, $\ldots$ and add, the equations of (4.47) become

$$C_{m+1} = u_{n+1} \cdots u_{m+1}[C_n + (C_n - u_n C_{n-1})\sigma_{m,n}] \tag{4.48}$$

and hence

$$\frac{A_{m+1}}{B_{m+1}} = \frac{A_n + (A_n - u_n A_{n-1})\sigma_{m,n}}{B_n + (B_n - u_n B_{n-1})\sigma_{m,n}} \tag{4.49}$$

where

$$\sigma_{m,n} = \frac{u'_{n+1}}{u_{n+1}} + \frac{u'_{n+1} u'_{n+2}}{u_{n+1} u_{n+2}} + \cdots + \frac{u'_{n+1} \cdots u'_{m+1}}{u_{n+1} \cdots u_{m+1}} \tag{4.50}$$

Now since we have $|u'_\nu / u_\nu| < M + \delta < 1$ for $\nu \geqslant n$, the series for $\sigma_{\infty,n}$

converges by comparison with a geometric series as

$$|\sigma_{\infty,n}| < (M + \delta) + (M + \delta)^2 + \cdots = (M + \delta)/(1 - M - \delta) \quad (4.51)$$

and so the continued fraction converges as long as

$$B_n + (B_n - u_n B_{n-1})\sigma_{\infty,n} \neq 0 \quad (4.52)$$

But this cannot always be zero, for when we begin the continued fraction at a_ν we have the corresponding $B_0 = B_1 = 1$ for (4.28), and, as close as we please, $u_n = u$, $\sigma_{\infty,n} = M/(1 - M)$. Thus (4.52) becomes as close as we please to

$$1 + (1 - u)M/(1 - M) = (1 - M + M - Mu)/(1 - M)$$

$$= (1 - u')/(1 - M) = u/(1 - M) \neq 0 \quad (4.53)$$

which completes the proof of the theorem. ∎

We will give one further general convergence theorem for continued fractions. It is due to Scott and Wall (1940).

Theorem 4.2 (parabola theorem). If all the a_n of the continued fraction

$$G = \cfrac{1}{1 + \cfrac{a_2}{1 + \cfrac{a_3}{1 + \cdots}}} \quad (4.54)$$

lie in truncated parabolic domains

$$|z| - R(z) \leqslant \zeta < \tfrac{1}{2} \quad (4.55)$$

$$|z| \leqslant M + \alpha n, \quad 0 < \alpha < 1, \quad (1 - 2\zeta)/\alpha > 1 \quad (4.56)$$

for some M, ζ, and α, where $R(z)$ is the real part of the complex variable z, then convergents A_n/B_n converge to a limit G. [The parabola (4.55) has its vertex at $z = -\tfrac{1}{4}$, its focus at $z = 0$, and opens to the right. See Fig. 4.1.]

Proof: An outline of the proof is as follows: We first show that the conditions in the theorem ensure that the a_n satisfy the fundamental inequalities. Then we relate the value of the continued fraction to the sum of a certain series. Finally, we are able to bound the sum of the series from the fundamental inequalities and thus establish the convergence of G.

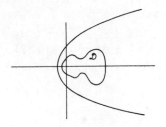

Fig. 4.1 The parabola of the parabola theorem. As originally stated by Scott and Wall (1940), the region containing the a_n was independent of the subscript ($\zeta = \frac{1}{2}$, $\alpha = 0$), and was closed and bounded, as illustrated by $\mathcal{D}$.

We wish to establish

$$r_1|1 + a_2| > |a_2|$$

$$r_2|1 + a_2 + a_3| > |a_3| \tag{4.57}$$

$$r_n|1 + a_n + a_{n+1}| > r_n r_{n-2}|a_n| + |a_{n+1}|, \qquad n = 3, 4, \ldots$$

which are the fundamental inequalities. The r_n will be shown to be uniformly less than a number less than unity. Now applying (4.55) to a_2, we get

$$|1 + a_2| \geqslant 1 + R(a_2) \geqslant 1 - \zeta + |a_2| > 1 - 2\zeta + |a_2| \tag{4.58}$$

Thus since by assumption (4.55) there must exist an r_1 such that

$$r_1|1 + a_2| > |a_2| \tag{4.59}$$

where

$$0 < r_1 < 1 - (1 - 2\zeta)/(M + 2\alpha) < 1 \tag{4.60}$$

Likewise

$$|1 + a_n + a_{n+1}| \geqslant 1 + R(a_n) + R(a_{n+1}) \geqslant 1 - 2\zeta + |a_n| + |a_{n+1}|$$

$$\tag{4.61}$$

Thus we can select an r_n such that

$$0 < r_n < 1 - (1 - 2\zeta)/[M + (n + 1)\alpha] \tag{4.62}$$

$$r_n(|1 + a_n + a_{n+1}| - |a_n|) > |a_{n+1}| \tag{4.63}$$

or, since $r_{n-2} < 1$,

$$r_n|1 + a_n + a_{n+1}| > r_n r_{n-2}|a_n| + |a_{n+1}| \tag{4.64}$$

By selecting $r_0 = 0$, we obtain (4.57) for r_2. Equation (4.62) is obeyed [by Eq. (4.60)] for all r_ζ.

We can now rewrite the nth convergent as

$$\frac{A_n}{B_n} = 1 + \sum_{j=2}^{n} \left(\frac{A_j}{B_j} - \frac{A_{j-1}}{B_{j-1}} \right) = 1 - \sum_{j=2}^{n} \frac{(-1)^n a_1 \cdots a_n}{B_j B_{j-1}} \qquad (4.65)$$

by Eq. (4.9), where $a_1 = 1$. If we define

$$\rho_n = -a_{n+1} B_{n-1} / B_{n+1}, \qquad n = 1, 2, \ldots \qquad (4.66)$$

then Eq. (4.36) becomes, since $B_0 = 1$,

$$A_n / B_n = 1 + \sum_{j=1}^{n-1} \left(\prod_{k=1}^{j} \rho_k \right) \qquad (4.67)$$

which re-expresses the nth convergent as a series. We now seek to show that

$$|\rho_k| \leqslant r_k \qquad (4.68)$$

First, from (4.2) and (4.16)

$$\rho_1 = -a_2 / B_2 = -a_2 / (1 + a_2) \qquad (4.69)$$

We thus have (4.68) for $k = 1$ directly from the first line of (4.57). Similarly, (4.68) for $k = 2$ follows from the second line of (4.57) by direct calculation as

$$\rho_2 = -a_3 B_1 / B_3 = -a_3 / (1 + a_2 + a_3) \qquad (4.70)$$

Now if we apply the recursion relations (4.5) to the even and odd parts of G [(4.16) and (4.17)] as though they were new continued fractions in their own right, we get for the original B_k

$$B_{k+2} = (1 + a_{k+1} + a_{k+2}) B_k - a_k a_{k+1} B_{k-2} \qquad (4.71)$$

Let us now suppose that (4.68) holds for all $k \leqslant n$. If $a_{n+2} \neq 0$, then from (4.71) we obtain

$$|\rho_{n+1}|^{-1} = \left| \frac{1 + a_{n+1} + a_{n+2}}{a_{n+2}} - \frac{a_{n+1}}{a_{n+2}} \frac{a_n B_{n-2}}{B_n} \right|$$

$$\geqslant \left| \frac{1 + a_{n+1} + a_{n+2}}{a_{n+2}} \right| - \left| \frac{a_{n+1}}{a_{n+2}} \right| r_{n-1} \geqslant \frac{1}{r_{n+1}} \qquad (4.72)$$

by the last line of (4.57). Thus we have shown (4.68) for all k, inductively.

By Cauchy's (1821) convergence criterion, it suffices to show that

$$|(A_n/B_n) - (A_m/B_m)| < \epsilon$$

for any $\epsilon > 0$ and all $n, m > N$ to prove that A_n/B_n tends to a limit as n tends to infinity. Now $m > n$,

$$\left| \frac{A_n}{B_n} - \frac{A_m}{B_m} \right| = \left| \sum_{j=n}^{m-1} \left(\prod_{k=1}^{j} \rho_k \right) \right|$$

$$\leqslant \sum_{j=n}^{m-1} \prod_{k=1}^{j} r_k < \sum_{j=n}^{m-1} \exp\left(-\sum_{k=1}^{j} \frac{1 - 2\zeta}{M + \alpha k} \right)$$

$$< \sum_{j=n}^{\infty} \exp\left(-\frac{1 - 2\zeta}{\alpha} \ln \frac{M + \alpha j}{M + \alpha} \right) = \sum_{j=n}^{\infty} \left(\frac{M + \alpha}{M + \alpha j} \right)^{(1-2\zeta)/\alpha}$$

$$< \frac{\alpha[M + \alpha(n-1)]}{1 - 2\zeta} \left[\frac{M + \alpha}{M + \alpha(n-1)} \right]^{(1-2\zeta)/\alpha} \tag{4.73}$$

which can be made smaller than any preassigned ϵ by choosing N large enough, since $[(1 - 2\zeta)/\alpha] > 1$ by assumption. Use has been made of $\ln(1 + x) \leqslant x$, and inequalities between sums and integrals. Thus the continued fraction converges. ∎

The scope of this theorem can be extended by means of more careful arguments and somewhat more complicated restrictions on the coefficients of the continued fraction. For example, if instead of our example (4.24), we choose

$$G = 1 + \cfrac{\frac{1}{2}}{1 - \cfrac{\frac{1}{4}}{1 - \cdots}} \tag{4.74}$$

then this continued fraction lies on the boundary of the parabolic region (4.27). We can again work out the general expressions for A_n and B_n for (4.74). They are

$$A_n = (2n + 1)(\tfrac{1}{2})^n, \qquad B_n = (n + 1)(\tfrac{1}{2})^n \tag{4.75}$$

so that A_n/B_n tends to a limit, two, but here the error is of the order $1/n$ instead of decaying geometrically as it does in the interior of the parabolic domain (4.55). The geometric decay is illustrated by (4.21) for our example (4.24). This example also lies on the boundary of the region covered by Theorem 4.1.

D. Relation of Continued Fractions to Taylor Series

After this brief introduction to continued fractions we are in a position to relate them to Padé approximants. This relation is most simply given from the form (4.11). Since we wish to have the continued fractions related to a Taylor series, we write

$$G(z) = b_0 + \cfrac{a_1 z}{1 + \cfrac{a_2 z}{1 + \cfrac{a_3 z}{1 + \cdots}}} \tag{4.76}$$

We observe by expansion of (4.76) for small z that

$$G(z) = b_0 + a_1 z - a_1 a_2 z^2 + O(z^3) \tag{4.77}$$

Hence a_2 affects only the coefficient of z^2 and perhaps higher terms and not at all the coefficient of z. At any stage we see that a_{n+1} only enters the Taylor series expansion at one higher power than does a_n. Thus by induction we conclude that if

$$G(z) = \sum_{n=0}^{\infty} g_n z^n \tag{4.78}$$

then g_n depends on $b_0, a_1, \ldots, a_n$, and that the $g_0, g_1, \ldots, g_n$ are reproduced exactly by the nth convergent to $G(z)$. The convergents are given by

$$\frac{A_0(z)}{B_0(z)} = \frac{b_0}{1}, \qquad \frac{A_1(z)}{B_1(z)} = \frac{b_0 + a_1 z}{1}, \qquad \frac{A_2(z)}{B_2(z)} = \frac{b_0 + a_1 z + b_0 a_2 z}{1 + a_2 z}$$

$$\tag{4.79}$$

and in general by (4.5) as

$$A_{n+1}(z) = A_n(z) + a_{n+1} z A_{n-1}(z), \qquad B_{n+1}(z) = B_n(z) + a_{n+1} z B_{n-1}(z)$$

$$\tag{4.80}$$

By the application of (4.80) to (4.79) as starting values, we can compute the degrees of the $A_n(z)$ and $B_n(z)$. They are

$$\text{degree}[A_{2n}(z)] = n \qquad\qquad \text{degree}[B_{2n}(z)] = n$$

$$\text{degree}[A_{2n+1}(z)] = n + 1 \qquad \text{degree}[B_{2n+1}(z)] = n \qquad (4.81)$$

Putting this information on the degrees together with the fact that the nth convergent reproduces the first $n + 1$ Taylor series coefficients of $G(z)$, we have proved, by the uniqueness theorem 1.1, the following theorem.

Theorem 4.2. The sequence of convergents of the continued fraction (4.76) occupies the stair step sequence $[0/0], [1/0], [1/1], [2/1], [2/2], \ldots$ in the Padé table for $G(z)$ [see Fig. 4.2].

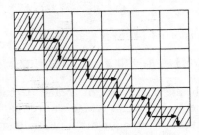

Fig. 4.2 The locations in the Padé table of the convergents to continued fraction (4.76).

The continued fraction (4.76) is called the *corresponding* (Perron, 1954) or *Stieltjes-type* (Wall, 1948) continued fraction to the power series (4.78). If we consider the even part of (4.76), we obtain the *associated* (Perron, 1954) or *Jacobi-type* (Wall, 1948) continued fraction of (4.78) as

$$G(z) = b_0 + \cfrac{a_1 z}{1 + a_2 z - \cfrac{a_2 a_3 z^2}{1 + (a_3 + a_4)z - \cfrac{a_4 a_5 z^2}{1 + (a_5 + a_6)z - \cdots}}}$$

$$(4.82)$$

The successive convergents of form (4.82) are given by the diagonal sequence $[0/0], [1/1], [2/2], \ldots$ of the Padé table.

The *equivalent* or *Euler-type* continued fractions based on Euler's identity

$$g_0 + g_1 x + g_2 x^2 + \cdots + g_n x^n$$

$$\equiv \cfrac{g_0}{1 - \cfrac{g_1 x / g_0}{1 + (g_1/g_0)x - \cfrac{g_2 x / g_1}{1 + (g_2/g_1)x - \cdots}}} \quad \cfrac{- g_n x / g_{n-1}}{1 + g_n x / g_{n-1}} \quad (4.83)$$

are related to Padé approximants, in as much as they are the partial sums of the Taylor series, disguised.

We can relate the parameters a_i of the continued fractions (4.76) and (4.82) directly to the g_i of the power series in the following way. By Theorem 4.2, the recursion relations (4.80) must be, except for normalization, simply the Frobenius three-term identities (3.24) and (3.25). We notice from (4.79) and (4.80) that the normalization implied by (4.80) is just $B_n(0) = 1$. Thus we can easily identify

$$a_{2n+1} = - \frac{C(n + 1/\quad n + 1)C(n/n - 1)}{C(n/n)C(n + 1/n)} \quad (4.84)$$

$$a_{2n} = - \frac{C(n + 1/n)C(n - 1/\quad n - 1)}{C(n/n - 1)C(n/n)} \quad n \geqslant 1 \quad (4.85)$$

by using the remark before Eq. (3.4) to renormalize S in (3.24) and (3.25) to conform with $B_n(0) = 1$.

E. Relation of Continued Fractions to Padé Approximants

By use of (3.34) we can re-express the Frobenius identities (3.37) and (3.38) in a form which brings out the relation between Padé approximants and the continued fraction (4.76). They are

$$\frac{S(L + 1/M)}{q_0(L + 1/M)} = \frac{S(L/M)}{q_0(L/M)}$$

$$+ \frac{P_{L+1}(L + 1/M)}{q_0(L + 1/M)} \frac{q_0(L/M - 1)}{P_L(L/M - 1)} x \frac{S(L/M - 1)}{q_0(L/M - 1)}$$

$$(4.86)$$

$$\frac{S(L/M + 1)}{q_0(L/M + 1)}$$

$$= \frac{S(L/M)}{q_0(L/M)} + \frac{q_{M+1}(L/M + 1)q_0(L - 1/M)}{q_0(L/M + 1)q_M(L - 1/M)} x \frac{S(L - 1/M)}{q_0(L - 1/M)}$$

$$(4.87)$$

Comparison with Eq. (4.80) leads to the identifications

$$a_{2n+1} = \frac{p_{n+1}(n + 1/n)q_0(n/n - 1)}{q_0(n + 1/n)p_n(n/n - 1)},$$

$$a_{2n} = \frac{q_n(n/n)q_0(n - 1/\quad n - 1)}{q_0(n/n)q_{n-1}(n - 1/\quad n - 1)} \qquad (4.88)$$

With these formulas we can, as a practical matter, readily compute the continued fraction from the Padé table, or compute the stair-step sequence of the Padé table from the continued fraction.

In summary, those continued fractions related to power series [defined by (4.4)] of the various types considered are the limits of particular sequences of entries in the Padé table: stair-step for the corresponding type, diagonal for the associated type, and vertical for the equivalent type.

F. Bigradients

Formula (4.80) converts a continued fraction into a sequence of convergents. The solution to the reverse problem of expanding a convergent, or more generally a quotient of two-power series, into the corresponding continued fraction as well as the calculation of the Padé approximants for that continued fraction can be given in terms of *bigradients*.

Suppose we consider

$$F(x) = C(x)/D(x) \qquad (4.89)$$

where

$$F(x) = \sum f_j x^j, \qquad C(x) = \sum c_j x^j, \qquad D(x) = \sum d_j x^j \qquad (4.90)$$

Bigradients are determinants of the form

$$
\Delta_{m,n} = \det
\begin{vmatrix}
d_0 & 0 & \cdots & 0 & 0 & \cdots & 0 & c_0 \\
d_1 & d_0 & \cdots & 0 & 0 & \cdots & c_0 & c_1 \\
\vdots & \vdots & & \vdots & \vdots & & \vdots & \vdots \\
d_{m+n-1} & d_{m+n-2} & \cdots & d_n & c_m & \cdots & c_{m+n-2} & c_{m+n-1}
\end{vmatrix}
$$

(4.91)

We can reduce $\Delta_{m,n}$ in the following way. Define

$$
E(x) = d_0/D(x) = \sum e_j x^j
$$

(4.92)

If we multiply the jth from the bottom row ($j \geqslant 2$) by e_j and add these products to the last row, we get

$$
0, \quad 0, \ldots, 0, \quad d_0 f_m, \ldots, d_0 f_{m+n-2}, \quad d_0 f_{m+n-1}
$$

(4.93)

by the properties of the reciprocal series $E(x)$. If we continue this process for the second, third from the bottom rows, we obtain

$$
\Delta_{m,n} = \det
\begin{vmatrix}
d_0 & 0 & \cdots & 0 & 0 & \cdots & 0 & d_0 f_0 \\
0 & d_0 & \cdots & 0 & 0 & \cdots & d_0 f_0 & d_0 f_1 \\
\vdots & \vdots & & \vdots & \vdots & & \vdots & \vdots \\
0 & 0 & \cdots & 0 & d_0 f_m & \cdots & d_0 f_{m+n-2} & d_0 f_{m+n-1}
\end{vmatrix}
$$

(4.94)

which can now, by column operations, be reduced to

$$
\Delta_{m,n} = d_0^{m+n} C(m/n)
$$

(4.95)

where C is defined by Eq. (2.1). Thus by means of (4.95) and (4.91) we can compute the $C(m/n)$ directly from the c_j and d_j and thus compute, via (4.84) and (4.85), the a_i of the corresponding continued fraction.

We remark that if $C(x)$ and $D(x)$ are polynomials of degree m and n exactly and if $\Delta_{m,n} = 0$, then by (4.95), $C(m/n) = 0$ and hence, by Padé's theorem (Theorem 2.3), the Padé approximant to $F(x)$ is not of full degree and thus $C(x)$ and $D(x)$ have a common divisor. This result is Trudi's theorem (Trudi, 1862).

We can give explicit formulas for the Padé approximants to the ratio of two series in terms of bigradients of the coefficients which are analogous to

Eq. (1.27), and reduce in a fairly clear way when $D(x) = 1.0$:

$$[L/M] = \frac{-\det\begin{vmatrix} d_0 & 0 & \cdots & 0 & 0 & \cdots & 0 & c_0 \\ d_1 & d_0 & \cdots & 0 & 0 & \cdots & c_0 & c_1 \\ \vdots & \vdots & & \vdots & \vdots & & \vdots & \vdots \\ d_{L+M} & d_{L+M-1} & \cdots & d_M & c_L & \cdots & c_{L+M-1} & c_{L+M} \\ 1 & x & \cdots & x^L & 0 & \cdots & 0 & 0 \end{vmatrix}}{\det\begin{vmatrix} d_0 & 0 & \cdots & 0 & 0 & \cdots & 0 & c_0 \\ d_1 & d_0 & \cdots & 0 & 0 & \cdots & c_0 & c_1 \\ \vdots & \vdots & & \vdots & \vdots & & \vdots & \vdots \\ d_{L+M} & d_{L+M-1} & \cdots & d_M & c_L & \cdots & c_{L+M-1} & c_{L+M} \\ 0 & 0 & \cdots & 0 & x^M & \cdots & x & m \end{vmatrix}}$$

$$= U(x)/V(x) \tag{4.96}$$

The simplest way to prove (4.96) is to consider

$C(x)V(x) - U(x)D(x)$

$$= \det\begin{vmatrix} d_0 & 0 & \cdots & 0 & 0 & \cdots & 0 & c_0 \\ d_1 & d_0 & \cdots & 0 & 0 & \cdots & c_0 & c_1 \\ \vdots & \vdots & & \vdots & \vdots & & \vdots & \vdots \\ d_{L+M} & d_{L+M-1} & \cdots & d_M & c_L & \cdots & c_{L+M-1} & c_{L+M} \\ D(x) & xD(x) & \cdots & x^L D(x) & x^M C(x) & \cdots & xC(x) & C(x) \end{vmatrix}$$

$$\tag{4.97}$$

By means of elementary row operations we can convert (4.97) into

$C(x)V(x) - U(x)D(x)$

$$= \sum_{k=1}^{\infty} \det\begin{vmatrix} d_0 & 0 & \cdots & 0 & 0 & \cdots & 0 & c_0 \\ d_1 & d_0 & \cdots & 0 & 0 & \cdots & c_0 & c_1 \\ \vdots & \vdots & & \vdots & \vdots & & \vdots & \vdots \\ d_{L+M} & d_{L+M-1} & \cdots & d_M & c_L & \cdots & c_{L+M-1} & c_{L+M} \\ d_{L+M+k} & d_{L+M+k-1} & \cdots & d_{M+k} & c_{L+k} & \cdots & c_{L+M+k-1} & c_{L+M+k} \end{vmatrix}$$

$$\times x^{L+M+k} \tag{4.98}$$

From Eq. (4.98) we see that (4.96) satisfies the Padé equations (1.13); thus, provided the appropriate determinants do not vanish, we have Eq. (4.96) by the uniqueness theorem 1.1.

Another useful result that we may obtain from (4.96) is Hadamard's (1892) formula for the relation between the C table for a power series and the C table for its reciprocal series. If we compute $\Delta'_{M,L}$ for $D(x)/1.0$ and $\Delta_{L,M}$ for $1.0/D(x)$, then from (4.91) the only difference is in the order of the columns. Yet by (4.95) we have

$$\Delta'_{M,L} = C'(M/L), \qquad \Delta_{L,M} = d_0^{L+M} C(L/M) \qquad (4.99)$$

where C' is the C table for the reciprocal series. One can transform $\Delta'_{M,L}$ into $\Delta_{L,M}$ by making $\frac{1}{2}(L+M)(L+M-1)$ column interchanges. Therefore we have shown Hadamard's formula

$$C'(M/L) = (-1)^{\frac{1}{2}(L+M)(L+M-1)} C(L/M)/f_0^{L+M} \qquad (4.100)$$

for the relation between the C table for $F(x)$ and the C table for its reciprocal series.

Throughout this chapter we have generally assumed that nothing vanished in such a way as to create difficulties. This assumption is equivalent, for continued fractions related to Taylor series, to assuming that the Padé table is normal. If the continued fraction (4.76) terminates, i.e., one of the $a_i = 0$, then we have simply a rational fraction. However, it might happen, that even though we compute from (4.84) that one of the a_i is zero, the truncated continued fraction is not equivalent to the power series. This case corresponds to a nonnormal Padé table. To deal with this case the *general corresponding* continued fraction is introduced with "gaps" in it. It is

$$C(z) = b_0 + \cfrac{a_1 z^{\alpha_1}}{1 + \cfrac{a_2 z^{\alpha_2}}{1 + \cfrac{a_3 z^{\alpha_3}}{1 + \cdots}}} \qquad (4.101)$$

where the α_i are positive integers. The theory of these continued fractions is related to the block-structure theory of the Padé table.

5

GAUSS'S HYPERGEOMETRIC FUNCTION

A. Gauss's Continued Fraction

One of the most widely occurring special mathematical functions is Gauss's hypergeometric function (1813). Both the Padé approximants and continued fractions can be derived explicitly for a wide variety of cases and many practical formulas given. An extensive collection is reported by Luke (1969).

The hypergeometric series is

$$_2F_1(\alpha, \beta; \gamma; x) = 1 + \frac{\alpha\beta}{1\cdot\gamma}x + \frac{\alpha(\alpha+1)\beta(\beta+1)}{1\cdot2\cdot\gamma(\gamma+1)}x^2$$

$$+ \frac{\alpha(\alpha+1)(\alpha+2)\beta(\beta+1)(\beta+2)}{1\cdot2\cdot3\gamma(\gamma+1)(\gamma+2)}x^3 + \cdots \quad (5.1)$$

where α and β are any complex constants and γ should not equal $0, -1, -2, \ldots$. The hypergeometric function is clearly a polynomial if α or β is a negative, real integer, and a constant if α or $\beta = 0$. Some special cases are

$$_2F_1(1, 1; 2; x) = -(1/x)\ln(1-x), \quad _2F_1(-k, \beta; \beta; x) = (1-x)^k$$

$$x\,_2F_1(\tfrac{1}{2}, \tfrac{1}{2}; \tfrac{3}{2}; x^2) = \sin^{-1}x, \quad x\,_2F_1(\tfrac{1}{2}, 1; \tfrac{3}{2}; -x^2) = \tan^{-1}x \quad (5.2)$$

The hypergeometric series obey identities which link any three series which are adjacent in the sense of having a difference in the parameters of unity. We have

$$_2F_1(\alpha, \beta+1; \gamma+1; x) - {_2F_1}(\alpha, \beta; \gamma; x)$$

$$= 1 + \frac{\alpha(\beta + 1)}{1 \cdot (\gamma + 1)} x + \frac{\alpha(\alpha + 1)(\beta + 1)(\beta + 2)}{1 \cdot 2(\gamma + 1)(\gamma + 2)} x^2 + \cdots$$

$$- 1 - \frac{\alpha \cdot \beta}{1 \cdot \gamma} x - \frac{\alpha(\alpha + 1)\beta(\beta + 1)}{1 \cdot 2(\gamma)(\gamma + 1)} x^2$$

$$= \frac{(\gamma - \beta)\alpha}{1 \cdot \gamma(\gamma + 1)} x + \frac{2(\gamma - \beta)\alpha(\alpha + 1)(\beta + 1)}{1 \cdot 2\gamma(\gamma + 1)(\gamma + 2)} x^2 + \cdots$$

$$= \frac{(\gamma - \beta)\alpha}{\gamma(\gamma + 1)} x \left[1 + \frac{(\alpha + 1)(\beta + 1)}{1 \cdot (\gamma + 2)} x \right.$$

$$\left. + \frac{(\alpha + 1)(\alpha + 2)(\beta + 1)(\beta + 2)}{1 \cdot 2(\gamma + 2)(\gamma + 3)} x^2 + \cdots \right] \tag{5.3}$$

or

$$_2F_1(\alpha, \beta + 1; \gamma + 1; x) - {}_2F_1(\alpha, \beta; \gamma; x)$$

$$= \frac{(\gamma - \beta)\alpha}{\gamma(\gamma + 1)} x \, {}_2F_1(\alpha + 1, \beta + 1; \gamma + 2; x) \tag{5.4}$$

Now since $_2F_1(\alpha, \beta; \gamma; x) = {}_2F_1(\beta, \alpha; \gamma; x)$ from Eq.(5.1), we have equally well derived

$$_2F_1(\alpha + 1, \beta; \gamma + 1; x) - {}_2F_1(\alpha, \beta; \gamma; x)$$

$$= \frac{(\gamma - \alpha)\beta}{\gamma(\gamma + 1)} x \, {}_2F_1(\alpha + 1, \beta + 1; \gamma + 2; x) \tag{5.5}$$

We can rewrite Eq. (5.4) as

$$\frac{_2F_1(\alpha, \beta + 1; \gamma + 1; x)}{_2F_1(\alpha, \beta; \gamma; x)}$$

$$= \frac{1}{1 - \dfrac{\alpha(\gamma - \beta)}{\gamma(\gamma + 1)} x \dfrac{_2F_1(\alpha + 1, \beta + 1; \gamma + 2; x)}{_2F_1(\alpha, \beta + 1; \gamma + 1; x)}} \tag{5.6}$$

and (replacing β by $\beta + 1$ and γ by $\gamma + 1$) Eq. (5.5) as

$$\frac{_2F_1(\alpha + 1, \beta + 1; \gamma + 2; x)}{_2F_1(\alpha, \beta + 1; \gamma + 1; x)}$$

$$= \frac{1}{1 - \dfrac{(\beta + 1)(\gamma - \alpha + 1)}{(\gamma + 1)(\gamma + 2)} x \dfrac{_2F_1(\alpha + 1, \beta + 2; \gamma + 3; x)}{_2F_1(\alpha + 1, \beta + 1; \gamma + 2; x)}} \tag{5.7}$$

which turns out to be just right to substitute into the denominator of (5.6). If we substitute (5.6) and (5.7) alternately, we obtain Gauss's continued fraction:

$$\frac{{}_2F_1(\alpha, \beta + 1; \gamma + 1; x)}{{}_2F_1(\alpha, \beta; \gamma; x)} = \cfrac{1}{1 + \cfrac{a_1 x}{1 + \cfrac{a_2 x}{1 + \cfrac{a_3 x}{1 + \cdots}}}} \tag{5.8}$$

where we have

$$a_{2n+1} = -\frac{(\alpha + n)(\gamma - \beta + n)}{(\gamma + 2n)(\gamma + 2n + 1)}, \qquad a_{2n} = -\frac{(\beta + n)(\gamma - \alpha + n)}{(\gamma + 2n - 1)(\gamma + 2n)} \tag{5.9}$$

Since the a_n tend to $-\frac{1}{4}$ as n goes to infinity, this continued fraction converges in the cut complex plane $1 \leqslant x \leqslant \infty$ by Theorem 4.1 to a meromorphic function. To see this result, we can apply Eq. (4.49) to (5.8) and note that A_n, B_n, A_{n-1}, and B_{n-1} are polynomials and $\sigma_{\infty, n}$ as the limit of a sequence of analytic functions in x is analytic (Copson, 1948). This region of convergence agrees precisely with the domain of analyticity of the hypergeometric function, and hence as the continued fraction is the ratio of two analytic functions, it is meromorphic in that domain.

B. Special Cases

One special case of particular interest is where $\beta = 0$. Since ${}_2F_1(\alpha, 0; \gamma; x) = 1$, Eq. (5.8) then becomes a direct expansion for the continued fraction itself. Thus

$$_2F_1(\alpha, 1; \gamma; x) = \cfrac{1}{1 + \cfrac{r_1 x}{1 + \cfrac{r_2 x}{1 + \cfrac{r_3 x}{1 + \cdots}}}} \tag{5.10}$$

where

$$r_{2n+1} = -\frac{(\alpha + n)(\gamma + n - 1)}{(\gamma + 2n - 1)(\gamma + 2n)},$$

$$r_{2n} = -\frac{n(\gamma + n - \alpha - 1)}{(\gamma + 2n - 2)(\gamma + 2n - 1)} \tag{5.11}$$

One can use (5.11), together with (4.87) and (4.88), to compute recursively the $[M/M]$ and $[M-1/M]$ Padé approximants. It turns out that a simple explicit expression results for the denominators. It is

$$Q_M^{(J)}(x) = q_0(L/M)\,{}_2F_1(-M, -\alpha - L; 1 - \gamma - L - M; x) \quad (5.12)$$

which is valid for $J = L - M \geqslant -1$. An explicit formula can also be given for the numerator polynomial, but in practice it is just as convenient to use the first L terms of

$$P_L^{(J)}(x) = Q_M^{(J)}(x)\,{}_2F_1(\alpha, 1; \gamma; x) + O(x^{L+M+1}) \quad (5.13)$$

Armed with these equations, we can work out quite a number of special cases. We will give several illustrations. One is

$$\log(1 + x) = x\,{}_2F_1(1, 1; 2; -x)$$

$$= \cfrac{x}{1 + \cfrac{\frac{1}{2}x}{1 + \cfrac{\frac{1}{6}x}{1 + \cfrac{\frac{3}{6}x}{1 + \cfrac{\frac{2}{10}x}{1 + \cfrac{\frac{3}{10}x}{1 - \cdots}}}}}} \quad (5.14)$$

the Padé approximants being given, of course, by (5.12) and (5.13). A related illustration is

$$\tanh^{-1} x = \frac{1}{2}\ln\left(\frac{1 + x}{1 - x}\right) = x\,{}_2F_1\left(\frac{1}{2}, 1; \frac{3}{2}; x^2\right)$$

$$= \cfrac{x}{1 - \cfrac{(1/3)x^2}{1 - \cfrac{(4/15)x^2}{1 - \cfrac{(9/35)x^2}{1 - \cdots}}}} \quad (5.15)$$

Equation (5.15) is an example of a C-type fraction, and the Padé table breaks up into 2×2 blocks for this, and for that matter, for any even or odd function. The $\tan^{-1} x$ is gotten by changing all the minus signs to plus in (5.15).

The following illustration terminates if k is a positive or negative integer or zero, but is otherwise infinite:

$$(1 - x)^k = {}_2F_1(-k, 1; 1; x)$$

$$= \cfrac{1}{1 + \cfrac{kx}{1 - \cfrac{\frac{1}{2}(k + 1)x}{1 + \cfrac{\frac{1}{6}(k - 1)x}{1 - \cfrac{\frac{1}{6}(k + 2)x}{1 + \cdots}}}}} \tag{5.16}$$

This fraction is useful for generating approximate expressions for irrational numbers. For example, $2^{1/3}$ is the value for $x = -1$, $k = \frac{1}{3}$. In this case (5.16) becomes

$$2^{1/3} = \cfrac{1}{1 - \cfrac{\frac{1}{3}}{1 + \cfrac{\frac{2}{3}}{1 + \cfrac{\frac{1}{9}}{1 + (7/18) \cdots}}}} \tag{5.17}$$

which has the initial convergents

$$1, \quad 3/2, \quad 5/4, \quad 24/19, \quad 131/104, \ldots$$
$$1, \quad 1.5, \quad 1.25, \quad 1.263, \quad 1.2596, \ldots$$

to $2^{1/3}$.

We also can derive

$$\int_0^x \frac{t^p \, dt}{1 + t^q} = \frac{x^{p+1}}{(p + 1)} \, {}_2F_1\left(\frac{p + 1}{q}, 1; \frac{p + 1}{q} + 1; - x^q\right) \tag{5.18}$$

The continued fraction (5.10) is given by

$$r_{2n+1} = - \frac{(qn + p)^2}{(2qn + p)(2qn + p + q)}, \tag{5.19}$$

$$r_{2n} = - \frac{q^2 n^2}{(2qn + p)(2qn + p - q)}$$

For the special case $p = 0$, $q = 2$, we get [Eq. (5.15)] $\tan^{-1} x$, by an elementary integration formula. Since $\tan^{-1} 1 = \frac{1}{4}\pi$ we have

$$\frac{\pi}{4} = \cfrac{1}{1 + \cfrac{(1/3)}{1 + \cfrac{(4/15)}{1 + (9/35) \cdot \cdot \cdot}}}$$

(5.20)

which yields the convergents

$$1, \quad 3/4, \quad 19/24, \quad 40/51, \ldots$$
$$0.75, \quad 0.791, \quad 0.7843$$

to the transcendental number $\frac{1}{4}\pi$.

We will illustrate Eq. (5.8) as well for

$$\frac{\sin^{-1} x}{(1 - x^2)^{1/2}} = \frac{x \, {}_2F_1(\frac{1}{2}, \frac{1}{2}; \frac{3}{2}; x^2)}{{}_2F_1(\frac{1}{2}, -\frac{1}{2}; \frac{1}{2}; x^2)}$$

$$= \cfrac{x}{1 - \cfrac{(2/3)x^2}{1 - \cfrac{(2/15)x^2}{1 - \cfrac{(12/35)x^2}{1 - (12/68)x^2 \cdot \cdot \cdot}}}}$$

(5.21)

and for

$$\frac{(1 + x)^k - (1 - x)^k}{(1 + x)^k + (1 - x)^k} = kx \frac{{}_2F_1(\frac{1}{2}(1 - k), \frac{1}{2}(2 - k); \frac{3}{2}; x^2)}{{}_2F_1(\frac{1}{2}(1 - k), -\frac{1}{2}k; \frac{1}{2}; x^2)}$$

$$= \cfrac{kx}{1 + \cfrac{(1/3)(k^2 - 1)x^2}{1 + \cfrac{(1/15)(k^2 - 4)x^2}{1 + (1/35)(k^2 - 9)x^2 + \cdot \cdot \cdot}}}$$

(5.22)

If we let $x = i \tan \phi$ in (5.22), we obtain

$$\frac{(1 + i \tan \phi)^k - (1 - i \tan \phi)^k}{(1 + i \tan \phi)^k + (1 - i \tan \phi)^k} = \frac{e^{ik\phi} - e^{-ik\phi}}{e^{ik\phi} + e^{-ik\phi}} = i \tan k\phi \quad (5.23)$$

Thus we get a continued fraction for

$$\tan k\phi = \cfrac{k \tan \phi}{1 - \cfrac{(1/3)(k^2 - 1)\tan^2 \phi}{1 - \cfrac{(1/15)(k^2 - 4)\tan^2 \phi}{1 - (1/35)(k^2 - 9)\tan^2 \phi + \cdots}}} \qquad (5.24)$$

which gives, when k is an integer, the trigonometric multiple angle formulas as (5.24) terminates. Another application is to note that we easily obtain by manipulation

$$\left(\frac{1 + x}{1 - x}\right)^k + 1 = \cfrac{2}{1 - \cfrac{kx}{1 + \cfrac{(1/3)(k^2 - 1)x^2}{1 + (1/15)(k^2 - 4)x^2 + \cdots}}} \qquad (5.25)$$

which when we note the identity

$$\tanh^{-1} x = \frac{1}{2}\ln\left(\frac{1 + x}{1 - x}\right) \qquad (5.26)$$

gives us a continued fraction expansion for

$$\exp(2k \tanh^{-1} x) \qquad (5.27)$$

By taking the limit

$$\lim_{k \to 0} \frac{\exp(2k \tanh^{-1} x) - 1}{2k} \qquad (5.28)$$

we recover the result (5.15).

C. Confluent Hypergeometric Function

We can use the foregoing results for the hypergeometric function to derive corresponding results for the confluent hypergeometric functions. First

$$_1F_1(\beta; \gamma; x) = \lim_{\alpha \to \infty} {}_2F_1(\alpha, \beta; \gamma; x/\alpha)$$

$$= 1 + \frac{\beta}{1 \cdot \gamma}x + \frac{\beta(\beta + 1)}{1 \cdot 2 \cdot \gamma(\gamma + 1)}x^2$$

$$+ \frac{\beta(\beta + 1)(\beta + 2)}{1 \cdot 2 \cdot \gamma(\gamma + 1)(\gamma + 2)}x^3 + \cdots \qquad (5.29)$$

represents a family of entire functions. Equations (5.8) and (5.9) become

$$\frac{{}_1F_1(\beta + 1; \gamma + 1; x)}{{}_1F_1(\beta; \gamma; x)} = \cfrac{1}{1 + \cfrac{a_1 x}{1 + \cfrac{a_2 x}{1 + \cfrac{a_3 x}{1 + \cdots}}}} \tag{5.30}$$

and

$$a_{2n+1} = -(\gamma - \beta + n)/[(\gamma + 2n)(\gamma + 2n + 1)] \tag{5.31}$$

$$a_{2n} = (\beta + n)/[(\gamma + 2n - 1)(\gamma + 2n)] \tag{5.32}$$

Equation (5.30) is a meromorphic function since it is the ratio of two entire functions. Since a_n tends to zero as n tends to infinity, by (5.31) and (5.32), the continued fraction (5.30) converges for all x not a zero of ${}_1F_1(\beta; \gamma; x)$, by Theorem 4.2. Again the special case $\beta = 0$ is of particular interest since it is closely related to the incomplete gamma function. We have

$$\gamma(\nu, x) = \int_0^x t^{\nu-1} e^{-t}\, dt = x^\nu e^{-x} \int_0^1 u^{\nu-1} e^{(1-u)x}\, du$$

$$= x^\nu e^{-x} \sum_{n=0}^\infty \frac{x^n}{\Gamma(n+1)} \int_0^1 u^{\nu-1}(1-u)^n\, du$$

$$= x^\nu e^{-x} \sum_{n=0}^\infty \frac{\Gamma(\nu)\Gamma(n+1)}{\Gamma(n+1)\Gamma(n+\nu)} x^n$$

$$= x^\nu e^{-x}\, {}_1F_1(1; 1+\nu; x) \tag{5.33}$$

by the standard β-integral (Peirce, 1910, No. 482), where $\Gamma(n)$ is the gamma function. The continued fraction is given by setting $\beta = 0$ in (5.30)–(5.32), where ${}_1F_1(0; \gamma; x) = 1$. The Padé denominators for ${}_1F_1(1; \gamma; x)$ are given from (5.12) by

$$Q_M^{(J)}(x) = q_0(L/M)\, {}_1F_1(-M; 1 - \gamma - L - M; -x) \tag{5.34}$$

for $J = L - M \geqslant -1$. The exponential function can be derived from (5.30) by setting $\beta = 0$, then $\gamma = 0$. Note that in (5.31) a_1 assumes $n = 0$ before γ or β goes to zero. Thus, as from (5.29), we see

$$e^x = {}_1F_1(\beta; \beta; x) \tag{5.35}$$

We can derive

$$e^x = \cfrac{1}{1 - \cfrac{x}{1 + \cfrac{(1 \cdot x/1 \cdot 2)}{1 - \cfrac{(1 \cdot x/2 \cdot 3)}{1 + \cfrac{(2 \cdot x/3 \cdot 4)}{1 - (2 \cdot x/4 \cdot 5) + \cdots}}}}} \qquad (5.36)$$

The explicit formula for the Padé approximants can be given very simply in this case since

$$e^x = 1/e^{-x} \qquad (5.37)$$

and we have by the uniqueness theorem 1.1

$$P_L^{(J)}(x)/Q_M^{(J)}(x) = Q_L^{(-J)}(x)/P_M^{(-J)}(x) \qquad (5.38)$$

Furthermore, since the exponential series extrapolated to negative powers of x automatically gives zero coefficients, Eq. (5.34) is valid for all $L, M \geqslant 0$; thus for e^x

$$[L/M] = {}_1F_1(-L; -L - M; x)/{}_1F_1(-M; -L - M; -x) \qquad (5.39)$$

which can be checked against Table 1.1. With this result we can easily obtain Padé's (1899) result for the limit as $L, M \to \infty$, while $M/L = \omega$. For fixed x, Eq. (5.39) goes term by term over to

$$[L/M] = \{\exp[x/(1 + \omega)]\}/\exp[-\omega x/(1 + \omega)] \qquad (5.40)$$

D. Bessel Functions

There are further interesting confluences of the hypergeometric function. In (5.29)–(5.32) let x be replaced by x/β and now let β go to infinity. Then we obtain

$$\begin{aligned} {}_0F_1(\gamma; x) &= \lim_{\beta \to \infty} {}_1F_1(\beta; \gamma; x/\beta) \\[2mm] &= 1 + \frac{x}{1 \cdot \gamma} + \frac{x^2}{1 \cdot 2\gamma(\gamma + 1)} + \frac{x^3}{1 \cdot 2 \cdot 3 \cdot \gamma(\gamma + 1)(\gamma + 2)} + \cdots \end{aligned}$$

$$(5.41)$$

From the definition of the Bessel function of the first kind we have

$$J_\nu(z) = (\tfrac{1}{2}z)^\nu \,_0F_1(1 + \nu; \, -\tfrac{1}{4}z^2)/\Gamma(\nu + 1) \qquad (5.42)$$

Thus we have obtained the result

$$\frac{J_\nu(z)}{J_{\nu-1}(z)} = \cfrac{\tfrac{1}{2}z/\nu}{1 - \cfrac{[\tfrac{1}{4}z^2/\nu(\nu + 1)]}{1 - \cfrac{[\tfrac{1}{4}z^2/(\nu + 1)(\nu + 2)]}{1 - [\tfrac{1}{4}z^2/(\nu + 2)(\nu + 3)] + \cdots}}} \qquad (5.43)$$

Since the limit in (5.43) of the a_n as n tends to infinity is zero, expansion (5.43) converges for all values of z that are not zeros of $J_{\nu-1}(z)$.

If we use the formulas for Bessel functions

$$J_{1/2}(z) = \left(\frac{2z}{\pi}\right)^{1/2} \frac{\sin z}{z}, \qquad J_{-1/2}(z) = \left(\frac{2z}{\pi}\right)^{1/2} \frac{\cos z}{z} \qquad (5.44)$$

then we obtain from (5.43) Lambert's continued fraction for $\tan z$:

$$\tan z = \cfrac{z}{1 - \cfrac{(z^2/3)}{1 - \cfrac{(z^2/15)}{1 - \cfrac{(z^2/35)}{1 - \cdots}}}} \qquad (5.45)$$

This result could also have been obtained from (5.24) by taking the limit $\phi \to 0$, with $k\phi = z$.

E. Divergent Series Derived by Confluence

We can also obtain results for a class of divergent series which are also asymptotically given by certain infinite integrals. In Eqs. (5.1), (5.8), and (5.9) let us replace x by $-\gamma x$ and let γ tend to infinity. Then we obtain

$$_2F_0(\alpha, \beta; \, -x) = \lim_{\gamma \to \infty} {}_2F_1(\alpha, \beta; \gamma; \, -\gamma x)$$

$$= 1 - \frac{\alpha\beta}{1!}x + \frac{\alpha(\alpha + 1)\beta(\beta + 1)}{2!}x^2$$

$$- \frac{\alpha(\alpha + 1)(\alpha + 2)\beta(\beta + 1)(\beta + 2)}{3!}x^3 + \cdots \qquad (5.46)$$

We can formally identify series (5.46) as

$$
{}_2F_0(\alpha, \beta; -x) = \frac{1}{\Gamma(\alpha)} \int_0^\infty \frac{t^{\alpha-1} e^{-t} \, dt}{(1 + tx)^\beta} = \frac{1}{\Gamma(\beta)} \int_0^\infty \frac{t^{\beta-1} e^{-t} \, dt}{(1 + tx)^\alpha} \quad (5.47)
$$

by expanding the integrands by the binomial theorem and integrating term by term. The continued fraction we obtain is

$$
\frac{{}_2F_0(\alpha, \beta + 1; -x)}{{}_2F_0(\alpha, \beta; -x)} = \cfrac{1}{1 + \cfrac{a_1 x}{1 + \cfrac{a_2 x}{1 + \cfrac{a_3 x}{1 + \cdot\,}}}} \quad (5.48)
$$

where

$$
a_{2n+1} = \alpha + n, \qquad a_{2n} = \beta + n \quad (5.49)
$$

Again, the special case where $\beta = 0$ is of interest, for it is just an integral rather than a ratio. In this case the Padé denominators are, from (5.12),

$$
Q_M^{(J)}(x) = q_0(L/M) \, {}_2F_0(-M, -\alpha - L; x) \quad (5.50)
$$

valid for $J = L - M \geqslant -1$. We have convergence of this continued fraction for real positive x by the parabola theorem 4.2 since by (5.49)

$$
\lim_{n \to \infty} (a_n / n) = \tfrac{1}{2} \quad (5.51)
$$

so that the parameters of the theorem can be chosen, even though the original series is divergent.

Several commonly occurring functions can be related to this integral. For example,

$$
{}_2F_0(\alpha, 1; -x) = \int_0^\infty \frac{\exp(-t) \, dt}{(1 + tx)^\alpha}
$$

$$
= x^{-\alpha} [\exp(x^{-1})] \int_{x^{-1}}^\infty u^{-\alpha} \exp(-u) \, du
$$

$$
= x^{-\alpha} [\exp(x^{-1})] \int_0^{\exp(-1/x)} \frac{dw}{(-\ln w)^\alpha} \quad (5.52)
$$

For $\alpha = 1$ we have the exponential integral. For $\alpha = \frac{1}{2}$ we have the error function as

$$\int_{x^{-1}}^{\infty} u^{-1/2} \exp(-u) \, du = 2 \int_{x^{-2}}^{\infty} \exp(-v^2) \, dv \qquad (5.53)$$

by a simple change of variables.

6

RECURSION RELATIONS

A. Classification of Problems

There are several methods available to actually calculate a Padé approximant. The first and most straightforward is to simply solve the Padé equations (1.16) directly by a standard elimination method for the coefficients of the numerator and denominator polynomials. If the Padé approximant exists, it can be found in this way. A second method is to evaluate the determinantal expression (1.27) or (4.96) for the answer. Although these expressions are very useful in many ways, unfortunately they do not form a very practical way to obtain the Padé approximants except where M is less than, say four. A third method is to apply the identities derived in Chapter 3 between the Padé approximants themselves to calculate them recursively. When the Padé table is normal these methods are, for Padé approximants of any appreciable size, by far the most efficient procedures.

In reality there are two distinct problems. The first one is to calculate the value only of a certain Padé approximant at a preselected value of its argument. The second problem is to compute the Padé approximant itself as a function of its argument—that is, to compute the coefficients of the numerator and denominator polynomials.

There is one other problem for which recursive Padé approximant methods are useful. This problem is the location of the smallest poles and zeros of a meromorphic function. It was in fact to the study of this very problem by D. Bernoulli that Cauchy (1821) referred in his foundational work. It turns out to be possible to develop recursion relations for quantities which converge directly to the location of the poles and zeros of smallest moduli.

Before discussing these problems we wish to draw attention (Shanks, 1955) to the relation of the first problem to the acceleration of the convergence of a sequence. Suppose we are given a sequence $\{S_p, p = 0, 1, \ldots \}$; we can write

$$S_p(\lambda) = S_0 + \sum_{n=1}^{p} (S_n - S_{n-1})\lambda^n \qquad (6.1)$$

which has the property $S_p(1) = S_p$. One way to construct a new sequence related to the $\{S_p\}$ is to Padé-approximate the series (6.1) and to evaluate the Padé approximants at $\lambda = 1$. This problem is then equivalent to the problem of evaluating a Padé approximant at only one point.

B. The Value Problem

Wynn (1956) gave a recursive procedure called the "epsilon algorithm" to construct Shank's transformations $e_m(S_p)$ of the sequence $\{S_p\}$ and hence a recursive procedure to evaluate a Padé approximant at a point ($\lambda = 1$). Later by eliminating certain auxiliary values from the epsilon algorithm Wynn (1966) gave a recursive procedure based on the identity (3.32). This identity relates the values of the Padé approximants in a pattern as illustrated in Fig. 2.1. His procedure, which we will now describe, seems to us to be as efficient and convenient as any for the calculation, either of the values in the upper left triangular portion of the Padé table, or for a single entry.

We first note that the $[L/0]$ Padé approximants to a Taylor series $A(x) = \sum_{n=0}^{\infty} a_n x^n$ are just the partial sums

$$[L/0] = \sum_{n=0}^{L} a_n x^n = S_l \qquad (6.2)$$

or in the case of (6.1) the sequence terms themselves. The idea is to generate successively Padé approximants with successively higher denominators. We solve (3.32) for the single Padé approximant with the largest degree denominator. Thus

$$[L/M + 1] = [L/M] + \left\{ ([L + 1/M] - [L/M])^{-1} \right.$$

$$+ ([L - 1/M] - [L/M])^{-1} - ([L/M - 1] - [L/M])^{-1} \left. \right\}^{-1} \qquad (6.3)$$

with the auxiliary conditions

$$[L/-1] = \infty, \qquad [-1/M] = 0 \tag{6.4}$$

to start Eq. (6.3) properly. With N terms of the power series we can compute the portion of the Padé table shown in Fig. 6.1. The algorithm breaks down if the Padé table is not normal.

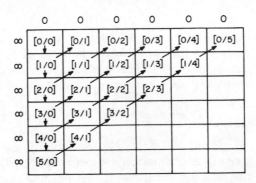

Fig. 6.1 The direction of computation of entries in the Padé table by the Wynn algorithm (6.3).

We can illustrate (6.3) by computing the [2/2] for $x = 1$ to Euler's divergent series $[{}_2F_0(1, 1; -x),$ Eq. (5.46)] which we found could be summed by Padé approximants in Chapter 5. The series is

$$A(x) = 1 - (1!)x + (2!)x^2 - (3!)x^3 + (4!)x^4 - \cdots$$

The partial sums are

$$S_0 = 1, \qquad S_1 = 0, \qquad S_2 = 2, \qquad S_3 = -4, \qquad S_4 = 20, \cdots$$

From these we compute, via (6.3) and (6.4),

$$[1/1] = \tfrac{2}{3}, \qquad [2/1] = \tfrac{1}{2}, \qquad [3/1] = \tfrac{4}{5}$$

We can also compute the $[0/1] = \tfrac{1}{2}$, but do not need to do so in order to get the [2, 2]. Finally, using (6.3) again, we get

$$[2/2] = 8/13$$

This result can be checked directly from Eqs. (5.48) and (5.49) for $\alpha = 1$ and $\beta = 0$, since it is just the convergent that stops with the contribution of a_4.

C. The Coefficient Problem

In order to calculate the coefficients of the Padé approximants, we use a recursion scheme (Baker, 1970) which follows the path in the Padé table shown in Fig. 6.2. It is based on the triangle identities (3.35) and (3.36), and is as efficient and convenient as any scheme we know of. It requires of the order of N^2 operations to derive an approximant. We introduce the sequence

$$\eta_{2j}(x)/\theta_{2j}(x) = [N - j/j], \qquad \eta_{2j+1}(x)/\theta_{2j+1}(x) = [N - j - 1/j]$$

(6.5)

Then we get the recursion relations

$$\frac{\eta_{2j}(x)}{\theta_{2j}(x)} = \frac{[\bar{\eta}_{2j-1}\eta_{2j-2}(x) - x\bar{\eta}_{2j-2}\eta_{2j-1}(x)]/\bar{\eta}_{2j-1}}{[\bar{\eta}_{2j-1}\theta_{2j-2}(x) - x\bar{\eta}_{2j-2}\theta_{2j-1}(x)]/\bar{\eta}_{2j-1}}$$

$$\frac{\eta_{2j+1}(x)}{\theta_{2j+1}(x)} = \frac{[\bar{\eta}_{2j}\eta_{2j-1}(x) - \bar{\eta}_{2j-1}\eta_{2j}(x)]/(\bar{\eta}_{2j} - \bar{\eta}_{2j-1})}{[\bar{\eta}_{2j}\theta_{2j-1}(x) - \bar{\eta}_{2j-1}\theta_{2j}(x)]/(\bar{\eta}_{2j} - \bar{\eta}_{2j-1})}$$

(6.6)

where $\bar{\eta}_j$ is the coefficient of the highest power of x in $\eta_j(x)$ (i.e., $x^{n-[(j+1)/2]}$). The starting values for these recursion relations are given by

$$\eta_0(x) = \sum_{k=0}^{n} a_k x^k, \qquad \theta_0(x) = 1.0$$

$$\eta_1(x) = \sum_{k=0}^{n-1} a_k x^k, \qquad \theta_1(x) = 1.0 \qquad (6.7)$$

The inclusion of the factors of $\bar{\eta}_{2j-1}$ and $\bar{\eta}_{2j} - \bar{\eta}_{2j-1}$ is to maintain the normalization condition $\theta_j(0) = 1$. We could have maintained the determinant normalization (3.34) by using $\bar{\theta}_{2j-1}$ and $\bar{\theta}_{2j}$ as divisors, respectively, in the first and second equations of (6.6). By $\bar{\theta}_j$ we denote the coefficient of the highest power of x in $\theta_j(x)$. Omitting these factors altogether, as in the original presentation of these relations, simply leads to a different normalization.

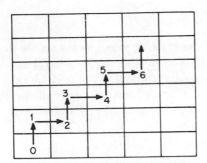

Fig. 6.2 The direction of calculation in Baker algorithm (6.6).

We will illustrate recursion (6.6) with Euler's divergent series and compute the [2/2] approximant. The starting values for $n = 4$ are

$$\eta_0(x) = 1 - (1!)x + (2!)x^2 - (3!)x^3 + (4!)x^4,$$

$$\theta_0(x) = 1.0, \quad \bar{\eta}_0 = 24$$

$$\eta_1(x) = 1 - x + 2x^2 - 6x^3, \qquad \theta_1(x) = 1.0, \quad \bar{\eta}_1 = -6$$

We will aim to compute $\eta_4(x)/\theta_4(x)$. First we obtain

$$\frac{\eta_2(x)}{\theta_2(x)} = \frac{[\bar{\eta}_1\eta_0(x) - x\bar{\eta}_0\eta_1(x)]/\bar{\eta}_1}{[\bar{\eta}_1\theta_0(x) - x\bar{\eta}_0\theta_1(x)]/\bar{\eta}_1}$$

$$= \frac{(-6 - 18x + 12x^2 - 12x^3)/(-6)}{(-6 - 24x)/(-6)}$$

$$= \frac{1 + 3x - 2x^2 + 2x^3}{1 + 4x}, \qquad \bar{\eta}_2 = 2$$

Next

$$\frac{\eta_3(x)}{\theta_3(x)} = \frac{[\bar{\eta}_2\eta_1(x) - \bar{\eta}_1\eta_2(x)]/(\bar{\eta}_2 - \bar{\eta}_1)}{[\bar{\eta}_2\theta_1(x) - \bar{\eta}_1\theta_2(x)]/(\bar{\eta}_2 - \bar{\eta}_1)}$$

$$= \frac{(8 + 16x - 8x^2)/8}{(8 + 24x)/8} = \frac{1 + 2x - x^2}{1 + 3x}, \qquad \bar{\eta}_3 = -1$$

Finally

$$\frac{\eta_4(x)}{\theta_4(x)} = \frac{[\bar\eta_3\eta_2(x) - x\bar\eta_2\eta_3(x)]/\bar\eta_3}{[\eta_3\theta_2(x) - x\bar\eta_2\theta_3(x)]/\bar\eta_3}$$

$$= \frac{(-1 - 5x - 2x^2)/(-1)}{(-1 - 6x - 6x^2)/(-1)} = \frac{1 + 5x + 2x^2}{1 + 6x + 6x^2} = [2/2]$$

If we evaluate the [2/2] approximant at $x = 1$, we recover the value $8/13$ quoted earlier. For large values of $N = L + M$ the method is clearly more efficient than the direct solution of the Padé equation (1.16), although the direct solution procedure is probably the best method when the Padé table is seriously non-normal. The difference in efficiency is, of course, not impressive for the [2/2] approximant.

In addition, we present Watson's algorithm (1973). It is as efficient as Baker's and more suitable for uses which require sequences of approximants that parallel the principal diagonal. We define the sequence

$$\eta_{2j}(x)/\theta_{2j}(x) = [L + j/j], \qquad \eta_{2j+1}(x)/\theta_{2j+1}(x) = [L + j + 1/j] \quad (6.8)$$

If we normalize $\theta_n(0) = 1.0$, then we can write, by (3.37) and (3.38),

$$\xi_{n+1}(x) = \xi_n(x) + g_n x \xi_{n-1}(x) \tag{6.9}$$

where

$$\xi_n(x) = F(x)\theta_n(x) + G(x)\eta_n(x) \tag{6.10}$$

with F and G arbitrary. To evaluate g_n, it is convenient to pick $F(x) = f(x)$, the function being approximated, and $G(x) = -1$. Then for x small, $\xi_n(x) = O(x^{L+n+1})$. The vanishing of the coefficients of x^ν, for $\nu = 0, \ldots, L + n$, on the right-hand side of (6.9) is automatic. The vanishing of the coefficient for $\nu = L + n + 1$ gives the following equation for g_n:

$$g_n = \left(\sum_m \theta_m^{(n)} f_{n-m+1}\right) / \sum_m \theta_m^{(n-1)} f_{n-m} \tag{6.11}$$

where the f_n are the coefficients of $f(x)$, and the $\theta_m^{(n)}$ are the coefficients of $\theta_n(x)$. The summations are over all m implied by the definitions (6.8) of $\theta_n(x)$, i.e., $m = 0, \ldots, \frac{1}{2}j$ for j even, and $m = 0, \ldots, \frac{1}{2}(j - 1)$ for j odd.

The Watson algorithm is then to start from the terms $n = 0, 1$ and generate g_1 by (6.11). Next use (6.9) to generate η_2 and θ_2. The process is then repeated for g_2, η_3, and θ_3, etc.

D. The Root Problem

We now turn to the problem of finding the poles and zeros of smallest modulus of an analytic function. The procedure is Rutishauser's (1954) quotient difference (Q-D) algorithm, of which we describe Gragg's (1972) variant. Suppose $A(z) = \Sigma a_j z^j$ is the ratio of two analytic (at least in a region) functions $B(z)/C(z)$; then it is meromorphic with poles at the zeros of $C(z)$ and zeros at the zeros of $B(z)$. In the Padé table for $A(z)$ the entries in any one column have denominators all of the same degree, but the numerators increase in degree. In the sense that successively higher and higher numbers of terms in the Taylor series are reproduced, the entries form better and better approximations. It is reasonable to suppose that the denominators will tend to a limiting polynomial which will vanish exactly at the poles of $A(z)$, that is, at the zeros of smallest modulus of $C(z)$. We will see in Part II that this conclusion usually holds and investigate the necessary conditions.

If we have a polynomial

$$R(x) = r_n x^n + r_{n-1} x^{n-1} + \cdots + r_0 = r_n \prod_{i=1}^{n} (x - x_i) \qquad (6.12)$$

we see from the factored form that

$$r_0 / r_n = (-1)^n \prod_{i=1}^{n} x_i \qquad (6.13)$$

where the x_i are the zeros of the polynomial. If we apply this argument to the $[L/M]$ Padé as $L \to \infty$ with M fixed, we expect (Hadamard, 1892; also see Chapter 11) in the notation of (3.33) and (3.34)

$$\lim_{L \to \infty} \frac{q_0(L - 1/M)}{q_M(L - 1/M)} = (-1)^M \prod_{i=1}^{M} u_i \qquad (6.14)$$

where the u_i are the roots of $C(z)$, with $|u_i| \leqslant |u_{i+1}|$, provided $|u_M| < |u_{M+1}|$. If, in addition, $|u_{M-1}| < |u_M|$, then we expect

$$\lim_{L \to \infty} \frac{q_M(L/M - 1)}{q_0(L/M - 1)} \frac{q_0(L - 1/M)}{q_M(L - 1/M)} = -u_M \qquad (6.15)$$

or, in terms of the entries in the C table,

$$\lim_{L \to \infty} \frac{C(L + 1/M - 1)C(L - 1/M)}{C(L/M - 1)C(L/M)} = u_M \qquad (6.16)$$

If we compute $D(z) = 1/A(z)$ and perform the same analysis on it as we have just done, then, if $C'(L/M)$ is the C table for the series $D(z)$, we derive the formula

$$\lim_{L \to \infty} \frac{C'(L + 1/M - 1)C'(L - 1/M)}{C'(L/M - 1)C'(L/M)} = v_M \qquad (6.17)$$

where v_i are the roots of $B(z)$, with $|v_i| \leqslant |v_{i+1}|$, and (6.17) is restricted to be valid only if $|v_{M-1}| < |v_M| < |v_{M+1}|$. By Hadamard's formula (4.100) we can re-express (6.17) in terms of the C table as

$$\lim_{M \to \infty} \frac{C(L - 1/M + 1)C(L/M - 1)}{C(L - 1/M)C(L/M)} = v_L \qquad (6.18)$$

A difference between (6.18) and (6.16) is that the size of the determinants in (6.16) remains fixed, while it increases as the limit is taken in (6.18).
In the light of (6.16) and (6.18), we define the variables

$$u(L/M) = \frac{C(L + 1/M - 1)C(L - 1/M)}{C(L/M - 1)C(L/M)},$$

$$v(L/M) = \frac{C(L - 1/M + 1)C(L/M - 1)}{C(L - 1/M)C(L/M)} \qquad (6.19)$$

so that (6.16) and (6.18) become

$$\lim_{L \to \infty} u(L/M) = u_M, \qquad \lim_{M \to \infty} v(L/M) = v_L \qquad (6.20)$$

We can derive the following identities among the $u(L/M)$ and the $v(L/M)$:

$$u(L/M)v(L/M) = u(L/M + 1)v(L + 1/M) \qquad (6.21)$$

$$u(L/M + 1) + v(L + 1/M) = u(L + 1/ M + 1) + v(L + 1/ M + 1)$$

$$(6.22)$$

These identities are called the *rhombus rules*. Equation (6.21) follows directly from the definition since

$$u(L/M)v(L/M)$$

$$= \frac{C(L+1/\quad M-1)C(L-1/M)C(L-1/\quad M+1)C(L/M-1)}{C(L/M-1)[C(L/M)]^2 C(L-1/M)}$$

$$= \frac{C(L+1/\quad M-1)C(L-1/\quad M+1)}{[C(L/M)]^2}$$

$$= \frac{C(L+1/M)C(L-1/\quad M+1)C(L/M+1)C(L+1/\quad M-1)}{C(L/M+1)[C(L/M)]^2 C(L+1/M)}$$

$$= u(L/M+1)v(L+1/M) \tag{6.23}$$

Equation (6.22) follows from the determinant identity (2.11). We write

$$u(L/M+1) + v(L+1/M)$$

$$= \frac{C(L+1/M)C(L-1/\quad M+1)}{C(L/M)C(L/M+1)} + \frac{C(L/M+1)C(L+1/\quad M-1)}{C(L/M)C(L+1/M)}$$

$$= \frac{\begin{aligned}&[C(L+1/M)]^2 C(L+1/\quad M+1)C(L-1/\quad M+1)\\&\quad + [C(L/M+1)]^2 C(L+1/\quad M-1)C(L+1/\quad M+1)\end{aligned}}{C(L/M)C(L/M+1)C(L+1/M)C(L+1/\quad M+1)}$$

$$= \frac{\begin{aligned}&[C(L+1/M)]^2\{[C(L/M+1)]^2 + C(L/M+2)C(L/M)\}\\&\quad + [C(L/M+1)]^2\{C(L+2/M)C(L/M) - [C(L+1/M)]^2\}\end{aligned}}{C(L/M)C(L/M+1)C\ (L+1/M)C(L+1/\quad M+1)}$$

$$= \frac{C(L+1/M)C(L/M+2)}{C(L/M+1)C(L+1/\quad M+1)} + \frac{C(L/M+1)C(L+2/M)}{C(L+1/M)C(L+1/\quad M+1)}$$

$$= v(L+1/\quad M+1) + u(L+1/\quad M+1) \tag{6.24}$$

which proves (6.22).

If we lay out our work in a u, v table as follows

$$
\begin{array}{ccccccc}
 & 0 & & 0 & & 0 & \\
0 & & v(1/1) & & v(1/2) & & v(1/3) \quad \cdots \\
 & u(1/1)\nearrow & & u(1/2)\nearrow & & u(1/3)\nearrow & \\
0 & \downarrow & v(2/1)\nearrow & & v(2/2) & & v(2/3)\;\nearrow \cdots \\
 & u(2/1)\nearrow & & u(2/2)\searrow\;(6.17) & & u(2/3)\nearrow & \\
0 & \downarrow & v(3/1)\nearrow & & v(3/2) & & v(3/3) \quad \cdots \\
 & u(3/1)\nearrow & & u(3/2) & & u(3/3) & \\
0 & \downarrow & v(4/1)\nearrow & & v(4/2)\searrow\;(6.18)\searrow & v(4/3) \quad \cdots \\
 & u(4/1)\nearrow & & u(4/2) & & u(4/3)\nearrow & \\
 & \vdots & \vdots & \vdots & \vdots & \vdots & \vdots
\end{array}
\qquad (6.25)
$$

then we can, using the starting values

$$
v(L/0) = 0, \qquad u(L/1) = a_{L-1}/a_L, \qquad u(0/M) = 0 \qquad (6.26)
$$

and Eq. (6.22) compute the first v column, $v(L/1)$. From the $v(L/1)$ and the $u(L/1)$ we can compute the $u(L/2)$ from (6.21). We can continue in this way, using the formulas (6.21) and (6.22) to recursively generate successive vertical columns of (6.25). Note that each calculation involves four terms arranged in a rhombus. We have shown this feature in (6.25) by arrows surrounding the appropriate equation number. If we have only a finite number of coefficients at our disposal, then we can only compute the upper left triangular portion of the table (6.25). The u's should converge as we move down vertical columns and the v's as we move across horizontal rows. If we had the values of the coefficients of the reciprocal series $D(z)$ instead of $A(z)$, we could have started with $v(1/M) = d_{M-1}/d_M$ and worked down the table instead of across, since the u, v table for the reciprocal is just the transpose of that for the series itself.

As an illustration of the use of the recursion method based on the rhombus rules, which is the quotient-difference algorithm, let us compute a few rows of the u, v table for

$$
A(x) = (1 - 3x + 2x^2)/(2 - 3x)
$$

Its reciprocal has the expansion

$$
D(x) = 1/A(x) = 2 + 3x + 5x^2 + 9x^3 + 17x^4 + 33x^5 + 65x^6 + \cdots
$$

Thus the table begins

0		0		0		0		0		
0		2/3		3/5		5/9		9/17		17/33 $\cdots$
	$-2/3$		1/15		2/45		4/153		8/561	$\cdots$
0		$-20/3$		9/10		17/18		33/34		$\cdots$

It will be observed that, as expected, the first row of v's is converging to the correct answer, $\frac{1}{2}$, and the second row to the correct answer, 1.

RELATION BETWEEN
ORTHOGONAL POLYNOMIALS AND
PADÉ APPROXIMANTS

A. Orthogonality Properties of Padé Denominators

In this chapter we ask, with Tschebycheff (1858): What are the properties of the polynomials $Q_M(x)$ which are the denominators of successive Padé approximants to a particular power series $A(x) = \Sigma a_j x^j$? We are led by this question very rapidly into the general theory of orthogonal polynomials, the Stieltjes (1889) expansion theorem, and the theory of Toeplitz forms [see Grenander and Szegö (1958)].

Our first step is to give a formal representation for the coefficients of the power series in terms of a moment problem. This step, at least in the early part of this chapter, is only a convenient device to guide our thinking. Suppose there exists a function $\varphi(u)$ such that

$$a_n = \int u^n \, d\varphi(u) \tag{7.1}$$

where Eq. (7.1) is the Stieltjes integral over some range of u. Let us next construct a family of orthogonal polynomials $\psi_m(u)$ which are orthogonal with respect to the weight function $d\varphi(u)$,

$$\int \psi_m(u)\psi_n(u) \, d\varphi(u) = 0, \qquad m \neq n \tag{7.2}$$

If these polynomials have the normalized expansions

$$\psi_m(u) = \sum_{\nu=0}^{m} \psi_{m,\nu} u^\nu, \qquad \psi_{m,m} = 1 \tag{7.3}$$

then Eq. (7.2) takes the form

$$\sum_{\nu=0}^{m} \sum_{\mu=0}^{n} \psi_{m,\nu} a_{\nu+\mu} \psi_{n,\mu} = 0, \qquad m \neq n \qquad (7.4)$$

Clearly, from (7.4), if we require $\psi_m(u)$ to be orthogonal to every $\psi_n(u)$, $n < m$, we may as well require instead that $\psi_m(u)$ be orthogonal to u^n for $n < m$. In that case we replace Eq. (7.4) by

$$\sum_{\nu=0}^{m} \psi_{m,\nu} a_{\nu+\mu} = 0, \qquad \mu = 0, 1, \ldots, m - 1 \qquad (7.5)$$

We can easily solve Eq. (7.5) for $\psi_m(u)$, by Cramer's rule. The result is

$$\psi_m(u) = \frac{\det \begin{vmatrix} a_0 & a_1 & \cdots & a_m \\ \vdots & \vdots & \ddots & \vdots \\ a_{m-1} & a_m & \cdots & a_{2m-1} \\ 1 & u & \cdots & u^m \end{vmatrix}}{\det \begin{vmatrix} a_0 & \cdots & a_{m-1} \\ \vdots & \ddots & \vdots \\ a_{m-1} & \cdots & a_{2m-2} \end{vmatrix}} \qquad (7.6)$$

If we compare (7.6) with the solution of the Padé equations (1.27), we find that

$$\psi_m(u) = u^m \mathcal{Q}_m^{(-1)}(u^{-1}) \qquad (7.7)$$

where $\mathcal{Q}_M^{(J)}(x)$ is the denominator of the $[L/M]$ Padé approximant satisfying $\mathcal{Q}_M^{(J)}(0) = 1.0$, where $J = L - M$. Clearly, the Padé approximant $\mathcal{Q}_M^{(J)}(x)$ must exist for $\psi_m(u)$ to exist, and vice versa.

Put another way, the denominators of the sequence of $[M - 1/M]$ Padé approximants for successive values of M are orthogonal to each other relative to the weight function which produces the sequence of a_j. If we repeat the argument with $u^{J+1} d\varphi(u)$ instead of $d\varphi(u)$ as the weight function, then we conclude that the denominators of the sequence of $[M + J/M]$ Padé approximants for successive values of n and $J \geqslant -1$ fixed are an orthogonal sequence also.

B. Examples from the Classical Orthogonal Polynomial Systems

A few examples of this result can easily be given if we select $d\varphi$ as an actual weight function. Suppose we pick $d\varphi(u) = du$ from -1 to $+1$; then

$$a_n = \int_{-1}^{+1} u^n \, du$$

$$A(x) = \int_{-1}^{+1} \sum u^n x^n \, du = \int_{-1}^{+1} \frac{du}{1 - ux} = \frac{1}{x} \ln\left(\frac{1 + x}{1 - x}\right)$$

$$= {}_2F_1\left(\frac{1}{2}, 1; \frac{3}{2}; x^2\right) \tag{7.8}$$

by Eq. (5.15). From the weight function we recognize at once that the solution for the denominators must be the Legendre polynomials, except for normalization (Morse and Feshbach, 1953), since they are orthogonal over the range -1 to $+1$ with uniform weight. The Frobenius identity (3.28) or (3.41) becomes the recursion relation between the Legendre polynomials. From our general formula (5.12) for the Padé denominators associated with Gauss's hypergeometric function we can write, $P_0 = 1$,

$$P_{2M+1}(x) \propto x^{2m+1} \mathcal{Q}_{2M+1}^{(-1)}(x^{-1})$$

$$= x^{2M+1} {}_2F_1\left(-M, -\tfrac{1}{2} - M; -\tfrac{1}{2} - 2M; x^{-2}\right)$$

$$P_{2M}(x) \propto x^{2M} \mathcal{Q}_{2M}^{(-1)}(x^{-1})$$

$$= x^{2M} {}_2F_1\left(-M, \tfrac{1}{2} - M; \tfrac{1}{2} - 2M; x^{-2}\right) \tag{7.9}$$

Another classical example is the Laguerre polynomials, which are the Padé denominators for Euler's divergent series

$$a_n = n! = \int_0^\infty u^n e^{-u} \, du, \quad A(x) = \int_0^\infty \frac{e^{-u} \, du}{1 - xu} = {}_2F_0(1, 1; x) \tag{7.10}$$

Again the Frobenius identities (3.28) or (3.41) are recursion relations for the Laguerre polynomials, which must be given by our general formula for Padé denominators (5.50) as

$$n! L_n(x) = x^n \mathcal{Q}_n^{(-1)}(x^{-1}) = x^n {}_2F_0(-n, -n; -x^{-1})$$

$$= n! {}_1F_1(-n; 1; x) \tag{7.11}$$

by the properties of the Gaussian hypergeometric functions.

The Hermite polynomials are orthogonal over the range $-\infty$ to $+\infty$ with the weight function $\exp(-u^2)$, and thus are the Padé denominators for (x complex here)

$$a_n = \Gamma(\tfrac{1}{2}n + \tfrac{1}{2}) = \int_{-\infty}^{\infty} u^n \exp(-u^2)\, du$$

$$A(x) = \int_{-\infty}^{\infty} \frac{\exp(-u^2)\, du}{1 - xu} = {}_2F_0\left(\frac{1}{2},\, 1;\, x^2\right)\Gamma\left(\frac{1}{2}\right) \qquad (7.12)$$

Once again the Frobenius identities give the recursion relations and our general formula (5.50) gives

$$H_{2M+1}(x) \propto x^{2M+1} \mathcal{Q}_{2M+1}^{(-1)}(x^{-1}) = x^{2M+1}\, {}_2F_0(-M,\, -\tfrac{1}{2} - M;\, -x^2)$$

$$H_{2M}(x) \propto x^{2M} \mathcal{Q}_{2M}^{(-1)}(x^{-1}) = x^{2M}\, {}_2F_0(-M,\, \tfrac{1}{2} - M;\, -x^{-2}) \qquad (7.13)$$

Finally, we can consider Fourier series in this way. Suppose we have the set $\{\cos n\theta,\, n = 0, 1, \ldots \}$. These functions are orthogonal over the range $0 \leqslant \theta \leqslant \pi$, which orthogonality integrals we can write as ($n \neq m$)

$$0 = \int_0^{\pi} \cos n\theta \cos m\theta\, d\theta = \int_{-1}^{+1} \cos n\theta(u) \cos m\theta(u) \frac{du}{(1 - u^2)^{1/2}} \qquad (7.14)$$

where we have changed variables to $\theta = \cos^{-1} u$. The multiple angle formulas (Peirce, 1910, No. 584)

$$\cos 2\theta = 2u^2 - 1, \qquad \cos 4\theta = 8u^4 - 8u^2 + 1$$

$$\cos 3\theta = 4u^3 - 3u, \qquad \cos 5\theta = 16u^5 - 20u^3 + 5u \qquad (7.15)$$

will thus be given by the Padé denominators to the series

$$a_n = \int_{-1}^{+1} u^n \frac{du}{(1 - u^2)^{1/2}}$$

$$A(x) = \int_{-1}^{+1} \frac{du}{(1 - ux)(1 - u^2)^{1/2}}$$

$$= [\Gamma(\tfrac{1}{2})]^2\, {}_2F_1(\tfrac{1}{2},\, 1;\, 1;\, x^2) \qquad (7.16)$$

The cosine recursion formula

$$\cos[(n+1)\theta] = 2 \cos \theta \cos n\theta - \cos[(n-1)\theta] \qquad (7.17)$$

becomes a consequence of the Frobenius identities. The general formula (5.12) for the Padé denominators yield

$$\cos[(2M+1)\theta] \propto x^{2M+1} \mathcal{Q}_{2M+1}^{(-1)}(x^{-1})$$

$$= x^{2M+1} {}_2F_1(-M, -\tfrac{1}{2}-M; -2M; x^{-2})$$

$$\cos(2M\theta) \propto x^{2M} \mathcal{Q}_{2M}^{(-1)}(x^{-1}) = x^{2M} {}_2F_1(-M, \tfrac{1}{2}-M; 1-2M; x^{-2})$$

$$(7.18)$$

C. Extremal Properties

It is also of interest to compute the normalization integral

$$\int \psi_m^2(u) \, d\varphi(u) = \int \sum_{\nu=0}^{m} \psi_{m,\nu} u^{\nu} \psi_m(u) \, d\varphi(u)$$

$$= \int \psi_{m,m} u^m \psi_m(u) \, d\varphi(u) = \int u^m \psi_m(u) \, d\varphi(u) \qquad (7.19)$$

by Eq. (7.5) and normalization (7.3). If we now substitute the solution (7.6) into (7.19) for $\psi_m(u)$ and use the definition (7.1) of $d\varphi(u)$, we obtain

$$\int \psi_M^2(u) \, d\varphi(u) = \frac{C(M/M+1)}{C(M-1/M)} \qquad (7.20)$$

in terms of the elements of the C-table determinants (2.1), where $C(-1/0) = 1$ for this purpose. From (7.6) and (7.20) it must be that $C(M-1/M) \neq 0$ if $\psi_M(u)$ is to exist.

We can derive Eq. (7.5) from another point of view and thereby find another property of the Padé denominators. Suppose we consider the normalizing integral and the subsidiary condition

$$N = \int \psi_m^2(u) \, d\varphi(u), \qquad \psi_{m,m} = 1 \qquad (7.21)$$

and seek to find a stationary solution such that $\delta(N) = 0$. If we differentiate (7.21) with respect to $\psi_{m,\nu}$, $\nu = 0, 1, \ldots, m-1$, we obtain exactly Eq. (7.5). Thus the polynomials $\psi_m(u)$ have the additional property that changes in their coefficients of order ϵ cause only changes of order ϵ^2

in N. If it should happen that the series corresponds to $d\varphi(u) \geqslant 0$, then $\psi_m(u)$ minimizes (7.21) subject to $\psi_{m,m} = 1$.

D. Decomposition of the Quadratic Form into a Sum of Squares

Although at first glance it would seem that one would have to re-solve for the minimizing polynomial ψ_m for each new value of m, we know from their relation to the Padé denominators (7.7) and the Frobenius identity (3.28) among the Padé denominators that they must satisfy a recursion relation. In fact, if we rewrite (3.28) in the notation of this chapter, it becomes

$$\psi_{M+1}(u) = \left\{ u + \left[\frac{C(L+1/\quad M+2)C(L/M-1)}{C(L/M)C(L+1/\quad M+1)} \right.\right.$$

$$\left.\left. - \frac{C(L+2/\quad M+1)C(L-1/M)}{C(L/M)C(L+1/\quad M+1)} \right] \right\} \psi_M(u)$$

$$- \frac{C(L+1/\quad M+1)C(L-1/\quad M-1)}{C^2(L/M)} \psi_{M-1}(u) \qquad (7.22)$$

where we need to set $L = M - 1$ for the principal sequence of orthogonal polynomials. Stieltjes (1889) gives an instructive method of relating the coefficients of the recursion relation (7.22) and thus the solution of the stationary value (7.21) to the formal decomposition of the bilinear form (7.4) into a sum of squares. To see his results, let us replace (7.4) by its infinite analog

$$Q(x, y) = \sum_{\nu, \mu} x_\nu a_{\mu+\nu} y_\mu \qquad (7.23)$$

Now to solve the stationary value problem, we had to solve (7.5). Instead of using Cramer's rule, we could have triangularized the coefficient matrix. In that procedure the first step is to transform the original equation (7.5) so that the coefficients of $\psi_{m,0}$ vanish in all equations except $\mu = 0$. This step is, of course, accomplished by subtracting appropriate multiples of the $\mu = 0$ equation from the others. The reduced equations have the same coefficients as the truncation of the new bilinear form

$$Q_1(x, y) = Q(x, y) - a_0 \left(\sum_\nu \frac{a_\nu}{a_0} x_\nu \right) \left(\sum_\mu \frac{a_\mu}{a_0} y_\mu \right) \qquad (7.24)$$

The new bilinear form is now independent of x_0 and y_0. In the solution for $m = 1$ we adjust $\psi_{1,0} = x_0 = y_0$. Referring to Eq. (7.5), we see that [truncating the sums in (7.24) at $\nu = \mu = 1$] the condition causes the sums in (7.24) to exactly vanish and the extremal value is just

$$\text{extreme}(Q) = Q_1(x_1 = y_1 = 1) = a_2 - (a_1^2/a_0) = \det\begin{vmatrix} a_0 & a_1 \\ a_1 & a_2 \end{vmatrix} / a_0$$

(7.25)

in accordance with Eq. (7.20).

By Eq. (7.24) we have reduced our problem for any m from an $(m + 1) \times (m + 1)$ quadratic form to an $m \times m$ quadratic form. If we let Eq. (7.20) keep track of the coefficient of $y_m x_m$ in the reduction, we obtain the Lagrange–Beltrami decomposition of a bilinear form

$$\sum_{\nu, \mu = 0} x_\nu a_{\mu + \nu} y_\mu = \sum_{m=0} E_M \left(\sum_{\nu = M} b_{M, \nu} x_\nu \right) \left(\sum_{\mu = M} b_{M, \mu} y_\mu \right)$$

(7.26)

where

$$b_{M, M} = 1, \qquad E_M = \frac{C(M/M + 1)}{C(M - 1/M)}, \qquad C(-1/0) = 1 \quad (7.27)$$

and the triangularized form of $(a_{\mu + \nu})$ is

$$\begin{aligned} B_{\mu, \nu} &= E_\mu b_{\mu, \nu}, & \nu \geqslant \mu \\ &= 0, & \nu < \mu \end{aligned}$$

(7.28)

From this triangularization we deduce the equation

$$b_{M-1, M-1}\psi_{M, M-1} + b_{M-1, M}\psi_{M, M} = 0$$

(7.29)

Using this equation and (7.27), we can identify the coefficients in (7.22) in terms of those of the decomposition (7.26) to give the Stieltjes expansion theorem

$$\psi_{M+1}(u) = (u + b_{M, M+1} - b_{M-1, M})\psi_M(u) - (E_M/E_{M-1})\psi_{M-1}(u)$$

(7.30)

which allows the recursive construction of the ψ's and thus of the Padé denominators from the solution of the decomposition problem.

One way to solve the decomposition problem is to re-express the fundamental bilinear form. Suppose

$$A(x) = \int_0^\infty B(xu)e^{-u}\,du = \sum a_j x^j, \qquad B(x) = \sum (a_j/j!)x^j. \quad (7.31)$$

Then we can write

$$Q(x,y) = \sum_{jk=0} x^j a_{j+k} y^k = [xA(x) - yA(y)]/(x - y) \qquad (7.32)$$

by direct expansion, or more conveniently for our present example,

$$\begin{aligned}
Q(x,y) &= \sum_{j,\,k=0} \frac{a_{j+k}}{j!k!}(j!\,k!\,x^j y^k) \\[2mm]
&= \sum_{j,\,k=0} \frac{a_{j+k}}{(j+k)!}\binom{j+k}{j}\int_0^\infty\int_0^\infty (ux)^j (vy)^k e^{-u-v}\,du\,dv \\[2mm]
&= \sum \frac{a_n}{n!}\int_0^\infty\int_0^\infty (ux + vy)^n e^{-u-v}\,du\,dv \\[2mm]
&= \int_0^\infty\int_0^\infty B(ux + vy)e^{-u-v}\,du\,dv \qquad\qquad\qquad (7.33)
\end{aligned}$$

where use has been made of the integral representation of $n!$.
   If we apply expression (7.33) to the example

$$A(x) = \int_0^\infty \operatorname{sech}^k(xu)e^{-u}\,du \qquad (7.34)$$

then we can solve the decomposition problem by use of

$$\begin{aligned}
&\operatorname{sech}^k(xu + yv) \\[2mm]
&\quad = (\cosh xu \cosh yv + \sinh xu \sinh yv)^{-k} \\[2mm]
&\quad = \operatorname{sech}^k(xu)\operatorname{sech}^k(yv)\sum_{M=0}^\infty \binom{-k}{M}\tanh^M(xu)\tanh^M(yv) \qquad (7.35)
\end{aligned}$$

by the binomial theorem. Thus using (7.33), we have

$$Q(x, y) = \int_0^\infty \int_0^\infty [\text{sech}^k(xu + yv)] e^{-u-v} \, du \, dv$$

$$= \sum_{M=0}^\infty \binom{-k}{M} \left\{ \int_0^\infty [\text{sech}^k(xu) \tanh^M(xu)] e^{-u} \, du \right\}$$

$$\times \left\{ \int_0^\infty [\text{sech}^k(yv) \tanh^M(yv)] e^{-v} \, dv \right\} \tag{7.36}$$

which yields at once

$$E_M = (-1)^M M! \left[\prod_{l=0}^{M-1} (k + l) \right] b_{M, M+1} = 0 \tag{7.37}$$

Thus the orthogonal polynomials for this case are given by (7.30) as

$$\psi_{M+1}(u) = u\psi_M(u) + M(k + M - 1)\psi_{M-1}(u),$$

$$\psi_0(u) = 1, \qquad \psi_1(u) = u. \tag{7.38}$$

which by relation (7.7) also gives the Padé denominators to (7.34).

E. The Eigenvalue Distribution for the Quadratic Form

In certain cases it is possible to deduce the limiting value of the normalization integral (7.20) directly from the properties of $d\phi(u)$, if it exists and is not just a formal device. The proof of these results will be found in Grenander and Szegö (1958) and here we will simply describe the ideas behind them. For definiteness let us suppose

$$a_n = \int_{-1}^{+1} v^n f(v) \, dv \tag{7.39}$$

where $f(v)$ is a real, nonnegative, continuous function of v. The idea is to transform this moment problem into a trigonometric moment problem which in turn is related to Toeplitz quadratic forms, for which an extensive body of results is known. To do this transformation, consider

$$\int_{-1}^{+1} \psi_m^2(v)f(v)\,dv$$

and substitute

$$v = \cos\theta = \tfrac{1}{2}(z + z^{-1}) \tag{7.40}$$

Thus we have the equivalent problem

$$\min\left\{\int_0^{+\pi} \psi_m^2(\cos\theta)f(\cos\theta)\sin\theta\,d\theta\right\}, \qquad \psi_{m,m} = 1 \tag{7.41}$$

If we express $\cos\theta$ in terms of $z = e^{i\theta}$, then

$$\psi_m^2(\cos\theta) = \left[\sum_{j=0}^{m} \psi_{m,j}\cos^j\theta\right]^2 = \left[\sum_{j=0}^{m} \psi_{m,j}(z + z^{-1})^j 2^{-j}\right]^2$$

$$= \left[\sum_{j=0}^{m} \psi_{m,j}(z^2 + 1)^j z^{m-j} 2^{-j}\right]\left[\sum_{j=0}^{m} \psi_{m,j}(z^{-2} + 1)^j z^{j-m} 2^{-j}\right]$$

$$= \left|\sum_{j=0}^{m} \psi_{m,j}(z^2 + 1)^j z^{m-j} 2^{-j}\right|^2 \tag{7.42}$$

since $z^{-1} = z^*$, since $|z| = 1$. (By z^* we denote the complex conjugate of z.) We note then that

$$\psi_m^2(\cos\theta) = \left|\sum_{j=0}^{2m} n_j z^j\right|^2 \tag{7.43}$$

where the η_j are real and satisfy

$$\eta_j = \eta_{2m-j}, \qquad \eta_0 = \eta_{2m} = 2^{-m} \tag{7.44}$$

The next step is to show that the extremum problem

$$T_m = (1/2\pi)\int_{-\pi}^{\pi} \left|\xi_0 + \xi_1 z + \cdots + \xi_{2m}z^{2m}\right|^2 f(\cos\theta)|\sin\theta|\,d\theta \tag{7.45}$$

subject only to $\xi_0 = 1$ is equivalent. We can see the equivalence by noting that as $|z| = 1$

$$\left|\xi_0 + \xi_1 z + \cdots + \xi_{2m} z^{2m}\right|^2$$

$$= (\xi_0 + \xi_1 z + \cdots + \xi_{2m} z^{2m})(\xi_0 + \xi_1 z^{-1} + \cdots + \xi_{2m} z^{-2m})$$

$$= (\xi_0 z^{2m} + \xi_1 z^{2m-1} + \cdots + \xi_{2m})(\xi_0 z^{-2m} + \xi_1 z^{1-2m} + \cdots + \xi_{2m})$$

$$= \left|\xi_0 z^{2m} + \xi_1 z^{2m-1} + \cdots + \xi_{2m}\right|^2 \tag{7.46}$$

Thus for any solution the symmetry property (7.44) must hold automatically and hence is no added restriction. Therefore, we have succeeded in transforming the original extremum problem into a trigonometric moment problem, except for a normalizing constant, $\eta_0 = 2^{-m}$.

We will now treat the problem (7.45). Temporarily, we will assume that $f(v) \geqslant m > 0$, and discuss the removal of this condition at a later stage. Before proceeding we will make the following observation about trigonometric polynomials on the unit circle $|z| = 1$. Suppose

$$\Xi_\eta(z) = \sum_{j=0}^{n} \xi_j z^j \tag{7.47}$$

vanishes for $z = z_0$, $|z_0| < 1$; then we have

$$\left|\Xi_n(z)\right| = \left|\frac{1 - z_0^* z}{z - z_0} \Xi_n(z)\right| \tag{7.48}$$

as a new polynomial which is equal to the original one, so far as problem (7.45) is concerned, and has the root at z_0 replaced by one at $(z_0^*)^{-1}$. Hence, we assume without loss of generality that the $\Xi_n(z)$ entering problem (7.45) do not vanish for $|z| < 1$.

By using the inequality between the arithmetic mean and the geometric mean [see Beckenbach and Bellman (1965)], i.e.,

$$(1/n) \sum_{j=1}^{n} \alpha_j \geqslant \left(\prod_{j=1}^{n} \alpha_j\right)^{1/n} \tag{7.49}$$

for all $\alpha_j > 0$, we can derive a lower bound for T_m. It is

$$T_m \geq \exp\left[(1/2\pi)\int_{-\pi}^{\pi} \ln|\xi_0 + \xi_1 z + \cdots + \xi_{2m} z^{2m}|^2 \, d\theta\right]$$

$$\times \exp\left\{(1/2\pi)\int_{-\pi}^{\pi} \ln[f(\cos\theta)] \, d\theta\right\}\exp\left\{(1/2\pi)\int_{-\pi}^{\pi} \ln[|\sin\theta|] \, d\theta\right\}$$

$$(7.50)$$

Let us examine

$$\frac{1}{2\pi}\int_{-\pi}^{\pi} \ln|\Xi_n(z)|^2 \, d\theta = \frac{1}{\pi}\int_{-\pi}^{\pi} \mathrm{Re}\{\ln\Xi_n(z)\} \, d\theta$$

$$= \mathrm{Re}\left\{\frac{1}{\pi i}\oint \ln\Xi_n(z) \, \frac{dz}{z}\right\} = 2\ln\xi_0 \quad (7.51)$$

by Cauchy's residue theorem, since $\Xi_n(z) \neq 0$ for $|z| < 1$. However, by our normalization $\xi_0 = 1$, so $\ln\xi_0 = 0$. Thus we have shown, using Peirce's tables (Peirce, 1910, No. 521) for the last integral in (7.50),

$$\mu_m \equiv \min T_m \geq \tfrac{1}{2}G(f(\cos\theta)) \quad (7.52)$$

where we define the geometric-mean function

$$G(q(\theta)) = \exp\left[(1/2\pi)\int_{-\pi}^{\pi} \ln[q(\theta)] \, d\theta\right] \quad (7.53)$$

The bound in (7.52) is uniform in m. Thus, since the minimum for problem (7.45) is nonincreasing as a function of m,

$$\mu_m \geq \lim_{m\to\infty} \mu_m \equiv \mu \geq \tfrac{1}{2}G(f(\cos\theta)) \quad (7.54)$$

We will now show that μ is actually equal to its lower bound. First we note that

$$G(r(\theta))G(s(\theta)) = G(r(\theta)s(\theta)), \qquad G(\lambda) = \lambda, \qquad \lambda \text{ a constant} \quad (7.55)$$

follow immediately from the properties of the logarithm. Now any $|\Xi_n(z)|^2$ can be obtained from the normalized one by multiplying by a constant.

Thus in light of property (7.55) we can write

$$T_m = \frac{1}{2\pi G(|\Xi_n(z)|^2)} \int_{-\pi}^{\pi} |\Xi_n(z)|^2 f(\cos\theta)|\sin\theta| \, d\theta \qquad (7.56)$$

without the restriction $\Xi_n(0) = 1$. Now, by Weierstrass's approximation theorem (Grenander and Szegö, 1958), we can pick a trigonometric poly-nomial to approximate uniformly within an error ϵ any continuous func-tion over the range $-\pi \leqslant \theta \leqslant \pi$. Let us select

$$\Xi_n(z) = z^n (f(\cos\theta)|\sin\alpha|)^{-1/2} + O(\epsilon) \qquad (7.57)$$

except for $|\sin\theta| < \epsilon$, where $\Xi_n(z) = z^n K + O(\epsilon)$, with K selected by continuity, by choosing n large enough to reduce the error to order ϵ. Equation (7.56) then becomes [by (7.55)]

$$T_m = G(f(\cos\theta))G(|\sin\theta|)\left[(1/2\pi)\int_{-\pi}^{\pi} d\theta\right] + O(\epsilon)$$

$$= \tfrac{1}{2}G(f(\cos\theta)) + O(\epsilon) \qquad (7.58)$$

But since $T_m \geqslant \mu_m$ and ϵ is as small as we please, we thus have shown, remembering our normalization factor in (7.44) and the value of μ_m from (7.20),

$$\lim_{M\to\infty}\left[2^{2M+1}\frac{C(M/M+1)}{C(M-1/M)}\right] = 2\pi \exp\left[\frac{1}{2\pi}\int_{-\pi}^{\pi} \ln f(\cos\theta)\, d\theta\right] \qquad (7.59)$$

where the coefficients are given by (7.39) and the elements of the C table are defined by (2.1).

The restriction, temporarily assumed, that $f(v) \geqslant m > 0$ can now be removed. Suppose only that $f(v) \geqslant 0$. First, the lower bound (7.52) on the limit μ is valid without regard to $f(v) \geqslant m > 0$, although with $f(v) \geqslant 0$ only, it may in fact be zero. The key step in obtaining an upper bound is the construction of the approximating polynomial (7.57), where use was made of the lower bound on $f(v)$ to bound the function being approxi-mated. Suppose we now choose, instead of (7.57),

$$\Xi_n(z) = z^n \min\left\{K, [f(\cos\theta)]^{-1/2}\right\} + O(\epsilon) \qquad (7.60)$$

Then Eq. (7.58) becomes

$$T_m = G(f(\cos\theta))\Big\{(1/2\pi)\int_{-\pi}^{+\pi}|\sin\theta|\,d\theta$$

$$-\int_{K^2f<1}[1 - K^2f(\cos\theta)]|\sin\theta|\,d\theta\Big\} + O(\epsilon) \qquad (7.61)$$

There are now two cases to distinguish. If the limit of the second integral as $K\to\infty$ is zero, then we have (7.59) since the corrections vanish for large enough m. On the other hand, if the limit of the second integral is not zero, then it necessarily follows that $G(f) = 0$, so that the upper bound tends to zero, and hence (7.59) is still correct in this case.

More general arguments can be given to relax the assumptions on $f(v)$ so that instead of Eq. (7.39) we could treat

$$a_n = \int_a^b v^n\,d\varphi(v) \qquad (7.62)$$

where $d\varphi \geq 0$, and a and b are finite. In this case (7.59) becomes

$$\lim_{M\to\infty}\left[\left(\frac{4}{b-a}\right)^{2M+1}\frac{C(M/M+1)}{C(M-1/M)}\right]$$

$$= 2\pi\exp\left\{\frac{1}{2\pi}\int_{-\pi}^{\pi}\ln\left[f\left(\frac{1}{2}(b-a)\cos\theta + \frac{1}{2}(a+b)\right)\right]d\theta\right\}$$

$$(7.63)$$

where now $f(v) = \varphi'(v)$, except at points of discontinuity of $\varphi(v)$.

In the case where $0 \leq a < b$ we can deduce the limit for ratios of the C-table elements for $J > -1$, as well as $J = -1$. To do this, we consider $v^J\,d\varphi(v)$ as a weight function. Then, by property (7.55) and the integral (Peirce, 1910, No. 523)

$$\frac{1}{2\pi}\int_{-\pi}^{\pi}\ln\left[\frac{1}{2}(a+b) + \frac{1}{2}(b-a)\cos\theta\right]d\theta = \ln\left[\frac{a+b+2(ab)^{1/2}}{4}\right]$$

$$(7.64)$$

we obtain

$$\lim_{M\to\infty} \left[\left(\frac{4}{b-a}\right)^{2M+1} \frac{C(M+J+1/M+1)}{C(M+J/M)} \right]$$

$$= 2\pi \left[\frac{a+b+2(ab)^{1/2}}{4} \right]^J$$

$$\times \exp\left\{ \frac{1}{2\pi} \int_{-\pi}^{\pi} \ln\left[f\left(\frac{1}{2}(b-a)\cos\theta + \frac{1}{2}(b+a)\right) \right] d\theta \right\}$$

(7.65)

where the elements of the C table [Eq. (2.1)] are computed from the a_j of (7.62), a_j of (7.62), and $f(v) = \varphi'(v)$ except at the points of discontinuity of $\varphi(v)$.

With these results, if an underlying distribution for a series is known, then the asymptotic behavior of the determinants and hence of the Padé approximants can be computed. For example, suppose we consider

$$a_n = \int_0^1 v^n \, dv = \frac{1}{n+1}, \qquad A(x) = -\frac{1}{x}\ln(1-x) = {}_2F_1(1, 1; 2; x)$$

(7.66)

Then, for large M, by (7.65),

$$\frac{C(M+J+1/M+1)}{C(M+J/M)} \approx 4^{-2M-J-1} \qquad (7.67)$$

since $\ln 1 = 0$. We can conveniently compare this result with the results of Chapter 5; Eq. (5.9) gives the limiting value of the continued fraction coefficients as $-\frac{1}{4}$ by use of Eq. (4.84), which gives the a_n in terms of ratios of C's. They yield

$$a_{2n+1} = -4^{-2n-1}4^{2n} = -\tfrac{1}{4}, \qquad a_{2n} = -4^{-2n}4^{2n-1} = -\tfrac{1}{4}$$

in exact agreement with the asymptotic results derived from the Gaussian hypergeometric function.

THE N-POINT PADÉ APPROXIMANT

A. General Padé Fitting Problem

Cauchy (1821) was the first to investigate fitting the values of a function at $n + m + 1$ points to a rational fraction $N(x)/D(x)$, where N is a polynomial of degree n and D is a polynomial of degree m. His formula is a generalization of Lagrange's famous polynomial interpolation formula and reduces to it in a straightforward way when $m = 0$. Jacobi (1846) gave a solution to this problem in terms of determinants, which he proved to be equivalent to Cauchy's solution. In addition, by taking the limit as all $n + m + 1$ points approached a single point, he obtained Eq. (1.27) for the Padé approximant to a function in terms of its Taylor series coefficients. We will pose a somewhat more general problem and give its solution.

Suppose we are given a set of n distinct points $\{z_i\}$ in the complex plane, and associated with each a representation of a function

$$\left\{ \sum_{j=0}^{m_i-1} u_{i,j}(z - z_i)^j + O[(z - z_i)^{m_i}] \right\} \tag{8.1}$$

We seek to find a rational fraction $P_L(z)/Q_M(z)$ such that it has properties (8.1) and the degrees of the numerator and denominator satisfy

$$L + M = N = \sum_{i=1}^{n} m_i - 1 \tag{8.2}$$

so that the number of independent coefficients exactly matches the number of conditions imposed. If there is only one point, then this problem is just the Padé problem. If all the $m_i = 1$, then this problem is just the Cauchy–Jacobi problem.

Before solving this problem it is convenient to give a generalized Lagrange interpolation formula. To this end, let us define the auxiliary function

$$f(z) = \prod_{i=1}^{n} (z - z_i)^{m_i} \tag{8.3}$$

Then we can verify that the polynomial

$$\varphi(z) = \sum_{i=1}^{n} f(z)[(m_i - 1)!]^{-1}\left(\frac{d}{da}\right)^{m_i-1}\left[\sum_{j=0}^{m_i-1} \frac{u_{ij}(a - z_i)^{m_i+j}}{(z - a)f(a)}\right]\Bigg|_{a=z_i} \tag{8.4}$$

has the property (8.1). Consider first the behavior of $\varphi(z)$ near $z = z_1$. Only the $i = 1$ term in the sum in (8.4) contributes because

$$f(z_1) = f^{(1)}(z_1) = \cdots = f^{(m_1-1)}(z_1) = 0$$

but the $i = 1$ term yields property (8.1) exactly near $z = z_1$. If we then consider every point in turn, we verify that (8.4) has property (8.1). From the form of (8.4) it is of degree at most N.

If we apply the general Lagrange interpolation formula (8.4) to x^p, $p \leqslant N$, then we have ($0 \leqslant p \leqslant N$)

$$z^p = \sum_{i=1}^{n} f(z)[(m_i - 1)!]^{-1}\left(\frac{d}{da}\right)^{m_i-1}\left[\frac{a^p(a - z_i)^{m_i}}{(z - a)f(a)}\right]\Bigg|_{a=z_i} \tag{8.5}$$

For the special case $z = 0$ we have

$$0 = \sum_{i=1}^{n} [(m_i - 1)!]^{-1}\left(\frac{d}{da}\right)^{m_i-1}\left[\frac{a^{p-1}(a - z_i)^{m_i}}{f(a)}\right]\Bigg|_{a=z_i}, \quad 0 < p \leqslant N$$

$$\tag{8.6}$$

The required equations to impose properties (8.1) are

$$P_L(z) - \left[\sum_{j=0}^{m_i-1} u_{i,j}(z - z_i)^j \right] Q_M(z) = O[(z - z_i)^{m_i}], \qquad i = 1, \ldots, n$$

$$Q_M(0) = 1 \tag{8.7}$$

A direct consequence of (8.7) is that

$$\left(\frac{d}{da} \right)^{m_i-1} \left\{ a^{p-1}(a - z_i)^{m_i} \frac{\left[P_L(a) - Q_M(a) \sum_{j=0}^{m_i-1} u_{ij}(a - z_i)^j \right]}{f(a)} \right\} \Bigg|_{a=z_i} = 0$$

$$\tag{8.8}$$

for any p. If we multiply Eq. (8.8) by the factor $[(m_i - 1)!]^{-1}$ and sum over i, then we derive

$$\sum_{i=1}^{n} [(m_i - 1)!]^{-1} \left(\frac{d}{da} \right)^{m_i-1}$$

$$\times \left\{ a^{p-1}(a - z_i)^{m_i} \frac{\left[P_L(a) - Q_M(a) \sum_{j=0}^{m_i-1} u_{ij}(a - z_i)^j \right]}{f(a)} \right\} \Bigg|_{a=z_i} = 0,$$

$$p = 1, \ldots, M. \tag{8.9}$$

Now by identity (8.6) the terms involving $P_L(z)$ vanish identically, and we are left with equations involving the $Q_M(z)$ coefficients alone. Let

$$Q_M(z) = \sum_{\mu=0}^{M} q_\mu z^\mu \tag{8.10}$$

Then (8.9) reduces to

$$\sum_{\mu=0}^{M} q_\mu v_{\mu+p} = 0, \qquad p = 1, \ldots, M \tag{8.11}$$

where we define

$$v_p = \sum_{i=1}^{n} [(m_i - 1)!]^{-1} \left(\frac{d}{da}\right)^{m_i-1} \left\{ \sum_{j=0}^{m_i-1} \frac{u_{ij}(a - z_i)^{m_i+j} a^{p-1}}{f(a)} \right\} \bigg|_{a=z_i} \qquad (8.12)$$

Equations (8.11) may be singular, which can imply that there are "unobtainable" values (Stoer, 1961).

B. Determinantal Solution

Equations (8.11) are of exactly the same form as the Padé equations (1.16). In fact, if we let $n = 1$, the Padé case, then (8.12) reduces ($z_i = 0$) to

$$v_p = u_{1, N+1-p} \qquad (8.13)$$

so that (8.11) becomes (1.16) exactly. In the other important special case, the Cauchy–Jacobi problem with all $m_i = 1$, (8.12) reduces to

$$v_p = \sum_{i=1}^{n} \frac{z_i^{p-1} u_{i,0}}{f'(z_i)} \qquad (8.14)$$

We can solve (8.11) as

$$Q_M(z) = \frac{\det \begin{vmatrix} v_{2M} & v_{2M-1} & \cdots & v_M \\ \vdots & \vdots & \ddots & \vdots \\ v_{M+1} & v_M & \cdots & v_1 \\ z^M & z^{M-1} & \cdots & 1 \end{vmatrix}}{\det \begin{vmatrix} v_{2M-1} & \cdots & v_M \\ \vdots & \ddots & \vdots \\ v_M & \cdots & v_1 \end{vmatrix}} \qquad (8.15)$$

In order to determine the numerator, it is convenient to use the generalized Lagrange interpolation formula (8.4). We can use Eq. (8.7) to work out the expansion parameters needed in terms of the $u_{i,j}$ of (8.1) and the q_μ of

(8.10). Thus we have

$$P_L(z) = \sum_{i=1}^{n} f(z)[(m_i - 1)!]^{-1}\left(\frac{d}{da}\right)^{m_i - 1}$$

$$\times \left\{\frac{(a - z_i)^{m_i}}{(z - a)f(a)}\left[\sum_{\mu=0}^{M} q_\mu a^\mu\right]\left[\sum_{j=0}^{m_i - 1} u_{ij}(a - z_i)^j\right]\right\}\Bigg|_{a = z_i} \qquad (8.16)$$

If we introduce the subsidiary polynomials

$$T_\mu(z) = \sum_{i=1}^{n} f(z)[(m_i - 1)!]^{-1}\left(\frac{d}{da}\right)^{m_i - 1}\left\{\frac{a^\mu(a - z_i)^{m_i}}{(z - a)f(a)}\sum_{j=0}^{m_i - 1} u_{ij}(a - z_i)^j\right\}\Bigg|_{a = z_i}$$

$$(8.17)$$

then, using (8.15) we can write

$$P_L(z)/Q_M(z) = \frac{\det\begin{vmatrix} v_{2M} & v_{2M-1} & \cdots & v_M \\ \vdots & \vdots & \ddots & \vdots \\ v_{M+1} & v_M & \cdots & v_1 \\ T_M & T_{M-1} & \cdots & T_0 \end{vmatrix}}{\det\begin{vmatrix} v_{2M} & v_{2M-1} & \cdots & v_M \\ \vdots & \vdots & \ddots & \vdots \\ v_{M+1} & v_M & \cdots & v_1 \\ z^M & z^{M-1} & \cdots & 1 \end{vmatrix}} \qquad (8.18)$$

In the Padé special case $n = 1$, $z_i = 0$, the polynomials (8.17) reduce to

$$T_\mu(z) = \sum_{j=\mu}^{N} u_{1, j-\mu} z^j \qquad (8.19)$$

which is identical to (1.27) when we remember (8.13), except that the degree of the numerator seems to be too large. However, by row manipulations in the numerator of (8.18) we can easily eliminate the extra powers of z and reduce the upper limit of (8.19) from N to L, in agreement with (1.27) for the solution of the Padé equations.

In the Cauchy–Jacobi special case, where all $m_i = 1$, the subsidiary polynomials (8.17) reduce to

$$T_\mu(z) = \sum_{i=1}^{\cdot n} \frac{z_i{}^\mu u_{i,0} f(z)}{(z - z_i) f'(z_i)} \tag{8.20}$$

Again, as in (8.19), the degree of (8.20) is N, but the linear combination formed in (8.18) reduces the degree of the numerator to L, by cancellation.

The computational part of the problem for $L = M = \frac{1}{2}N$, for example, can be simplified by recognizing that the solution of Eq. (8.11) can be obtained by the recursive method given in (6.5)–(6.7), which requires only of the order of M^2 operations. The construction of the derivatives by synthetic division and series manipulations can be performed in of the order of M^2 operations, as can the construction of the numerator from the T_μ. The computation of the T_μ takes of the order of M^2 operations in special cases (8.19) and (8.20).

C. Thiele's Reciprocal Difference Method

Just as we discussed in Chapter 6 for the Padé approximant problem, there are also particularly simple methods for computing the solution to certain of the Cauchy–Jacobi problems of fitting a rational fraction to a set of function values. One such method is Thiele's reciprocal-difference method [see, for example, Milne-Thompson (1951)], which produces a continued fraction type expansion and yields $L = M$ or $M - 1$. The procedure is as follows. We seek

$$\psi(z) = \cfrac{a_0}{1 + \cfrac{(z - z_0)a_1}{1 + \cfrac{(z - z_1)a_2}{1 + \cdot\cdot\cdot \quad \cfrac{(z - z_{p-1})a_p}{1 + (z - z_p)g_{p+1}(z)}}}} \tag{8.21}$$

Clearly, $\psi(z_0) = a_0$ since the corrections to this value vanish. If we define, and this procedure will use only the values at the points to be fitted,

$$g_0(z) = f(z), \qquad g_p(z) = \frac{g_{p-1}(z_{p-1}) - g_{p-1}(z)}{(z - z_{p-1})g_{p-1}(z)}, \qquad p \geqslant 1 \tag{8.22}$$

then we can easily find that

$$a_p = g_p(z_p) \tag{8.23}$$

by induction on form (8.21). We can then generate the function ψ_p for a fit on p points by means of the recursion formulas (4.5)

$$\psi_p = \frac{A_p}{B_p}, \qquad \frac{A_{n+1}}{B_{n+1}} = \frac{A_n + (z - z_n)a_{n+1}A_{n-1}}{B_n + (z - z_n)a_{n+1}B_{n-1}}$$

$$A_{-1} = 0, \qquad A_0 = a_0, \qquad B_{-1} = 1, \qquad B_0 = 1 \tag{8.24}$$

The ψ_{2p+1} will be of the form $P_p(z)/Q_{p+1}(z)$ and the ψ_{2p} of the form $P_p(z)/Q_p(z)$. Of the order of p^2 operations are required to produce a fit this way. Similar procedures (Wuytack, 1974) can be given along any stairs-step sequence parallel to the one described here.

In spite of the seemingly asymmetric manner in which ψ_p is constructed (a particular order of the points is chosen), it does not in fact depend on the order, since it is equivalent to the manifestly symmetric solution (8.14), (8.18), and (8.20).

D. An Orthogonality Property

Basdevant *et al.* (1969) have pointed out that the denominators in the Cauchy–Jacobi problem [or, more generally, for our problem (8.1)] have a kind of orthogonality similar to that discussed in Chapter 7 for the Padé denominators. To see this result, let us combine Eq. (8.9) by use of the coefficients of $Q_{M'}(z)$, $M' < M$. Then in a fashion similar to the reduction of (8.9) to (8.11) we have

$$\sum_{i=1}^{n} [(m_i - 1)!]^{-1} \left(\frac{d}{da}\right)^{m_i - 1} \left\{ \frac{Q_{M'}(a)Q_M(a) \displaystyle\sum_{j=0}^{m_i - 1} u_{ij}(a - z_i)^{m_i + j}}{f(a)} \right\} \Bigg|_{a = z_i}$$

$$= \sum_{\mu=0}^{M} \sum_{\nu=0}^{M'} q_\mu v_{\mu+\nu+1} q_\nu' = 0 \tag{8.25}$$

The same result would, of course, hold for any polynomial of degree less than M. The difference between this orthogonality property and that discussed in Chapter 7 is that here the weighting sequence $a_q = v_{2M+1-q}$ in general changes with M, whereas in Chapter 7 it does not.

E. The Nth Root Fitting Problem

In one important family of fitting problems we recover exactly the Padé denominators. Suppose we select as the points at which to fit, the roots of the equation

$$x^{N+1} = R^{N+1} \tag{8.26}$$

These roots are, of course, $x = R \exp[2\pi i j/(N + 1)], j = 0, \ldots, N$. If we fit the function

$$u(z) = \sum_{k=0}^{N} \alpha_k z^k \tag{8.27}$$

then, by (8.14), the v_p are

$$v_p = \sum_{j=0}^{N} \sum_{k=0}^{N} \alpha_k z_j^{p+k-1} / \left[(N + 1) z_j^N \right]$$

$$= \left[(N + 1) R^{N+1} \right]^{-1} \sum_{k=0}^{N} \alpha_k \sum_{j=0}^{N} z_j^{p+k} \tag{8.28}$$

But, by identity (8.6) specialized to this case

$$\sum_{j=0}^{N} z_j^l = 0, \qquad 0 < l \leqslant N \tag{8.29}$$

Thus, using (8.29) and (8.26) to reduce those cases where $p + k > N$, we have

$$v_p = \alpha_{N+1-p}, \qquad 1 \leqslant p \leqslant N + 1 \tag{8.30}$$

Reference to (8.13) and (8.15) shows that the solution to this problem gives us exactly the Padé denominator.

If we use (8.27), the Lagrange interpolation formula (8.4), and Eq. (8.26), we can evaluate the subsidiary polynomials (8.20) as

$$T_\mu(z) = \sum_{j=0}^{N-\mu} \alpha_j z^{j+\mu} + \sum_{j=N+1-\mu}^{N} \alpha_j R^{N+1} z^{j+\mu-N-1} \tag{8.31}$$

or from the fact that we have cancellation in formula (8.18) of powers of z higher than L, we need only consider the truncated polynomials

$$T_\mu(z) = \sum_{j=0}^{\min(L, N-\mu)} \alpha_j z^{j+\mu} + R^{N+1} \sum_{j=N+1-\mu}^{\min(L+N+1-\mu, N)} \alpha_j z^{j+\mu-N-1} \tag{8.32}$$

The first term in (8.32) is exactly the quantity which appears in the Padé approximant (1.27). The remainder of (8.32) arises from the difference between the N-point fitting problem and the N-derivative fitting problem. We observe that if we let $R \to 0$, that is, we fit on a circle which shrinks to the origin, the solutions become identical.

F. Acceleration of the Convergence of a Sequence

As an application of the Cauchy–Jacobi problem we will consider the following example (Basdevant, 1968). If instead of the scheme described in Chapter 6 by Eq. (6.1) et seq. to accelerate the convergence of a sequence we consider a given sequence $\{S_p, p = 0, 1, \ldots \}$ as a function of $(1/p)$, (Bulirsch and Stoer, 1964) as, for instance,

$$S_p = F(1/p) \tag{8.33}$$

then the limit of the sequence is

$$S_\infty = F(0)$$

and the idea is to compute $F(0)$ by extrapolating $F(1)$, $F(\frac{1}{2})$, $F(\frac{1}{3})$, etc. from a rational fraction fit to the S_p values.

Consider the sequence

$$S_p = \sum_{n=1}^{p} (1/n^2) \tag{8.34}$$

The limit of the S_p is well known to be $\pi^2/6 = 1.645 \ldots$. The first few terms are

$$S_1 = 1, \qquad S_2 = 1.25, \qquad S_3 = 1.36111, \ldots$$

If we use Thiele's reciprocal difference method [Eqs. (8.21)–(8.24)] we obtain by (8.22)

$$g_0(1) = 1.0, \qquad g_0(\tfrac{1}{2}) = 1.25, \qquad g_0(\tfrac{1}{3}) = 1.36111$$

$$g_1(\tfrac{1}{2}) = 2/5, \qquad g_1(\tfrac{1}{3}) = 39/98, \qquad g_2(\tfrac{1}{3}) = -2/65$$

or, using (8.23),

$$a_0 = 1, \qquad a_1 = 2/5, \qquad a_2 = -2/65$$

Finally, using (8.24) evaluated at $z = 0$, we have

$$\psi_0 = 1, \qquad \psi_1 = \frac{A_1}{B_1} = \frac{1}{1 - (2/5)} \cong 1.6667$$

$$\psi_2 = \frac{A_2}{B_2} = \frac{1 + (1/65)}{(3/5) + (1/65)} = \frac{33}{20} = 1.650$$

The convergence has been quite dramatically improved. The result ψ_2 depends only on the three values S_1, S_2, and S_3.

INVARIANCE PROPERTIES

A. Argument Transformations

In this chapter we pick up the thread begun in Chapter 1. We shall study groups of transformations that leave the process of Padé approximation invariant. Much of the ability of Padé approximant procedures to greatly extend the class of functions treatable directly from their Taylor series is rooted in their invariance properties.

The properties we will discuss will be characteristic of the principal diagonal sequence $[M/M]$ of Padé approximants. First we point out that any diagonal sequence $[M + J/M]$, with J fixed, is equivalent to a principal diagonal sequence of a closely related function. For, denoting the Padé approximant to $f(x)$ as

$$[L/M]_f, \tag{9.1}$$

then if $J < 0$, we have directly from Eq. (1.27)

$$[M + J/M]_f = [M/M]_g \tag{9.2}$$

where $g(x) = x^{-J}f(x)$. Likewise, if $J > 0$ and we define

$$g(x) = \left[f(x) - \sum_{j=0}^{J-1} f_j x^j \right] x^{-J} \tag{9.3}$$

then, by identity (3.45), Eq. (9.2) again holds.

We now prove that Padé approximation is invariant under Euler transformations (Baker *et al.*, 1961; Edrei, 1939).

Theorem 9.1. If $P_M(x)/Q_M(x)$ is the $[M/M]$ Padé approximant to $f(x)$, then $P_M(Ay/(1 + By))/Q_M(Ay/(1 + By))$ is the $[M/M]$ Padé approximant to $f(Ay/(1 + By))$.

Proof: If the numerator and denominator of $P_M(Ay/(1 + By))/Q_M(Ay/(1 + By))$ are multiplied by $(1 + By)^M$, then a general term becomes

$$[Ay/(1 + By)]^j (1 + By)^M = (Ay)^j (1 + By)^{M-j} \qquad (9.4)$$

a polynomial of degree M, and hence the fraction becomes of the form $p_M(y)/q_M(y)$. If we expand this rational fraction as a power series in y by expanding P_M/Q_M and then expanding $(1 + By)^{-j}$, we readily check that $p_M(y)/q_M(y)$ agrees with the expansion of $f(Ay/(1 + By))$ at least through terms in y^{2M}. Hence, by the uniqueness theorem 1.1, $p_M(y)/q_M(y)$ must be the required Padé approximant. ■

Corollary 9.1. If $P_{M+J}(x)/Q_M(x)$ is the $[M + J/M]$ Padé approximant to $f(x)$, then $(1 + By)^J P_{M+J}(Ay/(1 + By))/Q_M(Ay/(1 + By))$ is the $[M + J/M]$ Padé approximant to $(1 + By)^J f(Ay/(1 + By))$.

Proof: This result follows directly from Theorem 9.1 and Eqs. (9.2) and (9.3) after canceling a factor of $(Ay)^{-J}$ from both sides of the equation. ■

Corollary 9.2. If

$$T(y) = Ay^n/p_n(y), \qquad g(y) = f(T(y)) \qquad (9.5)$$

where $p_n(y)$ is a polynomial of degree at most n; then the $[nM + r/nM + s]$ Padé approximants, $r + s \leqslant n - 1$, to $g(y)$ are given by $P_M(T(y))/Q_M(T(y))$, where $P_M(x)/Q_M(x)$ is the $[M/M]$ Padé approximant to $f(x)$. The $[nM + r/nM + s](r + s \geqslant n, 0 \leqslant r \leqslant n - 1, 0 \leqslant s \leqslant n - 1)$ Padé approximants do not exist $(n > 1)$ and the Padé table is non-normal.

Proof: The proof is the same as for Theorem 9.1, which is a special case of this corollary $(n = 1)$. The multiplying factor is now $[p_n(y)]^M$, and the series expansion is good through order $y^{n(2M+1)-1}$. By application of the Padé block theorem (Theorem 2.3), the remainder of the theorem follows from the number of terms to which the Taylor series expansions agree. ■

B. Value Transformations

The theorems just proved relate to the transformation of the argument of the function. Since by Theorem 9.1 any point other than the origin can be mapped into, say $x = 1$, all points $x \neq 0$ are equivalent from the point of view of Padé approximation. Consequently, one expects the domain of convergence to be an intrinsic function of the analytic structure of the function being approximated, rather than artificially restricted to a circle centered at the origin, as is the case with Taylor series. An example of these theorems has been discussed in Chapter 1 [Eqs. (1.2)–(1.10)]. We will next discuss transformations of the value of the function. First we give the duality theorem which relates Padé approximants to a series and its reciprocal.

Theorem 9.2 (duality). If $P_L(x)/Q_M(x)$ is the $[L/M]$ Padé approximant to $f(x)$, then $Q_M(x)/P_L(x)$ is the $[M/L]$ Padé approximant to $1/f(x)$, provided $f_0 \neq 0$.

Proof: The simplest proof is to expand $1/f(x)$ and $Q_M(x)/P_L(x)$ in a Taylor series. Since by hypothesis and Taylor's theorem, they must agree with error $O(x^{M+L+1})$, the conclusion follows by the uniqueness theorem 1.1. This theorem also follows directly from the alternate bigradient representations (4.96) by use of the results for $f(x)/1$ and $1/[1/f(x)]$. ∎

The effect of this theorem is that the Padé table for $1/f(x)$ is the transpose of the table for $f(x)$ with the individual elements reciprocated. Next we see that the diagonal Padé approximants are invariant under linear fractional transformations of the function being approximated (Baker, 1965).

Theorem 9.3. If $P_M(x)/Q_M(x)$ is the $[M/M]$ Padé approximant to $f(x)$ and $C + Df(0) \neq 0$, then

$$\frac{A + B[P_M(x)/Q_M(x)]}{C + D[P_M(x)/Q_M(x)]} \tag{9.6}$$

is the $[M/M]$ Padé approximant to $\{A + Bf(x)\}/\{C + Df(x)\}$.

Proof: If the numerator and denominator of (9.6) are multiplied by $Q_M(x)$, then it becomes of the form $p_M(x)/q_M(x)$ as long as $q_M(0) = C + Df(0) \neq 0$. Since the series expansion of (9.6) is correct by the Padé equation (1.13) to $O(x^{2M+1})$, we have the conclusion of the theorem, by the uniqueness theorem 1.1. ∎

Needless to say, we may apply both Theorems 9.1 and 9.3 together. These theorems lead directly to, and are included as special cases of, the following result.

Theorem 9.4 (invariance theorem). If $P_M(x)/Q_M(x)$ is the $[M/M]$ Padé approximant to $f(x)$, and $C + Df(0) \neq 0$, then

$$
\frac{A + B\left[P_M\left(\dfrac{\alpha y}{1 + \beta y} \right) \Big/ Q_M\left(\dfrac{\alpha y}{1 + \beta y} \right) \right]}{C + D\left[P_M\left(\dfrac{\alpha y}{1 + \beta y} \right) \Big/ Q_M\left(\dfrac{\alpha y}{1 + \beta y} \right) \right]}
$$

is the $[M/M]$ Padé approximant to

$$
\{A + Bf[\alpha y/(1 + \beta y)]\}/\{C + Df[\alpha y/(1 + \beta y)]\}
$$

C. The Riemann Sphere

Since Padé approximation is invariant under the linear fractional group of transformations, it will be helpful to understand something about the structure of this group in order to preserve this symmetry of Padé approximation insofar as we are able (Baker, 1973). First, let us confine our attention to one-to-one transformations. If the pair of numbers (A, B) is simply proportional to the pair (C, D), then the result of the transformation

$$
w = T(z) = (Bz + A)/(Dz + C) \tag{9.7}
$$

is just $w = \mathrm{const}$, which is an uninteresting case. Thus we impose the condition $BC - AD = 1$. Any nonconstant, that is to say, nonsingular, transformation can be reduced to this case by multiplying A, B, C, and D all by the same factor, which does not change w.

It is well known, and can be easily checked by direct substitution, that the law of composition of two successive transformations is given by

$$
w = T_2(T_1(z)) = T_3(z), \quad \begin{pmatrix} B_3 & A_3 \\ D_3 & C_3 \end{pmatrix} = \begin{pmatrix} B_2 & A_2 \\ D_2 & C_2 \end{pmatrix}\begin{pmatrix} B_1 & A_1 \\ D_1 & C_1 \end{pmatrix}
$$

$$
\tag{9.8}
$$

where matrix multiplication is implied. Thus the group of nonsingular, linear, fractional transformations is isomorphic to the group of 2×2 matrices with unit determinant. Now, by the polar factorization theorem, we can factor any complex matrix $\mathbf{T}$

$$\mathbf{T} = \mathbf{UH}$$

where $\mathbf{U}$ is unitary, and $\mathbf{H}$ is Hermitian. That is if $\mathbf{A}^\dagger$ is the complex conjugate transpose of $\mathbf{A}$, then $\mathbf{U}^\dagger\mathbf{U} = \mathbf{I}$, and $\mathbf{H} = \mathbf{H}^\dagger$, with $\mathbf{I}$ the identity matrix. The factors $\mathbf{U}$ and $\mathbf{H}$ are uniquely determined by the requirement that all the eigenvalues (necessarily real) of $\mathbf{H}$ be greater than or equal to zero. Now any Hermitian matrix can be factored as

$$\mathbf{H} = \mathbf{V}^\dagger\mathbf{DV}$$

where $\mathbf{V}$ is unitary and $\mathbf{D}$ is a diagonal matrix. Thus we can write our transformation in the form

$$\mathbf{T} = \mathbf{UV}^\dagger\mathbf{DV} = \mathbf{U}_1\mathbf{DU}_2 \tag{9.9}$$

where $\mathbf{U}_1$ and $\mathbf{U}_2$ are unitary, since the product of two unitary matrices is again unitary.

In the diagonal matrix case the transformation (9.7) reduces to

$$w = (B/C)z \tag{9.10}$$

a simple change of scale. If we eliminate these scale changes from our consideration, then we have only the two-dimensional unitary group left. Namely, we have only $\mathbf{T}$ of the form

$$\mathbf{T} = \begin{pmatrix} b & a \\ -a^* & b^* \end{pmatrix}, \qquad |a|^2 + |b|^2 = 1 \tag{9.11}$$

If we digress into projective geometry, we can give a helpful geometric interpretation of the group of transformations (except for scale changes) under which the $[M/M]$ Padé approximants are invariant. Let us introduce Riemann's spherical representation of the complex plane. First imagine a sphere of unit radius having the unit disk of the z plane as its equatorial plane. If we draw a line from the north pole of the sphere to any point in the complex plane, it will cut the sphere in one and only one point (see Fig. 9.1 for a cross section of the Riemann sphere). The points $\hat{A}$ and $\hat{B}$ are then stereographic projections of the points A and B. The south pole S is a projection of the origin and the point at infinity is a projection of the north pole. Likewise, any line from the north pole of the sphere must, since the line is not parallel to the equatorial plane, intersect the equatorial or z plane in a single point. Thus the stereographic projection is a one-to-one mapping of the complex z plane and the point at infinity onto the Riemann sphere. If we introduce the three Cartesian coordinates (ξ, η, ζ) for a point on the sphere $(\xi^2 + \eta^2 + \zeta^2 = 1)$, then we can easily compute by trigonometry that

$$z = \frac{\xi + i\eta}{1 - \zeta}; \qquad \xi + i\eta = \frac{2z}{1 + |z|^2}, \qquad \zeta = \frac{|z|^2 - 1}{|z|^2 + 1} \qquad (9.12)$$

The opposite end of a diameter through (ξ, η, ζ) is at $(-\xi, -\eta, -\zeta)$, which corresponds to

$$z' = \frac{-\xi - i\eta}{1 + \zeta} = -\frac{1 - \zeta}{\xi - i\eta} = -\frac{1}{z^*} \qquad (9.13)$$

since $\xi^2 + \eta^2 = 1 - \zeta^2$. Thus the projections of opposite ends of a diameter on the Riemann sphere satisfy $z'z^* = -1$. Likewise one easily finds that the projections of reflection points (z, z'') in the equatorial plane satisfy $z''z^* = 1$.

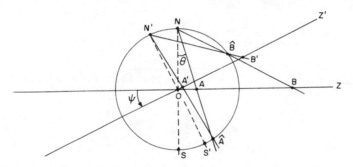

Fig. 9.1 Cross section of the Riemann sphere; the points $\hat{A}$ and $\hat{B}$ are the projections on the sphere of A and B. A rotation of the sphere (z' plane) leaves O fixed and A is mapped into A', B into B'. This mapping is a linear fractional transformation of the z plane into the z' plane.

Let us now consider what transformation of the z plane is induced by rotations of the Riemann sphere. First consider the special case of rotation about the ζ axis. Clearly in spherical coordinates, if $\xi = \cos \phi$ and $\eta = \sin \phi$, then a rotation ω gives $\xi' = \cos (\phi - \omega)$, $\eta' = \sin (\phi - \omega)$, $\zeta' = \zeta$. Referring to (9.12), we see $z' = e^{-i\omega}z$, or in the form (9.7),

$$z' = \frac{e^{-i\omega/2}z + 0}{0 \cdot z + e^{i\omega/2}} \qquad (9.14)$$

which is a unitary transformation

$$\mathbf{T}_\omega = \begin{pmatrix} e^{-i\omega/2} & 0 \\ 0 & e^{i\omega/2} \end{pmatrix} \tag{9.15}$$

Second let us consider the special case of rotation about the η axis. In Fig. 9.1 we have illustrated this rotation by an angle ψ. The points $\hat{A}$ and $\hat{B}$ are projected from a new north pole N' to new points A' and B' in the z' plane. The transformation is $A \rightarrow A'$ and $B \rightarrow B'$. To compute A' in terms of A and ψ, we note that the length $\overline{OA}$ is just $\tan \theta$. The corresponding angle for A' is, of course, $\theta - \psi$, and so the length $\overline{OA'}$ is $\tan(\theta - \psi)$. By the addition formula for $\tan \theta$ (Peirce, 1910, No. 593)

$$\overline{OA'} = \tan(\theta - \psi) = \frac{\tan \theta - \tan \psi}{\tan \psi \tan \theta + 1} = \frac{\overline{OA} - \tan \psi}{(\tan \psi)\,\overline{OA} + 1} \tag{9.16}$$

If $\eta \neq 0$, as in this example, one can easily verify that the general formula becomes

$$z' = \frac{z - \tan \psi}{(\tan \psi)z + 1} = \frac{(\cos \psi)z - \sin \psi}{(\sin \psi)z + \cos \psi} \tag{9.17}$$

as the direct complex extension of (9.16). Thus the transformation becomes, in form (9.7),

$$\mathbf{T}_\psi = \begin{pmatrix} \cos \psi & -\sin \psi \\ \sin \psi & \cos \psi \end{pmatrix} \tag{9.18}$$

which is again of the unitary form (9.11). But any rotation, say of angle ω_0 about an axis (θ_0, ϕ_0) in spherical coordinate notation, can be decomposed as a rotation of $-\phi_0$ about the ζ axis times a rotation of $-\theta_0$ about the new η axis, a rotation of ω_0 about the new ζ axis, a rotation of θ_0 about the new η axis, and finally a rotation of ϕ_0 about the new ζ axis. Since the product of unitary matrices is again unitary and we have any rotation decomposed as

$$\mathbf{T} = \begin{pmatrix} e^{i\phi_0/2} & 0 \\ 0 & e^{-i\phi_0/2} \end{pmatrix} \begin{pmatrix} \cos \theta_0 & -\sin \theta_0 \\ \sin \theta_0 & \cos \theta_0 \end{pmatrix} \begin{pmatrix} e^{i\omega_0/2} & 0 \\ 0 & e^{-i\omega_0/2} \end{pmatrix}$$

$$\times \begin{pmatrix} \cos \theta_0 & \sin \theta_0 \\ -\sin \theta_0 & \cos \theta_0 \end{pmatrix} \begin{pmatrix} e^{-i\phi_0/2} & 0 \\ 0 & e^{i\phi_0/2} \end{pmatrix} \tag{9.19}$$

it is explicitly unitary. Thus any rotation of the Riemann sphere corresponds to a unitary linear fractional transformation. By multiplying out (9.19), we obtain

$$
\mathbf{T} = \begin{pmatrix} \cos \tfrac{1}{2}\omega_0 + i\cos 2\theta_0 \sin \tfrac{1}{2}\omega_0 & i(\sin 2\theta_0)\,(\sin \tfrac{1}{2}\omega_0)e^{i\phi_0} \\ i(\sin 2\theta_0)\,(\sin \tfrac{1}{2}\omega_0)e^{-i\phi_0} & \cos \tfrac{1}{2}\omega_0 - i\cos 2\theta_0 \sin \tfrac{1}{2}\omega_0 \end{pmatrix}
$$

$$(9.20)$$

which by comparison with (9.11) allows us to solve for θ_0, ϕ_0, and ω_0 for any allowed a and b. Thus we have found that rotations of the Riemann sphere are equivalent to unitary, linear fractional transformations of the complex plane.

In the next part of this book we will be concerned with convergence. It seems most natural to introduce a distance between values of the Padé approximant which is invariant under rotations of the Riemann sphere. Such a distance could, for example, be the length of a chord between any two points projected onto the Riemann sphere. One can work out directly from (9.12) that the chord length on the Riemann sphere is

$$
D^2(z, w) = \frac{4|z - w|^2}{|1 + z^*w|^2 + |z - w|^2} = \frac{|2(z - w)|^2}{(1 + |z|^2)(1 + |w|^2)} \qquad (9.21)
$$

One can verify by direct substitution that a transformation (9.7) of the special unitary form (9.11) leaves the D of (9.21) invariant. Since from its definition as a chord length it is clear that $0 \leqslant D \leqslant 2$, we will be able to discuss convergence to the point at infinity with no more trouble, using (9.21) to define our distance, than convergence to any other point.

PART II

CONVERGENCE THEORY

10

NUMERICAL EXAMPLES

A. Regular Points

Before considering detailed proofs of the convergence properties of the Padé approximants it is instructive to consider a number of numerical examples to illustrate what we may reasonably expect to prove and what we may not. In light of the results of Chapter 9 on the invariance properties, we will confine our illustrations to the $[M/M]$ Padé approximants.

In Table 10.1 we list results for the values of Padé approximants at a regular point of the function considered.

The function $\exp[-x/(1+x)]$ is regular in the whole complex plane, except for an essential singularity $x = -1$. The increase in accuracy of the approximants at infinity is about two decimal places per unit increase in

TABLE 10.1 $[M/M]$ Evaluated at Infinity[a]

M	$\exp[-x/(1+x)]$	$[(1+2x)/(1+x)]^{1/2}$	$t(x)$	$w(x)$
1	1/3	7/5	—	−5/6
2	7/19	41/29	4/3	−0.7877462
3	71/193	239/169	5/4	−0.7896576
4	1001/2721	1393/985	—	−0.7895792
5	—	1.414213552	53/42	—
6	—	—	131/104	—
7	—	—	—	—
8	—	—	1.259978	—
Limit	0.367879441	1.414213562	1.259921	−0.7895822

[a]See Eq. (10.1) and (10.3), respectively, for $t(x)$ and $w(x)$.

M. The value given by the [4/4] approximant is off only one part in the eighth decimal place. The excellent results obtained in this case illustrate the fact that an essential singularity is, except in its neighborhood, almost identical with a pole of finite order. The poles of the Padé approximant tend, as one would expect, to cluster about the essential singularity. By the results of Chapter 9, this example is equivalent to e^{-x} at $x = 1$ and the general expression for the Padé approximant can be obtained from (5.39).

We have discussed the function $[(1 + 2x)/(1 + x)]^{1/2}$ extensively in Chapter 1. The convergence of the $[M/M]$ Padé approximants at infinity is again quite good. The error for $M = 5$ is only one part in the ninth place. All the poles of the Padé approximants lie on the ray between the branch point at $x = -\frac{1}{2}$ and the branch point at $x = -1$. Again, by the results of Chapter 9, this example is equivalent to $(1 + x)^{1/2}$ at $x = 1$ and the $[M/M]$ denominators can be deduced from (5.16) and (5.12).

The third example is

$$t(x) = \left\{ [(1 + x + x^2)(1 + 2x)]^{1/3} - 1 \right\}/x \tag{10.1}$$

and is interesting because the equations for the [1/1], [4/4],...,[3M + 1/3M + 1] Padé approximants are inconsistent and so those approximants do not exist. The function $t(x)$ has three branch points on the line $\mathrm{Re}(x) = -\frac{1}{2}$. The other Padé approximants exist and are clearly converging rapidly to $2^{1/3}$. The example proves that the entire sequence need not converge, or even exist. It is actually an example of Corollary 9.2, for if we rewrite Eq. (10.1) as

$$t(x) = 1 + \left\{ \left[1 + \left(\frac{x}{1 + x} \right)^3 \right]^{1/3} - 1 \right\} \frac{1 + x}{x} \tag{10.2}$$

then it follows that the Padé table (see Chapter 2) breaks up into 3×3 blocks with $[3l + 2/3m]$ ($0 \leqslant l, m < \infty$) their lowest-order approximants. Referring to Fig. 2.4, we see that the $[3M + 1/3M + 1]$ are inconsistent and that the other diagonal approximants do exist and are closely related to the $[M/M]$ and $[M + 1/M]$ Padé approximants to $(1 + y)^{1/3}$, which are known from Chapter 5.

The fourth example is

$$w(x) = \pi - 2 \tan^{-1}(5/12) - 2 \tan^{-1}(2.4 - 2.6x)$$

$$= 2 \sum_{n=1}^{\infty} [(\sin n\epsilon)/n](-x)^n \tag{10.3}$$

where $\sin \epsilon = 5/13$ and $\cos \epsilon = 12/13$. This function has branch points at

$$x = (12 \pm 5\sqrt{-1})/13. \tag{10.4}$$

All the poles of the Padé approximant lie on the arc of a circle through the origin and the two branch points. The convergence here is again good; the [4/4] is only off three parts in the sixth place.

B. Singular Points

In Table 10.2 we turn our attention to the convergence of Padé approximants at singular points. The functions $\operatorname{sech} x^{1/2}$ and $\tanh^2 x^{1/2}$ have essential singularities at infinity. From Table 10.2 we see that the error at infinity is decreasing by about a factor of minus five at each step. The closest poles are very accurately located by low-order Padé approximants. For instance, the [3/3] Padé approximant for $\operatorname{sech} x^{1/2}$ locates the closest pole to an accuracy of eight decimal places and the second nearest one to within one part in 200. The $\tanh^2 x^{1/2}$ has double poles. The Padé approximants simulate them by two very close single poles. For instance, the [4/4] approximant has its closest "double" pole at

$$x = -2.4674003 \pm 9.6396 \times 10^{-3}\sqrt{-1} \tag{10.5}$$

which may be compared with

$$x = -(\pi/2)^2 = -2.4674010 \tag{10.6}$$

The functions $a(x)$ and $b(x)$ in Table 10.2 are given by

$$a(x) = (1 - e^{-x})/x, \qquad b(x) = (1 - e^{-x}) \tag{10.7}$$

TABLE 10.2 [M/M] Evaluated at Infinity[a]

M	$\operatorname{sech} x^{1/2}$	$\tanh^2 x^{1/2}$	$a(x)$	$b(x)$	$(\tan^{-1}\sqrt{x})^2$	$1 - \dfrac{\tanh\sqrt{x}}{\sqrt{x}}$
1	$-1/5$	$3/2$	$-1/2$	2	1.5	$5/6$
2	$13/313$	$150/163$	$1/3$	0	1.8677	$14/15$
3	$-127/14615$	$7371/7244$	$-1/4$	2	2.0331	$27/28$
4	$17/9326$	0.99636	$1/5$	0	2.1271	$44/45$
5	-3.8254×10^{-4}	—	$-1/6$	2	2.1876	—
6	—	—	$1/7$	0	2.2299	—
7	—	—	$-1/8$	2	2.2611	—
8	—	—	$1/9$	0	2.2850	—
Limit	0	1	0	1	2.4674	1

[a] See Eq. (10.7) for $a(x)$ and $b(x)$.

From the entries in Table 10.2 it is very evident that the behavior of the Padé approximants to $a(x)$ and $b(x)$ is very different. The $[M/M]$ Padé approximants to $a(x)$ converge at the singular point $x = \infty$, although at a rate like only $1/(M + 1)$ instead of exponentially as we had for the first two examples, while the approximants to $b(\infty)$ do not converge at all but oscillate indefinitely.

The function $(\tan^{-1} x^{1/2})^2$ is regular in the complex plane cut from $x = -1$ to $-\infty$. The limit as $x \to \infty$ is $(\pi/2)^2$. The rate of convergence is similar, though not so obvious, as that for $a(x)$. The convergence of the $[M/M]$ Padé approximants is directly comparable to that of the well-known series [see Dwight (1947, No. 120), for example)

$$\left(\frac{\pi}{2}\right)^2 = \frac{3}{2}\left(1 + \frac{1}{2^2} + \frac{1}{3^2} + \frac{1}{4^2} + \cdots\right) \qquad (10.8)$$

The final example, $1 - [(\tanh x^{1/2})/x^{1/2}]$, has an essential singularity at $x = \infty$. The convergence is like $\{1 - [(2M + 1)(M + 1)]^{-1}\}$ at $x = \infty$. It would appear from these examples that the convergence at a singular point is determined by the rate and extent of the approach of the function to its limiting value. From considering $a(x)$ and $b(x)$ it would appear that the limit must be obtained by approach from a majority of the possible directions since $b(x)$ is continuous at infinity on any ray in the right half-plane save the imaginary axis, while $a(x)$ is continuous at infinity on any ray in the right half-plane, including the imaginary axis. Those sample functions that approach their limits exponentially fast show an exponential error decay, while those that approach only like a power have similarly an error decay which is a power of M. The relation between the powers is not evident since $a(x)$ approaches like $1/x$ with an error of $1/M$, while $1 - [(\tanh x^{1/2})/x^{1/2}]$ approaches like $x^{-1/2}$ with an error of $1/M^2$.

In cases where there is not a well-defined limit as x approaches the singular point we can specify which limit we intend by a sequence of points which approach the singular point along the desired path. If we use Padé approximants to obtain a convergent value at each point of this sequence, we can then estimate the desired limit. However, it would be less time consuming if we could simply pick our sequence of points so that we could compute $[M/M](x_M)$ without having to resort to a double limiting process. In Table 10.3 we illustrate this sort of procedure. The examples listed in this table have essential singularities at infinity, but no singularities in the finite plane. The $[M/M]$ Padé approximants have poles in the finite plane. We view the location of those poles as defining the neighborhood of infinity in which the Padé approximant breaks down. In Table 10.3 we list the magnitude of the smallest root (usually near the negative real axis) of the denominator. We also list the value of the Padé approximant at the magnitude of the smallest root.

TABLE 10.3 $[M/M]$ Evaluated at $|R|$

M	$\int_0^x e^{-y}\,dy$	$|R|$	$\left[\int_0^x \dfrac{\sin y}{y}\,dy\right]^2$	$|R|$	$\left[\int_0^x \left(\dfrac{\sin y}{y^3} - \dfrac{\cos y}{y^2}\right)^2 dy\right]^2$	$|R|$
1	1.0	2.0	4.5	3.0	0.0462963	2.73861
2	0.928203	3.46410	2.01456	4.50884	0.0402306	3.70084
3	1.0	4.64437	2.40654	5.53833	0.0437683	4.40956
4	0.994911	6.04653	1.80747	6.35405	0.0432710	4.99528
5	1.0	7.29348	2.50818	7.35452	0.0432724	4.95526[a]
6	0.999640	8.67206	2.31515	8.59101	0.0432865	5.66079
7	1.0	9.94356	—	—	—	—
Limit	1.0	—	2.46740	—	0.0438649	—

[a] Second smallest root.

For the function $b(x)$ the results are quite good, in contrast to those of Table 10.2. The odd approximants give exactly the correct value of the limit as $x \to +\infty$ and the even ones appear to converge rapidly. The smallest root of the denominator recedes to infinity in a manner almost linear with M.

The second function in Table 10.3 manifests a more typical behavior (we treat this function and the last one in this table as functions of x^2). Comparison with tabulated values of the sine integral (Jahnke and Emde, 1945) reveals that starting with the [3/3] approximant and going on, an accuracy of 0.2 is obtained and not improved upon. The magnitude of the smallest root in this case, as in the first example, appears to increase linearly with M. Since the radius within which the power series yields the function value to within a given accuracy also increases linearly with M in these two cases (though with a smaller coefficient for the same accuracy), it appears to be the case that we may proceed toward infinity a distance which is proportional to M.

The third function in Table 10.3 illustrates much the same behavior as the second. However, it also illustrates an important problem for convergence proofs for Padé approximants; namely, the appearance of a pole in a region which has already started to converge. The [4/4] Padé approximant very nearly fills a block (see Chapter 2) of order two in the Padé table. The smallest root of the denominator of the [5/5] is very nearly canceled by a root of the numerator.

root of numerator = 2.8852000,   root of denominator = 2.8851989

$$(10.9)$$

If we do not go as far as the first pole, but only to $x = \frac{1}{2}(M + 1)$, we find much improved convergence in these examples. In fact the [3/3] approximant gives the function value with an error of less than one part in 10^4, for at least the first two functions whose values are tabulated.

Another important problem connected with singular points arises when the location and precise nature are unknown, although they are known on general grounds to exist. Suppose we consider

$$f(x) = (1 - x/x_c)^{-\gamma}g(x) + h(x) \tag{10.10}$$

where the dominant singularity at $x = x_c$ is made manifest in (10.10); then for $x \approx x_c$ we would expect

$$\frac{d}{dx}\ln f(x) \approx \frac{-\gamma}{x - x_c} \tag{10.11}$$

We would expect that by forming Padé approximants to the logarithmic derivative we could, for this type of branch point singularity, locate it and determine its nature. We have tried these procedures on

$$f(x) = (1 - x)^{-1.5}(1 - \tfrac{1}{2}x)^{1.5} + e^{-x} \tag{10.12}$$

and summarize the results in Table 10.4. We observe from these results that, aside from occasional interruptions due to the occurrence of close poles and zeros, the procedure shows a steady convergence and that the accuracy in determining the location is noticeably higher than that in determining the nature.

TABLE 10.4 Poles and Residues of the [M/M] Approximants

M	Poles	Residues
2	0.96604	−1.25063
3	0.96526[a]	−1.24770[a]
4	1.00471	−1.60153
5	1.00170	−1.54827
6	1.00034	−1.51563
7	1.00035[a]	−1.51605[a]
8	1.00009	−1.50651
9	1.00005	−1.50443
Limit	1.00000	−1.50000

[a]Close pole and zero nearer to the origin.

C. Asymptotic Series

Let us now consider some examples of functions which are defined by asymptotic series and not by convergent series and try to evaluate them by Padé approximants at another singular point. We know already from Chapter 5, Section E that certain asymptotic series associated with confluent forms of the Gaussian hypergeometric functions have, for finite x, convergent diagonal Padé approximants which must converge at least as fast as a power of M. It will be interesting to see examples for them and other series as well.

Table 10.5 lists the values of the limit as x tends to infinity of the $[M/M]$ Padé approximants to several functions which tend to a constant on every ray except the negative real axis as x tends to infinity. The first function has the asymptotic expansion

$$\int_0^\infty \frac{e^{-t}\,dt}{1+xt} = 1 - (1!)x + (2!)x^2 - (3!)x^3 + \cdots \qquad (10.13)$$

which was first studied by Euler. This function, and the second one also, are special cases ($\alpha = 1$, $\frac{1}{2}$; $\beta = 1$) of (5.47) et seq. and so must converge for x finite. The value of the $[6/6]$ Padé approximant to the first function is 0.5968, compared to the exact value of 0.5963. The rate of convergence of the Padé approximants at $x = 1$ is essentially the same as at $x = \infty$; namely the error is proportional to $1/(M + 1)$. The limit at $x = \infty$ is approached by the first function like $(\ln x)/x$ and for the second, like $(\pi/x)^{1/2}$. In the case of the second function one can work out from Stirling's approximation to the gamma function that the error at infinity tends to zero like $M^{-1/2}$. The poles of the first two examples lie on the negative real axis, as they must. We shall see this result in Part III, since these functions are members of the class of series of Stieltjes.

TABLE 10.5 $[M/M]$ Evaluated at Infinity

M	$\displaystyle\int_0^\infty \frac{e^{-t}\,dt}{1+xt}$	$\displaystyle\pi^{-1/2}\int_0^\infty \frac{t^{-1/2}e^{-t}\,dt}{1+xt}$	$\displaystyle\int_0^\infty e^{-t}\left(\frac{1+2xt}{1+xt}\right)^{1/2}dt$	$4e^{2/x}[\mathrm{Erfc}\,(x^{-1/2})]^2$
1	1/2	2/3	6/5	1.0
2	1/3	$2\cdot4/3\cdot5$	376/297	1.39130
3	1/4	$2\cdot4\cdot6/3\cdot5\cdot7$	1.29935	1.61598
4	1/5	$2\cdot4\cdot6\cdot8/3\cdot5\cdot7\cdot9$	1.31982	1.76764
5	1/6	$2\cdot4\cdot6\cdot8\cdot10/3\cdot5\cdot7\cdot9\cdot11$	1.33377	1.87936
Limit	0	0	1.41421	3.14159

The third example is the reciprocal of a series of Stieltjes and the fourth is x over a series of Stieltjes. The third function appears to converge faster than $M^{-1/2}$ but more slowly than M^{-1}, and the fourth about like $M^{-1/2}$. For all the cases presented in Table 10.5 the coefficients of the power series ultimately diverge like $n!$ and the Padé approximants converge very slowly.

D. The Location of Cuts

A simple function like $f = (1 + x)^{-1/2}$ has the property that if we follow its value along a path starting at $(x = 0, f = 1)$ which encircles $x = -1$ and then returns to $x = 0$, we get $f = -1$. That is to say, f is not single-valued and so may not be equal to its analytic continuation. The Padé approximant, as a ratio of polynomials, must take on a single value at every point. We will now investigate the problem of Padé approximation to multivalued functions. Since

$$f = (1 + x)^{-1/2} = {}_2F_1(\tfrac{1}{2}, 1; 1; -x) \tag{10.14}$$

we have information on this example from Chapter 5, where Theorem 4.1 was used to prove that the diagonal Padé approximants converge in the whole complex plane except for $-1 \geqslant x \geqslant -\infty$. In this case we have a very reasonable result. The function f has two branch points $x = -1$ and $x = \infty$ such that if we introduce a cut between them, a single-valued function in the rest of the complex plane results. The diagonal Padé approximants then converge in just such a cut plane. Since f is discontinuous across the cut $[f(-2 + i\epsilon) \approx -i, f(-2 - i\epsilon) \approx +i]$, one wonders how the Padé approximants, which are continuous, except at poles, where they are infinite, can represent such a discontinuous function. The only way is to cluster poles along the cut. That a cut may be thought of as a line of poles can be seen from the standard integration formula (Peirce, 1910, No. 114)

$$(1 + x)^{-1/2} = \frac{1}{\pi} \int_1^\infty \frac{dz}{(x + z)(z - 1)^{1/2}} \tag{10.15}$$

We obtain an approximate representation of the integral by simply evaluating the integrand at a finite number of well-chosen points, which is nothing more, in this case, than a line of poles. Of course, a truly discontinuous function results only in the limit as the number of such poles becomes infinite.

Now as is well known, the cut, and by Cauchy's theorem the path of integration in Eq. (10.15), need not follow a ray to infinity. However, it is this particular cut that the Padé approximants have chosen.

In order to study the location of cuts implied by the Padé method, Baker *et al.* (1961) looked at, among others, the functions $t(x)$ [Eq. (10.1)], $w(x)$ [Eq. (10.3)], $u(x)$, $v(x)$, $s(x)$, $c(x)$, and $d(x)$, where

$$u(x) = (1 + x^2)^{1/2}/(1 + x)$$

$$v(x) = [(1 + x)/(1 + \tfrac{1}{4}x)]^{1/2} + \pi - 2\tan^{-1}(5/12)$$

$$- 2\tan^{-1}(2.4 - 2.6x)$$

$$s(x) = \pi - 2\tan^{-1}(2.4) - 2\tan^{-1}[(5 - 13x)/12]$$

$$+ (1 - x)^{-1/2} \tag{10.16}$$

$$c(x) = [(1 - x)/(1 - \tfrac{1}{4}x)]^{1/2} + \pi - 2\tan^{-1}(2.4)$$

$$- 2\tan^{-1}[(5 - 13x)/12]$$

$$d(x) = \left\{[(1 + x + x^2)(1 + x)]^{1/3} - 1\right\}/x$$

In Fig. 10.1 we have shown the branch points of these functions and connected by lines those branch points that may be connected by cuts to leave the function single valued in the cut plane. Earlier in this chapter we remarked that all the poles of $w(x)$ lie on the arc of a circle through the origin and the two branch points, Eq. (10.4). By using (5.15), a linear fractional transformation from Chapter 9, and the identity $\tan^{-1}x = -i \cdot \tanh^{-1}(ix)$, we can show that the results of Chapter 5 and 9 imply that the Padé approximants converge to $w(x)$ in the cut plane. The point is that a ray from a to b always maps into a circle through the origin and the transformed values of the two end points under a linear fractional transformation. Thus since the origin is a special point (we have the power series about that point), the arc of a circle (a straight line is here considered a degenerate circle) passing through the origin and any two points of the plane is a unique invariant path between them. By invariant we mean that the same prescription results whether or not we have made a linear fractional transformation. These circles project sterographically into small circles on the Riemann sphere introduced in Chapter 9.

In Table 10.6 we give some results for $u(x)$. By the arguments just given, for $u(x)$, $x = \infty$ is necessarily on a cut which runs from $x = +i$ through $x = \infty$ and back to $x = -i$. It is interesting to note that our expectations with regard to the location to which the poles converge is satisfied. That is, one pole converges to $x = -1$ and the rest to the cut. An interesting

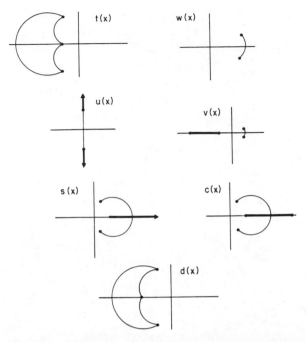

Fig. 10.1   The location of the singularities of the functions defined by Eq. (10.1), (10.3), and (10.16). Lines are drawn between those branch points that can be connected by cuts to leave a single-valued function.

behavior at $x = \infty$ (on the cut) is displayed in Table 10.6. The $[M/M]$ Padé approximants appear to converge to $\sqrt{2} \pm 1$ as M is even or odd. The sequence of $[M/M]$ Padé approximants do not converge to anything at $x = \infty$. In fact, although the sequences $[2M/2M]$ and $[2M + 1/2M + 1]$ converge separately, they do not converge to $x = \pm 1$, which are the values on each lip of the cut. This example suggests that Padé approximants do not necessarily converge on a cut.

In the cases of $v(x)$, $t(x)$, $s(x)$, $c(x)$, and $d(x)$, where there are more than two branch points, no such simple argument is known to give the exact location of the limiting cut. The numerical evidence of Baker *et al.* is somewhat inconclusive. It appears, however, that the limit set of poles of the $[M/M]$ Padé approximants still forms a simple curve. This curve seems always to lie in the interior of the largest region that can be formed by putting circular arcs through the origin and every pair of branch points in turn. In the case of $t(x)$ and $d(x)$, ϵ-shaped cuts were found, for the case of $c(x)$ a trident-shaped cut, and for $s(x)$ an infinitely long-handled, trident-shaped cut.

TABLE 10.6 $[M/M]$ Evaluated at Infinity

M	$u(x)$	$v(x)$
1	0.333333	−1.39793
2	2.333333	0.48732
3	0.411764	0.63483
4	2.411764	1.28677
5	0.414141	0.98855
6	2.414141	1.14742
7	0.414211	1.16828
8	2.414211	1.22103
Limit	±1	1.21042

In the case of $v(x)$ there is no linear fractional transformation which maps the point at infinity closer to the origin than all the singular points, as is true in the other examples. We see from Table 10.6, however, that the Padé approximants still seem to converge.

E. Noisy Series

As a final numerical example we will consider the effect of small, random variations in the coefficients. The particular example is due to Froissart (Basdevant, 1968). The Taylor series for

$$(1 - z)^{-1} = \sum_{n=0}^{\infty} z^n \qquad (10.17)$$

is exactly summed by the $[0/1]$ and $[1/1]$ Padé approximants. Suppose it is perturbed in a small, random way to

$$S(z) = \sum a_n z^n \qquad (10.18)$$

with

$$a_n = 1 + \epsilon r_n$$

where ϵ is a small number, and the r_n are independent, identically distributed random numbers satisfying

$$|r_n| \leqslant 1 \qquad (10.19)$$

Then the expected value of $S(z)$ is just $(1 - z)^{-1}$, at least for $|z| < 1$. However ($|z| < 1$)

$$E(|S(z)|^2) = |1 - z|^{-2} + \epsilon^2 \sigma^2 (1 - |z|^2)^{-1} \qquad (10.20)$$

where σ^2 is the expected value of $|r_n|^2$. From (10.20) we see that the expected value of $|S(z)|^2$ diverges as z approaches the unit circle. Thus, for almost all such functions the unit circle is a wall of poles, or a natural barrier. However, when diagonal Padé approximants are formed the following features are observed: (i) There is a pole near $z = 1$, which is very stable from order to order. (ii) There is a zero of large absolute value (roughly $1/\epsilon$). This zero simulates the fact that $(1 - z)^{-1}$ goes to zero at infinity, whereas the diagonal Padé approximants tend to a constant. (iii) Finally, all the other poles and zeros group themselves into pole–zero pairs with a separation of the order ϵ when the pair is near $|z| = 1$. The spacing decreases as the pair approaches the origin and increases as it recedes. The location of these pairs is very unstable. As an example we give one [4/4] for $\epsilon = 0.05$ in Fig. 10.2. The large zero is off the plot at $z \approx -55 + 5i$. In this case $|z_0 - 1| \sim 3 \times 10^{-3}$. In another case where $\epsilon = 10^{-3}$ a typical result is $|z_0 - 1| \sim 10^{-5}$, where z_0 is the location of the pole approximating the real pole at $z = 1$. The results are rather nice in this example. Even if we compute the Padé approximants for large $|z|$, outside the unit circle we get $(1 - z)^{-1}$ within an error of ϵ, usually. The Padé approximants appear to have damped the effect of random fluctuations in the power series coefficients rather than amplifying it.

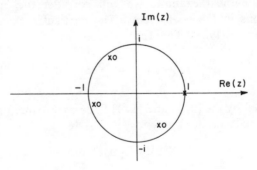

Fig. 10.2 Sample [4/4] Padé approximant to the series for $(1 - z)^{-1}$ subject to random noise. $\times$, poles; $\bigcirc$, zeros.

11

CONVERGENCE OF VERTICAL AND

HORIZONTAL SEQUENCES

A. Convergence on the Sphere

In the study of the convergence properties of the Padé approximant, it is desirable to have a definition of convergence which reflects the ability of the Padé approximant to accurately reproduce the behavior of a function near a polar singularity. For example, as we discussed in the beginning of the preceding chapter, rapidly improving estimates of the location of the nearest pole of sech x are obtained by low-order Padé approximants. With the usual definition of convergence, however, we would conclude, nonetheless, that the error made by any finite-order Padé approximant at that pole is indefinitely large because all such Padé approximants are finite there and the correct value is infinite. However, if we consider reciprocals, then the reciprocal to the Padé approximants converges uniformly in the neighborhood of the pole to the reciprocal of the function. In light of the invariance properties of Padé approximants with respect to reciprocation (Theorem 9.2) and more general transformations (Theorem 9.3), and the interpretation of these results in terms of rotations and reflections of the Riemann sphere, we introduce the following definition, based on the chord length formula (9.21).

Definition. A sequence of complex numbers w_n is said to *converge on the sphere* if for any $\epsilon > 0$ there exists an N such that

$$D^2(w_n, w_m) = \frac{4|w_n - w_m|^2}{|1 + w_n^* w_m|^2 + |w_n - w_m|^2} \leqslant \epsilon^2 \qquad (11.1)$$

for all $n, m \geqslant N$. [Ostrowski (1925) appears to have introduced the use of the chordal distance in the study of meromorphic functions.]

We remark that the definition (11.1) can easily be seen to be satisfied if the w_n are tending to infinity, since we can rewrite (11.1) as

$$D^2(w_n, w_m) = \frac{4|w_m^{-1} - w_n^{-1}|^2}{|w_n^{-1*}w_m^{-1} + 1|^2 + |w_m^{-1} - w_n^{-1}|^2}$$

(11.2)

which clearly tends to zero over unity.

B. Functions with Only Polar Singularities

We will first present in this chapter a theorem due to Montessus de Ballore (1902, 1905). His original result was stated in terms of ordinary convergence rather than convergence on the sphere as we shall do. Before going on to the theorem we will need the following two lemmas.

Lemma 11.1. If the constants a_n can be represented in the form

$$a_n = \sum_{j=1}^{s} A_j y_j^n$$

(11.3)

where the y_j are all distinct and nonzero, then if

$$\mathbf{B} = \begin{pmatrix} a_{r-s+1} & a_{r-s+2} & \cdots & a_r \\ \vdots & \vdots & \ddots & \vdots \\ a_r & a_{r+1} & \cdots & a_{r+s-1} \end{pmatrix}$$

(11.4)

we have

$$\mathbf{B}^{-1} = \begin{pmatrix} (-1)^{s-1}\sigma_{s-1}(y_1)/f'(y_1) & \cdots & (-1)^{s-1}\sigma_{s-1}(y_s)/f'(y_s) \\ (-1)^{s-2}\sigma_{s-2}(y_1)/f'(y_1) & \cdots & (-1)^{s-2}\sigma_{s-2}(y_s)/f'(y_s) \\ \vdots & \ddots & \vdots \\ 1/f'(y_1) & \cdots & 1/f'(y_s) \end{pmatrix}$$

$$\times \begin{pmatrix} A_1^{-1}y_1^{s-1-r} & 0 & \cdots & 0 \\ 0 & A_2^{-1}y_2^{s-1-r} & \cdots & 0 \\ \vdots & \vdots & \ddots & \vdots \\ 0 & 0 & \cdots & A_s^{-1}y_s^{s-1-r} \end{pmatrix}$$

$$\times \begin{pmatrix} (-1)^{s-1}\sigma_{s-1}(y_1)/f'(y_1) & (-1)^{s-2}\sigma_{s-2}(y_1)/f'(y_1) & \cdots & 1/f'(y_1) \\ \vdots & \vdots & \ddots & \vdots \\ (-1)^{s-1}\sigma_{s-1}(y_s)/f'(y_s) & (-1)^{s-2}\sigma_{s-2}(y_s)/f'(y_s) & \cdots & 1/f'(y_s) \end{pmatrix}$$

(11.5)

where

$$f(x) = \prod_{j=1}^{s} (x - y_j) \tag{11.6}$$

and the $\sigma_j(y_k)$ are the elementary symmetric polynomials on the set $\{y_l | l = 1, 2, \ldots, s; \; l \neq k\}$. That is to say, the σ_j are given by

$$\sum_{j=0}^{s-1} (-1)^j \sigma_j(y_k) x^{s-j-1} = f(x)/(x - y_k) \tag{11.7}$$

Proof: The results of this lemma depend on the following factorization of **B**:

$$\mathbf{B} = \begin{pmatrix} 1 & 1 & \cdots & 1 \\ y_1 & y_2 & \cdots & y_j \\ \vdots & \vdots & \ddots & \vdots \\ y_1^{s-1} & y_2^{s-1} & \cdots & y_j^{s-1} \end{pmatrix}$$

$$\times \begin{pmatrix} A_1 y_1^{r-s+1} & 0 & \cdots & 0 \\ 0 & A_2 y_2^{r-s+1} & \cdots & 0 \\ \vdots & \vdots & \ddots & \vdots \\ 0 & 0 & \cdots & A_s y_s^{r-s+1} \end{pmatrix}$$

$$\times \begin{pmatrix} 1 & y_1 & \cdots & y_1^{s-1} \\ 1 & y_2 & \cdots & y_2^{s-1} \\ \vdots & \vdots & \ddots & \vdots \\ 1 & y_s & \cdots & y_s^{s-1} \end{pmatrix} \tag{11.8}$$

That this factorization is correct can be readily seen by multiplication starting from right to left and comparison with definitions (11.3) and (11.4). The conclusion of the lemma, (11.5), can be verified directly by forming $\mathbf{BB}^{-1}$. We note that except on the diagonals of the product of the last matrix factor in (11.8) with the first factor in (11.5), the elements are, by (11.7), just $f(y_j)/[(y_j - y_k)f'(y_k)] = 0$, while the diagonals are just $f'(y_j)/f'(y_j) = 1$. This product is then just the identity matrix. Continuing in this way, we complete the verification of the conclusion of this lemma. ∎

Lemma 11.2. If the constants a_n can be represented in the form

$$(-1)^n a_n = \left[\sum_{j=1}^{t} \sum_{k=1}^{m_j} A_{j,k} \binom{-k}{n} \right] y_j^n \qquad (11.9)$$

where

$$\sum_{j=1}^{t} m_j = s \qquad \text{and} \qquad A_{j,m_j} \neq 0 \qquad (11.10)$$

then for **B** given by (11.4) we have

$$B^{-1} = \mathbf{C}^T \mathbf{D} \mathbf{C} \qquad (11.11)$$

where

$$
\mathbf{C} = \begin{pmatrix}
(-1)^{s-1}\sigma_{s-1}(y_1, 1) & (-1)^{s-2}\sigma_{s-2}(y_1, 1) & \cdots & \sigma_0(y_1, 1) \\
\vdots & \vdots & & \vdots \\
(-1)^{s-1}\sigma_{s-1}(y_1, m_1) & (-1)^{s-2}\sigma_{s-2}(y_1, m_1) & \cdots & \sigma_0(y_1, m_1) \\
(-1)^{s-1}\sigma_{s-1}(y_2, 1) & (-1)^{s-2}\sigma_{s-2}(y_2, 1) & \cdots & \sigma_0(y_2, 1) \\
\vdots & \vdots & \ddots & \vdots \\
(-1)^{s-1}\sigma_{s-1}(y_t, m_t) & (-1)^{s-2}\sigma_{s-2}(y_t, m_t) & \cdots & \sigma_0(y_t, m_t)
\end{pmatrix}
$$

$$(11.12)$$

We use the notation

$$g(x) = \prod_{j=1}^{t} (x - y_j)^{m_j}, \qquad \sigma_j(y_k, l) = 0, \quad j < l - 1$$

$$\sum_{j=l-1}^{s-1} (-1)^{j-l+1} \sigma_j(y_k, l) x^{s-j-1} = m_k! \frac{g(x)}{(x - y_k)^l}, \qquad j \geq l-1. \quad (11.13)$$

D is all zero except for t $m_j \times m_j$ diagonal blocks $\mathbf{D}_j$ of the form

$$
\mathbf{D}_j = \begin{pmatrix}
\dfrac{1}{(m_j-1)!}\left(\dfrac{d}{d\zeta}\right)^{m_j-1}\alpha_j(\zeta) & \dfrac{1}{(m_j-2)!}\left(\dfrac{d}{d\zeta}\right)^{m_j-2}\alpha_j(\zeta) & \cdots & \dfrac{d}{d\zeta}\alpha_j(\zeta) & \alpha_j(\zeta) \\
\dfrac{1}{(m_j-2)!}\left(\dfrac{d}{d\zeta}\right)^{m_j-2}\alpha_j(\zeta) & \dfrac{1}{(m_j-3)!}\left(\dfrac{d}{d\zeta}\right)^{m_j-3}\alpha_j(\zeta) & \cdots & \alpha_j(\zeta) & 0 \\
\vdots & \vdots & & \vdots & \vdots \\
\alpha_j(\zeta) & 0 & \cdots & 0 & 0
\end{pmatrix}
$$

$$(11.14)$$

evaluated at $\zeta = y_j$ using the notation

$$\alpha_j(\zeta) = \frac{\zeta^{s-1-r}}{\left[\Sigma_{k=1}^{m_j} A_{j,k}(\zeta - y_j)^{m_j - k}\right]\Pi'_{l \neq j}\left[(\zeta - y_l)^{m_l}(y_j - y_l)^{m_l}\right]} \tag{11.15}$$

Proof: This lemma can be derived by taking the appropriate limit of Lemma 11.1 and rearranging the results so that all the singularities cancel out. First we observe that form (11.9) is the binomial expansion of the function (in powers of z^{-1})

$$h(z) = \sum_{j=1}^{t}\left[\sum_{k=1}^{m_j}\frac{A_{j,k}}{(z - y_j)^k}\right] \tag{11.16}$$

If we use the Lagrange interpolation formula, as discussed in Chapter 8, for s distinct points with the properties

$$z_{j,k} \approx y_j, \quad k = 1, \ldots, m_j; \quad j = 1, \ldots, t \tag{11.17}$$

then rewriting $h(z)$ as

$$h(z) = \sum_{j=1}^{t}\frac{\left[\Sigma_{k=1}^{m_j} A_{j,k}(z - y_j)^{m_j - k}\right]\Pi'_{l \neq j}(z - y_l)^{m_l}}{\Pi'_{l=1}(z - y_l)^{m_l}} \tag{11.18}$$

we can re-express [by (8.4)] the numerator of (11.18) exactly and approximate the denominator as

$$h(z) \approx \frac{\phi(z)}{f(z)}$$

$$= \sum_{\nu=1}^{t}\sum_{\mu=1}^{m_\nu}\frac{\left\{\Sigma_{j=1}^{t}\left[\Sigma_{k=1}^{m_j} A_{j,k}(z_{\nu,\mu} - y_j)^{m_j - k}\right]\Pi'_{l \neq j}(z_{\nu,\mu} - y_l)^{m_l}\right\}}{(z - z_{\nu,\mu})f'(z_{\nu,\mu})} \tag{11.19}$$

Here we use

$$f(z) = \prod_{j=1}^{t}\left[\prod_{k=1}^{m_j}(z - z_{j,k})\right] \tag{11.20}$$

a polynomial of degree s. The representation (11.19) reduces exactly to form (11.18) in the limit as the approximate equality sign in (11.17) becomes an equality sign. That this result is correct follows because $\phi(z)$ in (11.19) is exactly equal for all $\{z_{\nu,\mu}\}$ to the numerator in (11.18), and $f(z)$ in (11.19) goes directly to the denominator of (11.18) by construction.

The idea of the proof of this lemma is to apply Lemma 11.1 to the coefficients of the series generated by (11.19) and take the limit as the $z_{j,k}$ approach y_j. To accomplish this end, we need to factor further the rightmost factor of (11.5). It will suffice to show how this is done for the first $m_1 + 1$ rows. We obtain Eq. (11.21),

$$
\mathbf{LM} =
\begin{pmatrix}
\dfrac{1}{f'(z_{1,1})} & 0 & 0 & \cdots & 0 & 0 \\[2ex]
\dfrac{1}{f'(z_{1,2})} & \dfrac{z_{1,2}-z_{1,1}}{f'(z_{1,2})} & 0 & \cdots & 0 & 0 \\[2ex]
\dfrac{1}{f'(z_{1,3})} & \dfrac{z_{1,3}-z_{1,1}}{f'(z_{1,3})} & \dfrac{(z_{1,3}-z_{1,1})(z_{1,3}-z_{1,2})}{f'(z_{1,3})} & \cdots & 0 & 0 \\[2ex]
\vdots & \vdots & \vdots & \ddots & \vdots & \vdots \\[2ex]
\dfrac{1}{f'(z_{1,m_1})} & \dfrac{z_{1,m_1}-z_{1,1}}{f'(z_{1,m_1})} & \dfrac{(z_{1,m_1}-z_{1,1})(z_{1,m_1}-z_{1,2})}{f'(z_1 m_1)} & \cdots & \dfrac{\Pi_{k=1}^{m_1-1}(z_{1,m_1}-z_{1,k})}{f'(z_{1,m_1})} & 0 \\[2ex]
0 & 0 & 0 & \cdots & 0 & \dfrac{1}{f'(z_2,1)}
\end{pmatrix}
$$

$$
\times
\begin{pmatrix}
(-1)^{s-1}\sigma_{s-1}(z_{1,1}) & (-1)^{s-2}\sigma_{s-2}(z_{1,1}) & \cdots & -\sigma_1(z_{1,1}) & 1 \\[1.5ex]
(-1)^{s-2}\sigma_{s-2}(z_{1,1},z_{1,2}) & (-1)^{s-2}\sigma_{s-3}(z_{1,1},z_{1,2}) & \cdots & 1 & 0 \\[1.5ex]
\vdots & \vdots & \ddots & \vdots & \vdots \\[1.5ex]
(-1)^{s-m_1}\sigma_{s-m_1}(z_{1,k},\, k=1,m_1) & (-1)^{s-m_1}\sigma_{s-m_1}(z_{1,k},\, k=1,m_1) & \cdots & 0 & 0 \\[1.5ex]
(-1)^{s-1}\sigma_{s-1}(z_{2,1}) & (-1)^{s-1}\sigma_{s-1}(z_{2,1}) & \cdots & -\sigma_1(z_{2,1}) & 1
\end{pmatrix}
$$

$$(11.21)$$

where we define for $\{w_\tau,\ \tau=1,n\}$ a subset of the $z_{j,k}$,

$$
\sum_{j=0}^{s-n}(-1)^j\sigma_j(\{w_\tau\})x^{s-n-j} = \frac{f(x)}{\Pi_{\tau=1}^n(x-w_\tau)} \tag{11.22}
$$

The first factor is a lower left triangular matrix with zeros outside the t $m_j \times m_j$ diagonal blocks. This factor becomes singular as the $z_{j,k}$ approach y_j. The second factor simply tends directly to $\mathbf{C}$ as given by (11.12), in this limit. That this factorization is correct can be verified by multiplying the factors together and repeatedly reducing the result by means of the identity

$$\sigma_u(w_a) + \sigma_{u-1}(w_a, w_b)w_a \equiv \sigma_u(w_b) + \sigma_{u-1}(w_a, w_b)w_b \qquad (11.23)$$

which follows immediately from the definitions.

By Lemma 11.1 we will thus have proven this lemma if we can show that the limit as $z_{j,k}$ approach y_j of $\mathbf{L}^T\mathbf{NL}$ is $\mathbf{D}$ where, by (11.5), (11.8), and (11.19), we have for the diagonal matrix $\mathbf{N}$ the diagonal elements

$$N(\nu, \mu) = \frac{f'(z_{\nu, \mu})z_{\nu, \mu}^{s-1-r}}{\sum_{j=1}^{t}\left[\sum_{k=1}^{m_j}A_{j,k}(z_{\nu, \mu} - y_j)^{m_j-k}\right]\prod_{l \neq j}^{t}(z_{\nu, \mu} - y_l)^{m_l}} \qquad (11.24)$$

in our customary order.

If we now construct $\mathbf{L}^T\mathbf{NL}$, we see that the factors $f'(z_{\nu, \mu})$ are removed when $\mathbf{NL}$ is multiplied, leaving a lower triangular matrix. The final multiplication by $\mathbf{L}^T$ yields formulas which are difference approximations to various derivatives of a single function. When their limit is taken as the $z_{j,k}$ approach the y_j, Eq. (11.14) and (11.15) result. The other terms in the summation in (11.24) vanish to higher order and do not contribute to (11.15). The lower right triangular part forms the same pattern, except that here integrals replace derivatives which, when evaluated over zero range, give zero. Use has also been made of the formula

$$\frac{d}{dx}(xF(x)) = x\frac{d}{dx}(F(x)) + F(x)$$

in the final derivation. ∎

We now give the theorem, as extended by Wilson (1928a), of Montessus de Ballore (1902, 1905), which was an application of the results of Hadamard (1892). We have restated it in terms of convergence on the sphere.

Theorem 11.1 (Montessus). Let $f(z)$ be regular inside the circle $|z| < R$, except for poles of total multiplicity M. Then in the limit as L goes to infinity $[L/M]$ converges uniformly to $f(z)$ on the sphere in $|z| \leqslant \rho$, for any $\rho < R$.

Proof: Let the poles of $f(z)$, $|z| < R$, be denoted by r_i

$$|r_1| \leqslant |r_2| \leqslant \cdots \leqslant |r_n| \qquad (11.25)$$

with corresponding multiplicities

$$m_1, m_2, \ldots, m_n, \qquad \sum_{i=1}^{n} m_i = M \qquad (11.26)$$

Define the polynomial

$$\tau(z) = 1 + \sum_{j=1}^{M} \tau_j z^j = \prod_{i=1}^{n} (1 - z/r_i)^{m_i} \tag{11.27}$$

By the assumptions of the theorem, the coefficients of

$$B(z) = \tau(z)f(z) = \sum_{i=1}^{\infty} b_j z^j \tag{11.28}$$

satisfy the Cauchy inequalities

$$|b_j| < K/S^j \tag{11.29}$$

for any $S < R$ and some K. By multiplying the rows of the determinants in (1.27) by the coefficients τ_j and adding to the last row, we can rewrite Eq. (1.27) as

$$[L/M] = \frac{\det \begin{vmatrix} f_{L-M+1} & \cdots & f_L & b_{L+1} \\ \vdots & \ddots & \vdots & \vdots \\ f_L & \cdots & f_{L+M-1} & b_{L+M} \\ \sum_{j=M}^{L} f_{j-M} z^j & \cdots & \sum_{j=1}^{L} f_{j-1} z^j & \sum_{j=0}^{L} b_j z^j \end{vmatrix}}{\det \begin{vmatrix} f_{L-M+1} & \cdots & f_L & b_{L+1} \\ \vdots & \ddots & \vdots & \vdots \\ f_L & \cdots & f_{L+M-1} & b_{L+M} \\ z^M & \cdots & z & \tau(z) \end{vmatrix}}$$

$$\tag{11.30}$$

If we expand the numerator and denominator determinants along the last row and column, we have by the standard formula for the matrix inverse, when we cancel a factor of $C(L/M)$ [Eq. (2.1)] from the numerator and the denominator

$$[L/M] = \frac{\sum_{j=0}^{L} b_j z^j + \mathbf{f}_L^{\mathrm{T}} \mathbf{U}^{-1}(L/M) \mathbf{b}_L}{\tau(z) + \mathbf{z}^{\mathrm{T}} \mathbf{U}^{-1}(L/M) \mathbf{b}_L} \tag{11.31}$$

where

$$U(L/M) = \begin{pmatrix} f_{L-M+1} & \cdots & f_L \\ \vdots & \ddots & \vdots \\ f_L & \cdots & f_{L+M-1} \end{pmatrix}, \qquad \mathbf{b}_L^T = (b_{L+1}, \ldots, b_{L+M})$$

$$\mathbf{f}_L^T = \left(\sum_{j=M}^{L} f_{j-M} z^j, \ldots, \sum_{j=1}^{L} f_{j-1} z^j \right), \qquad \mathbf{z}^T = (z^M, \ldots, z) \quad (11.32)$$

The cancellation of $C(L/M)$ is allowed if it does not vanish. Since for L large enough, by (11.29), $U(L/M)$ tends to the form given in Lemma 11.2, which has an explicit, nonsingular inverse since $A_{j, m_j} \neq 0$, then this step is justified for all sufficiently large L. We can consider the behavior of $U^{-1}(L/M)$ as $L \to \infty$, by writing it, in analogy to Lemma 11.2, as

$$U^{-1}(L/M) = C^T \Omega^{-1} C \qquad (11.33)$$

where

$$\Omega = CU(L/M)C^T \qquad (11.34)$$

If we compute directly from the definitions (11.12) and (11.32), we find

$$\Omega = D + O \begin{pmatrix} b_{L-M+1} & \cdots & b_L \\ \vdots & \ddots & \vdots \\ b_L & \cdots & b_{L+M-1} \end{pmatrix} \qquad (11.35)$$

From (11.35), since $\sum b_j |r_i|^j$ converges geometrically for all i, it follows that Ω^{-1} converges geometrically to D^{-1}. From the form of the inverse (11.33) and (11.35) it follows that $\mathbf{f}_L^T U^{-1}(L/M)$ is uniformly bounded for $|z| \leq \rho < S$, since the polar singularity terms are transformed by C^T to match their analogous reciprocal in D. Thus, using (11.14) and (11.29), we have

$$[L/M] = \frac{\sum_{j=0}^{L} b_j z^j + K_N \max\left(1, (z/S)^L\right)(y_{M-1}/S)^L L^{m_n - 1}}{\tau(z) + K_D (y_{M-1}/S)^L L^{m_n - 1}} \qquad (11.36)$$

where K_N and K_D are L and z dependent, but are uniformly bounded in L

and z. If we now compute $D^2(f, [L/M])$ from Eq. (11.1), we have

$$D^2(f(z), [L/M])$$

$$= 4|\sigma(L/M, z)|^2 \left\{ \left| \, |\tau(z)|^2 + |B(z)|^2 + \tau^*(z) K_D(y_{M-1}/S)^L L^{m_n-1} \right.\right.$$

$$+ B^*(z)\left[-\sum_{j=L+1}^{\infty} b_j z^{j+1} + K_N \max(1, (x/S))(y_{M-1}/S)^L L^{m_n-1} \right] \left| \vphantom{\sum} \right.^2$$

$$\left. + |\sigma(L/M, z)|^2 \right\}^{-1} \tag{11.37}$$

where

$$\sigma(L/M, z) = B(z)\left[\tau(z) + K_D(y_{M-1}/S)^L \right]$$

$$- \tau(z)\left[\sum_{j=0}^{L} b_j z^j + K_N \max\left(1, (z/S)^L\right)(y_{M-1}/S)^L \right]$$

$$= B(z)Q_M(z) - \tau(z)P_L(z) = O(z^{L+M+1}) \tag{11.38}$$

by the fundamental Padé equation (1.13). Since (i) these terms arise from the product BQ and (ii) by (11.33) Q is uniformly bounded, we deduce from (11.29) that $|\sigma(L/M, z)|$ is uniformly bounded by $\bar{K}|z/S|^{L+M+1}$ for some $\bar{K}$. Since by hypothesis τ and B do not vanish together, $|\tau|^2 + |B|^2$ has a nonzero minimum in $|z| \leqslant S$. Since the rest of the terms combined with $|\tau|^2 + |B|^2$ tend to zero, these results suffice to prove that D^2 can be made arbitrarily small over any disk $|z| \leqslant \rho < S < R$. Since for any $\rho < R$ we can always pick such an S, the theorem is proven. ∎

It will be observed that we have established that $D(f, [L/M])$ tends to zero like $K'|z/S|^{L+M+1}$ as L tends to infinity. This behavior is exactly the same as the convergence displayed by the partial sums of a Taylor series to a regular function in the usual sense. The case of the behavior of the $[L/M]$ Padé approximants on the bounding circle as L tends to infinity was discussed by Wilson (1927) and was found to be identical, as one might surmise from (11.33), to that of the Taylor series expansion to $B(z)$ [see Dienes (1957)]. As an example illustrating this theorem, consider the

$[L/1]$ approximation to

$$f(x) = (1 - x)^{-1} + (1 - 2x)^{-1}$$

$$[L/1] = \sum_{j=0}^{L-1} (2^j + 1)x^j + (2^L + 1)x^L \left(1 - 2\frac{1 + 2^{-L-1}}{1 + 2^{-L}} x\right)^{-1}$$

$$f(x) - [L/1] = \frac{x^{L+2}/(1 + 2^{-L})}{(1 - x)(1 - 2x)[1 - 2(1 + 2^{-L-1})x/(1 + 2^{-L})]}$$

which shows in specific detail the results of the theorem.

C. Functions with Polar Singularities and "Smooth" Nonpolar Singularity

Wilson (1928a, 1930) considered a number of special cases in which convergence can be proven for the $[L/M + \mu]$ Padé approximants to functions which are regular in a circle of radius R, except for n poles of total multiplicity M. The cases he considered include a multiple pole at distance R from the origin of order greater than μ, an algebraic branch point, a logarithmic singularity, and a general algebraic logarithmic singularity. The characteristic feature of these cases is that their Taylor series representations are *smooth* in a certain sense. We will summarize and extend these results in the following theorem. For the special cases considered by Wilson the determinant condition (11.39) given in the theorem can be readily verified.

Theorem 11.2 (Wilson). If $f(z)$ is regular inside $|z| < R$, except for n interior poles r_i with corresponding multiplicities m_i, $\sum_{i=1}^{n} m_i = M$; then if the coefficients of

$$B(z) = \sum_{j=0}^{\infty} b_j z^j = f(z) \prod_{i=1}^{n} (1 - z/r_i)^{m_i} \tag{11.39}$$

satisfy the condition for $n = L - M - \mu + 1$ to $L + M + \mu$

$$b_n = \Gamma^{-n}\beta(L)\left[\sum_{j=0}^{2\mu-2} \alpha_j(L)\left(\frac{n}{L} - 1\right)^j + \gamma(n, L)\left(\frac{M + \mu}{L}\right)^{2\mu-1}\right] \tag{11.40}$$

where $|\Gamma| = R$ and $\alpha_j(L)$ and $\gamma(n, L)$ are uniformly bounded as $L \to \infty$,

and

$$\lim_{L\to\infty} \det \begin{vmatrix} (2\mu - 2)! \, \alpha_{2\mu-2}(L) & \cdots & (\mu - 1)! \, \alpha_{\mu-1}(L) \\ \vdots & \ddots & \vdots \\ (\mu - 1)! \, \alpha_{\mu-1}(L) & \cdots & \alpha_0(L) \end{vmatrix} \qquad (11.41)$$

exists and is not zero, then in the limit as L goes to infinity, $[L/M + \mu]$ converges uniformly to $f(z)$ on the sphere in $|z| \leqslant \rho$ for any $\rho < R$.

Proof: In our proof we will employ the same sort of algebraic reductions which led to Lemma 11.2, to reduce the problem to a form similar to that obtained in the proof of Montesuss's theorem. We wish to construct

$$\Omega = \mathbf{C}\mathbf{U}(L/M + \mu)\mathbf{C}^{\mathbf{T}} \qquad (11.42)$$

in the notation of (11.12) and (11.32). From the property

$$b_{n+1} - \Gamma^{-1}b_n = \Gamma^{-n-1}\beta(L)\left[\sum_{j=0}^{2\mu-3} \alpha_{j+1}(L) \sum_{k=0}^{j} \binom{j+1}{k+1}\left(\frac{n}{L} - 1\right)^{j-k} L^{-k-1} \right.$$

$$\left. + \gamma'(n, L)\left(\frac{M + \mu}{L}\right)^{2\mu-1} \right] \qquad (11.43)$$

and the definition of $\mathbf{C}$ we can easily write down, to leading order in L^{-1}, the last $\mu \times \mu$ diagonal block of Ω. It is

$$\frac{[f^{(\mu)}(\Gamma^{-1})]^2 \beta(L)}{(-\Gamma)^{L+M+1-\mu}}$$

$$\times \begin{pmatrix} (2\mu - 2)! \, \alpha_{2\mu-2}(L)(\Gamma L)^{2-2\mu} & \cdots & (\mu - 1)! \, \alpha_{\mu-1}(L)(\Gamma L)^{1-\mu} \\ \vdots & \ddots & \vdots \\ (\mu - 1)! \, \alpha_{\mu-1}(L)(\Gamma L)^{1-\mu} & \cdots & \alpha_0(L)(\Gamma L)^0 \end{pmatrix}$$

$$(11.44)$$

plus terms of lower order in L in every element. We define

$$f(z) = (z - 1/\Gamma)^\mu \prod_{i=1}^{n} (z - 1/r_i)^{m_i} \tag{11.45}$$

The effects of the pole terms cancel exactly here. Because of the assumption $|\Gamma| > |r_i|$ for all i, the rest of the matrix elements are dominated by the pole terms and we have, to leading order, the structure for Ω of a series of $m_i \times m_i$ diagonal blocks containing the elements

$$[f^{(m_i)}(r_i^{-1})]^2(-r_i)^{M-L-1} \begin{pmatrix} 0 & \cdots & & A_{i,\,m_i} \\ \vdots & \ddots & & \vdots \\ & & & \\ A_{i,\,m_i} & \cdots & \sum_{l-1}^{m_i} A_{i,\,m_i-l+1} \begin{pmatrix} -l \\ L-M+1 \end{pmatrix} \end{pmatrix} \tag{11.46}$$

where use has been made of the binomial identity

$$\begin{pmatrix} -l \\ n \end{pmatrix} + \begin{pmatrix} -l \\ n-1 \end{pmatrix} = \begin{pmatrix} -l+1 \\ n \end{pmatrix} \tag{11.47}$$

in obtaining (11.46), which is expressed in the notation of (11.9). Form (11.46) is a lower right triangular matrix, which is explicitly inverted by $\mathbf{D}_j$ of Eq. (11.14).

If we now define, as in Eq. (11.28) in the proof of Montessus's theorem

$$\tau(z) = (1 - z/\Gamma)^\mu \prod_{i=1}^{n} (1 - z/r_i)^{m_i} \tag{11.48}$$

and

$$H(z) = \tau(z)f(z) = \sum_{j=0}^{\infty} h_j z^j \tag{11.49}$$

then in the range $n = L - M - \mu + 1$ to $L + M + \mu$ we have the h_n represented [using property (11.43)] to leading order in L^{-1} by

$$h_n = L^{-\mu}\Gamma^{-n}\beta(L) \left[\sum_{j=\mu}^{2\mu-2} \alpha_j(L) \frac{j!}{(j-\mu)!} \left(\frac{n}{L} - 1 \right)^{j-\mu} \right] \tag{11.50}$$

If we now reduce the Padé approximant as we did in (11.30) and (11.31), we obtain

$$[L/M + \mu] = \frac{\sum_{j=0}^{L} h_j z^j + \mathbf{f}_L^T \mathbf{U}^{-1}(L/M + \mu)\mathbf{h}_L}{\tau(z) + \mathbf{z}^T \mathbf{U}^{-1}(L/M + \mu)\mathbf{h}_L} \qquad (11.51)$$

where $\mathbf{h}_L$ is defined in analogy to $\mathbf{b}_L$ of (11.32). We can write, by (11.42),

$$\mathbf{U}^{-1}(L/M + \mu) = \mathbf{C}^T \mathbf{\Omega}^{-1} \mathbf{C} \qquad (11.52)$$

Again the cancellation of the divergences due to the polar contributions follows, as we argued in (11.33) to (11.35). For large L, $\mathbf{D}$ [Eq. (11.11)] is dominated by the boundary singularity contributions, which, by (11.41) and (11.44), contribute to the last diagonal blocks terms of order

$$\frac{\Gamma^L}{\beta(L)} \begin{pmatrix} L^{2\mu-2} & L^{2\mu-3} & \cdots & L^{\mu-1} \\ L^{2\mu-3} & L^{2\mu-4} & \cdots & L^{\mu-2} \\ \vdots & \vdots & \ddots & \vdots \\ L^{\mu-1} & L^{\mu-2} & \cdots & L^0 \end{pmatrix} \qquad (11.53)$$

If we use property (11.43) and the structure of $\mathbf{C}$, we can compute

$$\mathbf{Ch}_L = \frac{\beta(L)}{\Gamma^L} \begin{pmatrix} L^{1-2\mu}\gamma(n, L) \\ L^{2-2\mu}\alpha_{2\mu-2}(L) \\ \vdots \\ L^{-\mu}\alpha_\mu(L) \end{pmatrix} \qquad (11.54)$$

to leading order in L^{-1}. If we combine (11.53) and (11.54), we obtain for fixed $|z| < S < R$, by using the convergence of $H(z)$ for this value of z, to leading order in small quantities,

$$[L/M + \mu] = \frac{\sum_{j=0}^{L} h_j z^j + K_N/L}{\tau(z) + K_0/L} \qquad (11.55)$$

If we substitute (11.55) into (11.1) to give $D^2([L'/M + \mu], [L/M + \mu])$, we obtain results quite analogous to those obtained in the proof of Montessus's theorem in Eqs. (11.34) and (11.35). Again D goes to zero like $K'|z/S|^{L+M+\mu+1}$ uniformly in $|z| \leqslant S < R$. In this case, however, the denominator does not converge geometrically but only like L^{-1}. Since D converges geometrically in the interior of the region, so too must the roots of the Padé denominators to the r_i. Thus it is the convergence of the Padé denominator roots to the point Γ which go like L^{-1}. Since $\tau(z)$ and $H(z)$ do not vanish simultaneously in $|z| \leqslant S < R$, we conclude the proof of this theorem just as we did Theorem 11.1. ∎

As an example of this theorem, we consider the $[L/1]$ Padé approximants to

$$f(x) = (1 - x)^{-2},$$

$$[L/1] = \sum_{j=0}^{L-1} (1 + j)x^j + (L + 1)x^L\left(1 - \frac{L + 2}{L + 1}x\right)^{-1}$$

$$f(x) - [L/1] = \frac{-x^{L+2}/(L + 1)}{(1 - x)^2[1 - x - x/(L + 1)]}$$

which illustrate the conclusions of the theorem. In the error term an extra convergence factor of $(L + 1)^{-1}$ will be noted over what was established in the theorem.

Again Wilson (1928a, 1930) has treated the boundary behavior and again it parallels exactly the behavior of the partial sums of the Taylor series to $H(z)$, Eq. (11.49).

D. Several "Smooth" Boundary Circle Singularities

Next we will discuss the convergence properties when there are several singularities on the boundary circle $|z| = R$ (Hadamard, 1892; Wilson, 1928b, 1930; Parlett, 1968). The simplest case occurs when all the singularities are simple poles. We will first illustrate this case by an example. Let

$$f(x) = (1 - x)^{-1} + (1 - e^{i\theta}x)^{-1}$$

$$[L/1] = \sum_{j=0}^{L-1} (e^{ij\theta} + 1)x^j + (e^{iL\theta} + 1)x^L\left(1 - \frac{e^{i(L+1)\theta} + 1}{e^{iL\theta} + 1}x\right)^{-1}$$

$$\tag{11.56}$$

Then the denominator vanishes at

$$x_L = e^{-i\theta/2}\{[\cos(\tfrac{1}{2}L\theta)]/\cos[\tfrac{1}{2}(L + 1)\theta]\} \tag{11.57}$$

Clearly, since x_L can be significantly less than unity, the radius of convergence of $f(x)$, we cannot expect to prove convergence on the sphere for the whole sequence of $[L/1]$ Padé approximants in this case. What one can show here is that there exists a subsequence of $[L/1]$ for which $|x_L| \geqslant 1$, and for which $[L/1]$ converges to $f(x)$ for $|x| < 1$. Other results on *convergence in the mean on the sphere* will be discussed in a later chapter. What can be said in general about convergence on the sphere uniform over the disk $|z| \leqslant \rho < R$ in the case of polar boundary singularities has already been said by Montessus's theorem (Theorem 11.1), that is, we have such convergence when the number of poles exactly coincides with the number of poles counted according to their multiplicity enclosed within $|z| < R$, and not otherwise.

Next suppose that instead of obeying (11.40), the coefficients of $B(z)$ [Eq. (11.39)] satisfy

$$b_n = \sum_{v=1}^{V} \Gamma_v^{-n} \beta_v(L) \left[\sum_{j=0}^{2\mu_v - 1} \alpha_{v,j}(L) \left(\frac{n}{L} - 1 \right)^j + \gamma_v(n, L) \left(\frac{M + \mu_v}{L} \right)^{2\mu_v} \right]$$

(11.58)

where $|\Gamma_v| = R$. The approximation procedure will, with the number of poles available, try to represent the dominant behavior. To select an appropriate $\tau(z)$ [Eq. (11.48)], we recognize that the most sensitive part of the calculation is (11.44). It has a background error which behaves asymptotically, as L goes to infinity in proportion to

$$\sum_{v=0}^{V} \beta_v(L) L^{-2\mu_v}$$

(11.59)

after the inclusion of $\mu = \sum_{v=1}^{V} \mu_v$ poles. For the previous line of argument to go through, (11.59) must be smaller than $\beta_v(L) L^{2-2\mu_v}$ for every v $(0 \leqslant v \leqslant V)$. Put another way, the total μ is to be partitioned into the $\{\mu_v\}$ in such a way as to minimize (11.59) as L goes to infinity (Saff, 1969). If this partition is not unique, as in our example (11.56), then no progress can be expected in general. Suppose therefore that the partition is unique. We seek to bound again the error in (11.51). The argument given before (11.54) again holds for any particular value of v, say v_0, but we must also consider the other terms for different values of $v \neq v_0$. Since factors of $(1 - z/\Gamma_v)^{\mu_v}$ appear in $\tau(z)$, these same arguments also improve the convergence in L to roughly the same rate as for $v = v_0$. Consequently the argument goes through and the conclusion, (11.55), is slightly modified to read

$$[L/M + \mu] = \frac{\sum_{j=0}^{L} h_j z^j + K_N \phi(L)}{\tau(z) + K_D \phi(L)} \tag{11.60}$$

where $\phi(L)$ goes to zero. That $\phi(L)$ goes to zero and goes no faster than L^{-1} follows because each term in (11.51) varies by integral powers of L^{-1} and so the spread in rate of decay can be no more than one power of L; thus $\phi(L)$ must go to zero, and will not go faster than our result in Theorem 11.2. Consequently we conclude:

Theorem 11.3. Let $f(z)$ satisfy the conditions of Theorem 11.2, except that (11.58) replaces (11.40) and there exists a unique partition of $\mu = \{\mu_v\}$ so as to minimize (11.59) as L goes to infinity; then in the limit as L goes to infinity, $[L/M + \mu]$ converges uniformly to $f(z)$ on the sphere in $|z| \leqslant \rho$ for any $\rho < R$. Condition (11.41) is now assumed for the $\alpha_{v,j}$ for each $v = 1, \ldots, V$ separately.

E. "Smooth" Entire Functions

There is one more general theorem for vertical sequences of Padé approximants to "smooth" power series. It could happen that a function is entire, except for a finite number of poles in the complex plane. Then, for the degree of the denominator greater than the number of poles, there is nothing obvious for the rest of the roots of the denominator to converge to except infinity. The following theorem treats this case and has the added refinement of specifying the rate of approach of those "extra poles" to infinity (Baker, 1973).

Theorem 11.4. Let $f(z)$ be regular for all finite z, except for n poles r_i with corresponding multiplicities m_i, $\sum_{i=1}^{n} m_i = M$, and let the coefficients of Eq. (11.39),

$$B(z) = \sum_{j=0}^{\infty} b_j z^j = f(z) \prod_{i=1}^{n} (1 - z/r_i)^{m_i}$$

satisfy the condition, for $n = L - M - \mu + 1$ to $L + M + \mu$,

$$b_n = \frac{\Lambda^{-n}}{(n!)^{\theta}} \beta(L) \left[\sum_{j=0}^{2\mu-2} \alpha_j(L) \left(\frac{n}{L} - 1 \right)^j + \gamma(n, L) \left(\frac{M + \mu}{L} \right)^{2\mu-1} \right] \tag{11.61}$$

where $\alpha_j(L)$ and $\gamma(n, L)$ are uniformly bounded as $L \to \infty$, and $\theta > 0$, $|\Lambda| \neq 0$. Then in the limit as L goes to infinity, $[L/M + \mu]$ converges uniformly to $f(z)$ on the sphere in $|z| \leq \rho$ for any $\rho < \infty$.

Proof: The plan of the proof is the same as for Theorem 11.2, with some modifications in detail. First we need to re-express $n!$ as

$$n! = L! \, L^{n-L} \prod_{k=L+1}^{n} (k/L) = L! \, L^{n-L} \exp\left[\sum_{k=L+1}^{n} \ln(k/L) \right]$$

$$= L! \, L^{n-L} \exp\left[- \sum_{j=1}^{\infty} j^{-1} \sum_{l=1}^{n-L} (-l/L)^j \right] \tag{11.62}$$

Since the sum of l^j is a polynomial in $n - L$ of degree $j + 1$, we can write

$$n! = L! \, L^{n-L} \exp\left[-\delta_1(L)\left(\frac{n}{L} - 1 \right) - L \sum_{j=2}^{\infty} \delta_j(L)\left(\frac{n}{L} - 1 \right)^j \right] \tag{11.63}$$

which, although we assumed $n \geqslant L$, is also valid for $n < L$ as well. Using (11.63), we can rewrite

$$b_n = \frac{(\Lambda L^\theta)^{-n} \beta(L)}{(L!)^\theta L^{-\theta L}} \exp\left[\theta \delta_1(L)\left(\frac{n}{L} - 1 \right) + \theta L \sum_{j=2}^{\infty} \delta_j(L)\left(\frac{n}{L} - 1 \right)^j \right]$$

$$\times \left[\sum_{j=0}^{2\mu-2} \alpha_j(L)\left(\frac{n}{L} - 1 \right)^j + \gamma(n, L)\left(\frac{M + \mu}{L} \right)^{2\mu-1} \right] \tag{11.64}$$

On expanding the exponential series in (11.56) in powers of $n - L$, we see that the dominant terms (for large L) arise from the $j = 2$ and $j = 3$ terms. If we define the factor

$$\bar{\Lambda} = \Lambda\{\exp[-\theta\delta_1(L)/L]\}[1 - \alpha_1 L^{-1}/(\alpha_0 - \alpha_1)] \tag{11.65}$$

then to leading order (for each power of $n - L$) in L^{-1} we have

$$
b_n = \frac{(\overline{\Lambda} L^{\theta})^{-n} \beta(L)[\alpha_0(L) - \alpha_1(L)]}{(L!)^{\theta} L^{-\theta L} \exp[\theta \delta_1(L)]}
$$

$$
\times \sum_{k=0}^{\infty} \frac{1}{k!} \left\{ -\frac{1}{2} \theta \left[\frac{(n - L)^2}{L} - \frac{1}{3} \frac{(n - L)^3}{L^2} \right] \right\}^k \quad (11.66)
$$

where explicit asymptotic values $\delta_2 = \frac{1}{2}$, $\delta_3 = -\frac{1}{6}$ have been used. Form (11.66) implies in turn the values of the expansion (to leading order in L^{-1})

$$
b_n = \frac{(\overline{\Lambda} L^{\theta})^{-n} \beta(L)[\alpha_0(L) - \alpha_1(L)]}{(L!)^{\theta} L^{-\theta L} \exp[\theta \delta_1(L)]}
$$

$$
\times \sum_{k=0}^{\infty} \frac{\theta^k}{2^k k!} \left[\frac{(n - L)^{2k}}{(-L)^k} + \frac{(n - L)^{2k+1}}{(-L)^{k+1}} \right] \quad (11.67)
$$

Except for two differences, we have now converted the form (11.61) into the form (11.40), for which Theorem 11.2 is valid. The differences are: (i) the location of the apparent singularity is now L dependent and (ii) the coefficient of the jth power of $n - L$ now decreases like $L^{-[(j+1)/2]}$ instead of L^{-j} as in (11.40). We can simply repeat, line by line, the proof of Theorem 11.2; however, now we need to keep track of the powers of $\overline{\Lambda} L^{\theta}$. Also, the counting up of the powers of L will be slightly different. We now define

$$
\tau(z) = \left(1 - z/\overline{\Lambda} L^{\theta}\right)^{\mu} \prod_{i=1}^{n} \left(1 - z/r_i\right)^{n_i} \quad (11.68)
$$

and use (11.49):

$$
H(z) = \tau(z) f(z) = \sum_{j=0}^{\infty} h_j z^j
$$

We can, as in the proof of Theorem 11.2, reduce the Padé approximant to the form (11.51):

$$
[L/M + \mu] = \frac{\sum_{j=0}^{L} h_j z^j + \mathbf{f}_L^{\mathsf{T}} U^{-1}(L/M + \mu) \mathbf{h}_L}{\tau(z) + \mathbf{z}^{\mathsf{T}} U^{-1}(L/M + \mu) \mathbf{h}_L}
$$

The analysis of the terms of $U^{-1}(L/M + \mu)$ is slightly different. From (11.67) we deduce the $\alpha_j(L)$ and from (11.44) we deduce by calculating the order of the determinant and its minors, for the last $\mu \times \mu$ diagonal block of Ω^{-1} [Eq. (11.42)] terms of the order

$$\frac{(L!)^\theta (\Lambda L^\theta)^{L+M-\mu+1}}{\beta(L) L^{\theta L}}$$

$$\times \begin{pmatrix} (\overline{\Lambda} L^\theta)^{2\mu-2} L^{\mu-1} & (\overline{\Lambda} L^\theta)^{2\mu-3} L^{\mu-2} & \cdots & (\overline{\Lambda} L^\theta)^{\mu-1} L^{[(\mu-1)/2]} \\ \vdots & \vdots & \ddots & \vdots \\ (\overline{\Lambda} L^\theta)^{\mu-1} L^{[(\mu-1)/2]} & (\overline{\Lambda} L^\theta)^{\mu-2} L^{[(\mu-2)/2]} & \cdots & (\overline{\Lambda} L^\theta)^0 L^0 \end{pmatrix} \tag{11.69}$$

We can compute, also from (11.54), directly, by keeping track of the $\overline{\Lambda} L^\theta$ factors

$$\mathbf{Ch}_L = O\left(\frac{\beta(L) L^{\theta L}}{(L!)^\theta (\Lambda L^\theta)^{L+M}} \begin{pmatrix} (\Lambda L^\theta)^{-\mu} L^{-\mu} \\ (\Lambda L^\theta)^{1-\mu} L^{1-\mu} \\ \vdots \\ (\Lambda L^\theta)^{-1} L^{-(\mu+1)/2} \end{pmatrix} \right) \tag{11.70}$$

Combining (11.69) and (11.70) and doing the multiplication by C^T, we get ($\mu \neq 0$)

$$\mathbf{z}^T U^{-1} \mathbf{h}_L = O\left\{ L^{-1} \left[\sum_{j=1}^{\mu} (z/\overline{\Lambda} L^\theta)^j \right] \left(\sum_{j=0}^{M} z^j \right) \right\} \tag{11.71}$$

This form follows because there must necessarily be a factor of $(\Lambda L^\theta)^{\mu+1-j}$ in the first μ rows of C^T. The corrections due to the polar terms are of much higher order and go to zero in the same way that the Taylor series partial sums of $H(z)$ converge to $H(z)$, evaluated at the various r_i. Result (11.71), together with arguments similar to those used previously, allows us to establish form (11.55) for the Padé approximant. Convergence now follows as in Theorem 11.2.

We have only one final condition to check, and that is one analogous to the determinant condition (11.41). We implicitly assumed that the determinant of (11.44), which we inverted to obtain estimate (11.69), did not vanish to leading order in L^{-1}. If we define

$$v_k = (-1)^k [(2k)!/2^k k!] = -(-2)^{k-1} \cdot \tfrac{1}{2} \cdot \tfrac{3}{2} \cdots (k - \tfrac{1}{2}) \quad (11.72)$$

then by multiplying the rows of (11.44) by $(\bar{\Lambda} L^\theta L)^{(\mu-j)/2}$ and the columns by $(\bar{\Lambda} L^\theta L)^{(\mu-k)/2}$, where $j = 1, \ldots, \mu$ and $k = 1, \ldots, \mu$ are the row and column indices, respectively, we can pick out the leading order determinant of (11.44) as

$$\det \propto \begin{vmatrix} v_{\mu-1} & 0 & v_{\mu-2} & 0 & \cdots \\ 0 & v_{\mu-2} & 0 & v_{\mu-3} & \cdots \\ v_{\mu-2} & 0 & v_{\mu-3} & 0 & \cdots \\ \vdots & \vdots & \vdots & \vdots & \ddots \end{vmatrix} \quad (11.73)$$

which by rearranging rows and columns becomes

$$\det \propto \det \begin{vmatrix} v_0 & v_1 & \cdots & v_{[(\mu-1)/2]} \\ v_1 & v_2 & \cdots & v_{[(\mu-1)/2]+1} \\ \vdots & \vdots & \ddots & \vdots \\ v_{[(\mu-1)/2]} & v_{[(\mu-1)/2]+1} & \cdots & v_{2[(\mu-1)/2]} \end{vmatrix}$$

$$\times \det \begin{vmatrix} v_1 & \cdots & v_{[\mu/2]} \\ \vdots & \ddots & \vdots \\ v_{[\mu/2]} & \cdots & v_{2[\mu/2]-1} \end{vmatrix} \quad (11.74)$$

However, these determinants are just those which occur in the $[[\tfrac{1}{2}(\mu - 1)]/[\tfrac{1}{2}(\mu - 1)] + 1]$ and $[[\tfrac{1}{2}\mu]/[\tfrac{1}{2}\mu]]$ Padé approximants to $_2F_0(\tfrac{1}{2}, 1; -x)$. We discussed this case in Chapter 5, [Eq. (5.46) et seq.] and it follows from that discussion that these determinants are not zero. Thus we have justified our estimate (11.69) and thereby completed the proof of this theorem. ■

As an example of the results of this theorem we can rewrite the denominator of the $[L/\mu]$ Padé approximant to e^x as

$$Q_\mu(x) = 1 + \sum_{k=1}^{\mu} \binom{\mu}{k} \prod_{l=0}^{k-1} \left(\frac{-x}{L + \mu - l} \right) \quad (11.75)$$

by (5.39). From (11.75) it is clear that

$$Q_\mu(x) = (1 - x/L)^\mu + O\left[L^{-1} \sum_{i=1}^{\mu} (x/L)^j \right] \tag{11.76}$$

as expected from Theorem 11.4. For large L, by solving (11.75) or (11.76) for its roots, we find them to be $x_i = L(1 + O(L^{-1/\mu}))$. Thus the results of Theorem 11.4 give the dominant rate of convergence and the first-order correction to that rate correctly for this example.

It is possible to generalize Theorem 11.4 in the same way as Theorem 11.3 generalized Theorem 11.2. One would not be interested in considering different θ's since for large L one set of terms would be asymptotically negligible, just as one was unconcerned with different $|\Gamma_v|$ in Theorem 11.3. If we consider the possibility of coefficients b_n of the form of the sum of several terms of the form (11.61), then unless $|\Lambda|$ is the same for all such terms, they will not all contribute asymptotically to the same order. If, on the other hand, we have

$$b_n = \sum_{v=1}^{V} \frac{\Lambda_v^{-n}}{(n!)^\theta} \beta_v(L) \left[\sum_{j=0}^{2\mu_v - 1} \alpha_{v,j}(L) \left(\frac{n}{L} - 1 \right)^j + \gamma_v(n, L) \left(\frac{M + \mu_v}{L} \right)^{2\mu_v} \right] \tag{11.77}$$

where $|\Lambda_v| = R$, then, just as Theorem 11.3 follows from Theorem 11.4, we have the result:

Corollary 11.1. Let $f(z)$ satisfy the conditions of Theorem 11.4, except that (11.77) replaces (11.61) and there exists a unique partition of $\mu = \{\mu_v\}$ so as to minimize

$$\sum_{v=0}^{V} \beta_v(L) L^{-\mu_v} \tag{11.78}$$

as L goes to infinity. Then in the limit as L goes to infinity, $[L/M + \mu]$ converges uniformly to $f(z)$ on the sphere in $|z| \leqslant \rho$ for any $\rho < \infty$.

F. Nonvertical Sequences

Wallin (1972) has observed that these results can easily be extended to nonvertical sequences of Padé approximants.

Corollary 11.2 (Wallin). If $f(z)$ satisfies the conditions of Theorems 11.2–11.4 or Corollary 11.1 for all μ, then there exists an infinite sequence of $[L/M]$ Padé approximants which converges on the sphere uniformly in the domain D specified by the respective theorem or corollary. This sequence satisfies the property

$$L_i \to \infty \quad \text{and} \quad M_i \to \infty \quad \text{as} \quad i \to \infty \quad (11.79)$$

Proof: By the relevant theorem or corollary we can show that for all $[L/m]$ Padé approximants with $L \geq L(m)$ we have

$$D^2(f(z), [L/m]) < \epsilon_m^2 \quad (11.80)$$

everywhere in D. If we select the sequence ϵ_m so that ϵ_m tends to zero as m tends to infinity, then $[L(m)/m]$ satisfies the requirements of this corollary.

A further simple corollary follows from the duality theorem 9.2.

Corollary 11.3. Let $f(z)$ satisfy the conditions of Theorems 11.1–11.4 or Corollary 11.1, except that the specifications of conditions concerning poles be replaced by specifications on zeros. Then the dual conclusions hold; i.e., we have convergence on the sphere for the $[l/M]$ Padé approximants as M goes to infinity with l fixed.

G. General Entire Functions

In the absence of "smoothness" assumptions on the series coefficients, no theorems of the above character are known, nor can they be expected. In example (11.56) we saw that the entire $[L/1]$ sequence did not converge. This problem was further studied in Theorem 11.3 and Corollary 11.1. Perron (1954) has given the following interesting example. First select a sequence $\{r_n\}$ of complex numbers which are distributed over the complex plane in any desired fashion. Then select the coefficients, for $n = 0, 1, \ldots$,

$$a_{3n} = r_n/(3n + 2)!, \qquad a_{3n+1} = 1/(3n + 2)!,$$

$$a_{3n+2} = 1/(3n + 2)! \quad \text{if} \quad |r_n| \leq 1 \quad (11.81)$$

or

$$a_{3n} = 1/(3n + 2)!, \qquad a_{3n+1} = 1/(3n + 2)!,$$

$$a_{3n+2} = 1/[r_n(3n + 2)!] \quad \text{if} \quad |r_n| > 1 \quad (11.82)$$

Then the function

$$A(x) = \sum_{n=0}^{\infty} a_n x^n \quad (11.83)$$

has the properties that it is entire, by comparison with the exponential series, and that either the $[3n/1]$ or $[3n + 1/1]$ approximant has a pole at r_n as $|r_n|$ is $\leqslant 1$ or > 1, respectively, since

$$[L/1] = \sum_{n=0}^{L-1} a_n x^n + a_L x^L / (1 - a_{L+1} x / a_L) \qquad (11.84)$$

Thus Perron's example can be used to produce functions such that *every point* of the complex plane is a limit point of poles. Nevertheless, the pole of the $[3n + 2/1]$ Padé approximants necessarily lies outside the region $|x| \leqslant (3n + 5)(3n + 4)(3n + 3)$, and from this fact and (11.84) we can easily prove that the subsequence $[3n + 2/1]$ converges to $A(x)$ at every point of the complex plane. We will now prove that, at least under certain conditions, such convergent subsequences always exist (Beardon, 1968b; Baker, 1973). First we will treat the simple case of the $[L/1]$ sequence of Padé approximants.

Theorem 11.5 (Beardon). If $f(z)$ is analytic in $|z| \leqslant 1/\rho$, then there exists an infinite subsequence of the $[L/1]$ Padé approximants which converge uniformly in any disk $|z| \leqslant R$, for $R\rho < 1$, to the function defined by the power series.

Proof: First suppose $f(z)$ is a polynomial of degree l; then all the $[L/1]$ Padé approximants with $L \geqslant l$ are identical to $f(z)$, by Theorem 2.2, so the theorem is true in this case. For the rest of the proof we will assume that $f(z)$ has an infinite number of nonzero coefficients. We next observe that for the coefficients $f(z) = \Sigma f_j z^j$

$$\lim_{L \to \infty} \inf |f_{L+1}/f_L| \leqslant \rho \qquad (11.85)$$

For, suppose the contrary. Then there must exist an L_0 such that for all $l \leqslant L_0$

$$|f_{l+1}/f_l| \geqslant \sigma > \rho \qquad (11.86)$$

It is no loss of generality to assume $f_{L_0} \neq 0$. Thus by (11.72) we have

$$|f_l| \geqslant |f_{L_0}| \sigma^{l - L_0} \qquad (11.87)$$

If we now apply Cauchy's nth root test for the radius of convergence R, we get

$$R = \left[\lim_{l \to \infty} \sup(f_l)^{1/l} \right]^{-1} \leqslant \sigma^{-1} \qquad (11.88)$$

But this result contradicts the hypothesis that $f(z)$ has radius of convergence at least $1/\rho$. Thus (11.85) must hold. However, (11.85) implies that there exists a subsequence of values for L for which $|f_{L+1}/f_L|$ tends to a limit less than or equal to ρ. If we select this subsequence for L, then from (11.84) we have for any fixed z, $|z| \leq R$, convergence of the sum term by the definition. Since $f_L z^L$ tends to zero and the denominator is uniformly bounded away from zero for $|z| \leq R$, the maximum contribution of the pole term tends to zero. ∎

We can now deduce as a corollary to Theorem 11.5 the corresponding result (Baker, 1973) for an entire function. All that is required is to let $\rho \to 0$ in Theorem 11.5.

Corollary 11.4. If $f(z)$ is an entire function, then there exists an infinite subsequence of the $[L/1]$ Padé approximants which converge uniformly in any closed, bounded region of the complex plane to the entire function defined by the power series.

We conclude this section with a theorem (Baker, 1973) on the convergence for the $[L/2]$ vertical subsequence. Before presenting it, it is convenient first to give two lemmas which will be required in the proof.

Lemma 11.3. Let $f(z)$ be an entire function such that

$$f(z) = \sum_{j=0}^{\infty} f_j z^j, \qquad |f_j| \leq K/(j!)^\theta \tag{11.89}$$

for K and θ greater than zero. In addition, suppose that an infinite number of $f_j \neq 0$. Then for $0 < \zeta < \theta$

$$\lim_{L \to \infty} \inf \max\{L^\zeta |f_{L+1}/f_L|, L^\zeta |f_{L+2}/f_L|^{1/2}, \ldots, L^\zeta |f_{L+k}/f_L|^{1/k}\} = 0 \tag{11.90}$$

Proof: Assume the contrary; then for all L greater than or equal to some L_0 and for some $\sigma > 0$ it must be that

$$\max\{L^\zeta |f_{L+1}/f_L|, \ldots, L^\zeta |f_{L+k}/f_L|^{1/k}\} \geq \sigma > 0 \tag{11.91}$$

It is no loss of generality to assume $f_{L_0} \neq 0$. We now start with f_{L_0} and select L_1 as any $L_0 < L_0 + k$ for which

$$|f_{L_1}/f_{L_0}| \geq (\sigma/L_0^\zeta)^{L_1 - L_0} \tag{11.92}$$

which is possible by (11.91). We continue in this way to pick $L_1 < L_2$

$\leqslant L_1 + k$, and so on, so that we can select an infinite subsequence of L_i such that, by (11.92),

$$|f_{L_i}| \geqslant |f_{L_0}|\sigma^{(L_i - L_0)}\left(\frac{L_0!}{L_i!}\right)^{\varsigma}, \tag{11.93}$$

where use has been made of the properties of the factorial function in writing (11.93). Now if L_i is taken large enough, then the lower bound (11.93) must exceed the upper bound (11.89) since $\varsigma < \theta$. Hence we have a contradiction and therefore (11.90) must hold, since every term in (11.90) is nonnegative. ∎

Lemma 11.4. If that for $\rho > 0$, $\varsigma > 0$ we have

$$\max\{L^{\varsigma}|a_{L+1}/a_L|, L^{\varsigma}|a_{L+2}/a_L|^{1/2}, \ldots, L^{\varsigma}|a_{L+k}/a_L|^{1/k}\} > \rho \tag{11.94}$$

and

$$\max\{(L+1)^{\varsigma}|a_{L+2}/a_{L+1}|, (L+1)^{\varsigma}|a_{L+3}/a_{L+1}|^{1/2}, \ldots,$$

$$(L+1)^{\varsigma}|a_{L+k+1}/a_{L+1}|^{1/k}\} \leqslant \rho, \tag{11.95}$$

then

$$|a_{L+1}/a_L| > \rho L^{-\varsigma}. \tag{11.96}$$

Proof: Suppose the first term in (11.94) is the maximum one; then we have (11.96) directly. If one of the other terms $L^{\varsigma}|a_{L+j}/a_L|^{1/j}$ is the maximum one, then we can write, using (11.94) and (11.95),

$$\rho^{(j-1)/j} \geqslant (L+1)^{(j-1)\varsigma/j}|a_{L+j}/a_{L+1}|^{1/j}$$

$$= (L+1)^{(j-1)\varsigma/j}|a_{L+j}/a_L|^{1/j}|a_L/a_{L+1}|^{1/j}$$

$$> (L+1)^{(j-1)\varsigma/j}|a_L/a_{L+1}|^{1/j}\rho/L^{\varsigma} \tag{11.97}$$

or, rewriting (11.97), we obtain

$$|a_{L+1}/a_L| > \left[(L+1)^{j-1}/L^j\right]^{\varsigma}\rho > \rho L^{-\varsigma}$$

since $(L+1)/L > 1$ and $\varsigma > 0$, which provides the conclusion of the lemma for the remainder of the cases. ∎

Theorem 11.6. If $f(z)$ is an entire function satisfying condition (11.89) with $\theta > 1$, then there exists an infinite subsequence of $[L/2]$ Padé approximants which converge uniformly in any closed, bounded region of the complex plane to the entire function defined by the power series.

Proof: If $f(z)$ has only a finite number of nonzero coefficients, then by Theorem 2.2, beyond a certain point all the $[L/2]$ Padé approximants are identical with the function and thus the theorem follows immediately in this case. For the rest of the proof we will assume that there are an infinite number of $f_j \neq 0$.

We can therefore apply Lemma 11.3 for $k = 2$ and $\zeta > 1$ and show that there exists an infinite number of L's for which expression (11.91) is less than any $\epsilon > 0$ we choose. If every such term with L greater than a prescribed J is less than or equal to ϵ, select a smaller ϵ so that at least some terms are greater than ϵ. Pick now the first $L > J$ such that

$$\max\{ L^\zeta |f_{L+1}/f_L|, L^\zeta |f_{L+2}/f_L|^{1/2}\} \leqslant \epsilon \qquad (11.98)$$

and

$$\max\{ (L - 1)^\zeta |f_L/f_{L-1}|, (L - 1)^\zeta |f_{L+1}/f_L|^{1/2}\} > \epsilon \qquad (11.99)$$

We can clearly pick such an L, from the conditions we have imposed. Applying Lemma 11.4, we have

$$|f_L/f_{L-1}| > \epsilon/(L - 1)^\zeta \qquad (11.100)$$

At this point it is convenient to write out explicitly from (1.27) the Padé denominator. Let

$$Q(x) = 1 + \alpha x + \beta x^2 \qquad (11.101)$$

then

$$\alpha = \left[(f_{L+2}f_{L-1}/f_L^2) - (f_{L+1}/f_L) \right] / \left[(f_{L+1}f_{L-1}/f_L^2) - 1 \right]$$

$$\beta = \left[(f_{L+2}/f_L) - (f_{L+1}/f_L)^2 \right] / \left[(f_{L+1}f_{L-1}/f_L^2) - 1 \right] \qquad (11.102)$$

With (11.102) as a guide as to what needs to be bounded, we compute

$$|f_{L-1}/f_L| \, |f_{L+1}/f_L| - 1 \leqslant (1 - L^{-1})^\zeta - 1 \leqslant -\zeta/L \qquad (11.103)$$

since $\zeta > 1$. Further, (11.98) and (11.100), we have

$$|f_{L-1}/f_L| \, |f_{L+2}/f_L| \leqslant \epsilon(L - 1)^\zeta L^{-2\zeta} < \epsilon L^{-\zeta} \qquad (11.104)$$

Now, applying (11.98), (11.103), and (11.104) to (11.102), we have

$$|\alpha| < (2\epsilon/\zeta)L^{1-\zeta}, \qquad |\beta| < (2\epsilon^2/\zeta)L^{1-2\zeta} \tag{11.105}$$

but we can make ϵ as small as we please and L as large as we like. Hence (11.105) implies that $|\alpha|$ and $|\beta|$ tend to zero. Consequently, the roots of (11.101) recede to infinity. From the facts (i) $|\alpha|$ and $|\beta|$ tend to zero, (ii) the convergence, by definition, of the partial sums of the Taylor series to $f(z)$, and (iii) the Padé numerator $P(z)$ is a truncation of $Q(z)f(z)$ to a polynomial of degree L, we can readily establish that $P(z)$ converges to $f(z)$ in any closed, bounded region of the complex plane. The result completes the proof. ∎

We remark that since $|f_{L+1}/f_L|$ tends to zero for the selected subsequence, then for the same subsequence of L's the $[L/1]$ Padé approximants also converge. We know of no reason to suppose that Theorem 11.6 is the best possible result for that case, as have been, in some sense, the other results of this chapter.

H. The N-Point Padé Approximants

Our goal is to generalize Theorem 11.1 to the N-point case. As a preliminary to discussing the convergence of vertical columns in the Padé table, we first find it convenient to rewrite the Lagrange interpolation formula (8.4) by means of the calculus of residues (Copson, 1948) in Hermite's form. To this end, using $f(z)$ defined by (8.3),

$$f(z) = \prod_{i=1}^{n} (z - z_i)^{m_i}$$

we can consider the polynomial

$$\psi(z) = \frac{1}{2\pi i} \int_{\Gamma} \frac{f(t) - f(z)}{t - z} \frac{U(t)}{f(t)} \, dt \tag{11.106}$$

which is of degree N by (8.2), since $t - z$ is a factor of $f(t) - f(z)$. We choose the contour Γ so that it encircles all of the points z_i and none of the singular points of $U(t)$. We can evaluate $\psi(z)$ by means of the residue theorem. Let $U(t)$ have the property (8.1),

$$U(z) \approx \left\{ \sum_{j=0}^{m_i-1} u_{i,j}(z - z_i)^j + O\big((z - z_i)^{m_i}\big) \right\}$$

for z near z_i, $i = 1, \ldots, n$. From the calculus of residues we know that the only contributions to (11.106) come from the terms $a_i/(z - z_i)$ after a partial fraction expansion has separated out the higher-order pole contributions. Thus, summing over the different singularities, we have

$$\psi(z) = f(z) \sum_{i=1}^{n} [(m_i - 1)!]^{-1} \left(\frac{d}{dt}\right)^{m_i - 1}$$

$$\times \left[\frac{U(t)(t - z_i)^{m_i}}{(z - t)f(t)} \right] \Bigg|_{t = z_i} \tag{11.107}$$

which is exactly, by (8.4), the Lagrange interpolating polynomial $\varphi(z)$. Thus (11.106), Hermite's formula, is an equivalent expression of Lagrange's interpolation formula. In this form the distinction between specifying a higher-order fit of a value and $\nu - 1$ derivatives at one point and specifying a value fit at ν coincident points disappears. It is convenient for us to simply treat all fitting problems as value fits in which the points need not be distinct.

In order to discuss the limiting behavior as N goes to infinity of the N-point Padé approximant, we must first specify the limiting behavior of points at which the fit is being made. We will denote the sequence of points as

$$\beta_1^{(0)}$$
$$\beta_1^{(1)}, \quad \beta_2^{(1)}$$
$$\cdots \tag{11.108}$$
$$\beta_1^{(n)}, \quad \beta_2^{(n)}, \ldots, \beta_{n+1}^{(n)}$$
$$\cdots \qquad \cdots$$

where the superscript denotes the degree of fit and the subscript distinguishes the point. Of course if we are treating only one sequence, the value $\beta_i^{(j)}$ would be independent of j, but we do not need to make that restriction. Since we will use the Hermite formula (11.106), it will be necessary to have control of the limiting behavior of $f_n(z)$

$$f_n(z) = \prod_{i=1}^{n+1} (z - \beta_i^{(n)}) \tag{11.109}$$

We shall therefore make the following assumptions on the $\beta_i^{(j)}$:

(i)   We can select a closed, bounded point set E whose complement K in the complex plane (including the point at infinity) is connected and contains no limit point of the $\beta_i^{(j)}$.

(ii)

$$\lim_{n \to \infty} \frac{1}{n} \sum_{i=1}^{n+1} \ln|z - \beta_i^{(n)}| = G(z) \qquad (11.110)$$

uniformly on each closed, bounded subset of K, where $G(z)$ tends to infinity as z tends to infinity.

As an example of $\beta_i^{(j)}$ that satisfy these conditions, we can choose $\beta_i^{(j)} = 0$, the one-point Padé case. Then E can be selected as the set $\{z = 0\}$. In this case $G_a(z) = \ln|z|$. For the two-point Padé case we could choose for each j, $\beta_i^{(j)} = 0$ for $i \leqslant [\frac{1}{2} j]$ and $\beta_i^{(j)} = 1$ for $i > [\frac{1}{2} j]$. Then we set $E = \{z = 0, 1\}$ and $G_b(z) = \frac{1}{2} \ln|z(z - 1)|$. As a third example, we pick $\beta_k^{(j)} = \exp[2\pi i k/(j + 1)]$; then we can select E as the set $\{|z| \leqslant 1\}$. By use of a known integral (Peirce, 1910, No. 523), we readily derive that $G_c(z) = \ln|z|$ for $|z| > 1$, and $G_c(z) = 0$ for $|z| \leqslant 1$. We have illustrated these cases in Fig. 11.1 by drawing in the level contours of $G(z)$ (light lines) and the set E (heavy dots and lines). The function $G(z)$ can be thought of as the "two-dimensional" electrostatic potential of a charge of $1/n$ on each $\beta_i^{(n)}$.

With these preliminaries we are now in a position to prove Saff's theorem (1972), which describes the natural region of convergence of the N-point Padé approximant. It makes plain the separate roles played by the

(a)

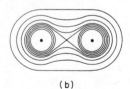

(b)

Fig. 11.1 Level contours of $G(z)$ (light lines) and the set E (heavy dots and lines) for (a) $\{ \beta_i^{(j)} = 0 \}$, (b) $\{ \beta_i^{(j)} = 0$ for $i \leqslant [\frac{1}{2}j]$, $\beta_i^{(j)} = 1$ for $i > [\frac{1}{2}j]\}$, (c) $\{ \beta_k^{(j)} = \exp[2\pi i k/(j + 1)]\}$.

(c)

location of the N points at which the function is fit, as well as the singularity structure.

Theorem 11.7 (Saff). Let there be given a sequence $\beta_i^{(j)}$ ($i = 1, \ldots, j + 1$); $j = 0, 1, \ldots$ and a closed, bounded set E whose complement K (with respect to the extended complex plane) is connected and contains no limit point of $\{\beta_i^{(j)}\}$, let the limit (11.110) exist, and define a $G(z)$ in K which goes to infinity as z goes to infinity. Let Γ_σ denote the locus $G(z) = \ln \sigma$ and let E_σ denote the interior of Γ_σ. Let $F(z)$ be regular in E_ρ except for poles of total multiplicity M. Then the $(L + M + 1)$-point Padé approximant $P_L(x)/Q_M(x)$ derived from fitting at $\beta_i^{(L+M)}$ converges uniformly on the sphere to $F(z)$ interior to and on Γ_σ for any $\sigma < \rho$.

Proof: First we treat the case where all the polar singularities of $F(z)$ are simple poles. Let them be located at $\alpha_1, \ldots, \alpha_M$. We define

$$T_0(z) = 1, \qquad T_k(z) = \prod_{i=1}^{k} (z - \alpha_i), \qquad 1 \leqslant k \leqslant M \quad (11.111)$$

and

$$t_L(z) = \sum_{k=1}^{M} a_k^{(L)} T_{k-1}(z) + T_M(z) \qquad (11.112)$$

Using (11.106), we obtain the interpolating polynomial of degree $L + M$ to the analytic function $t_L(z) T_M(z) F(z)$ as

$$\psi_L(z) = \frac{1}{2\pi i} \int_{\Gamma_R} \left[1 - \frac{f_{L+M}(z)}{f_{L+M}(v)} \right] \frac{t_L(v) T_M(v) F(v)}{v - z} \, dv \quad (11.113)$$

where $f_{L+M}(z)$ is defined by (11.109). We select $\sigma < R < \rho$ such that E_R contains all M of the poles of $F(z)$. The idea is to pick the $a_k^{(L)}$ such that $T_M(z)$ is a factor of $\psi_L(z)$. This requirement holds if and only if $\psi_L(\alpha_i) = 0$, $i = 1, \ldots, M$. It then follows by (11.113) that the $a_k^{(L)}$ must satisfy the equations

$$\sum_{k=1}^{M} c_{jk}^{(L)} a_k^{(L)} = d_j^{(L)}, \qquad j = 1, \ldots, M \qquad (11.114)$$

where

$$c_{jk}^{(L)} = \frac{1}{2\pi i} \int_{\Gamma_R} \left[1 - \frac{f_{L+M}(\alpha_j)}{f_{L+M}(v)} \right] \frac{T_{k-1}(v) T_M(v) F(v)}{v - \alpha_j} \, dv$$

$$(11.115)$$

$$d_j^{(L)} = -\frac{1}{2\pi i} \int_{\Gamma_R} \left[1 - \frac{f_{L+M}(\alpha_j)}{f_{L+M}(v)} \right] \frac{[T_M(v)]^2 F(v)}{v - \alpha_j} \, dv$$

From (11.110) and since all α_j lie inside Γ_R, we deduce as $L \rightarrow \infty$ that $|f_{L+M}(\alpha_j)/f_{L+M}(v)| \rightarrow 0$ uniformly on Γ_R. Hence Eqs. (11.115) become

$$\lim_{L \to \infty} c_{jk}^{(L)} = c_{jk} = \frac{1}{2\pi i} \int_{\Gamma_R} \frac{T_{k-1}(v) T_M(v) F(v)}{v - \alpha_j} \, dv$$

$$\lim_{L \to \infty} d_j^{(L)} = d_j = \frac{-1}{2\pi i} \int_{\Gamma_R} \frac{[T_M(v)]^2 F(v)}{v - \alpha_j} \, dv \tag{11.116}$$

By Cauchy's integral theorem (Copson, 1948) we have

$$c_{jk} \begin{matrix} = 0, & k > j \\ \neq 0, & k = j \end{matrix}, \qquad d_j = 0 \tag{11.117}$$

Thus

$$\lim_{L \to \infty} \det[c_{jk}^{(L)}] = \prod_{l=1}^{M} \frac{1}{2\pi i} \int_{\Gamma_R} \frac{T_{l-1}(v) T_M(v) F(v)}{v - \alpha_l} \, dv \neq 0 \tag{11.118}$$

We conclude therefore that for L large enough we can solve (11.114) for the $a_k^{(L)}$. Since the right-hand side tends to zero, it follows directly that

$$\lim_{L \to \infty} a_k^{(L)} = 0, \qquad 1 \leqslant k \leqslant M \tag{11.119}$$

and hence from (11.112)

$$\lim_{L \to \infty} t_L(z) = T_M(z) \tag{11.120}$$

interior to and on $G(z) = \ln \sigma$, for example.

Next we remove the restriction that the α_j are distinct. We will find that the conclusion (11.120) continues to hold. The only variant in the argument occurs in the definition of the coefficients $c_{jk}^{(L)}$ and $d_j^{(L)}$ in Eq. (11.114). We illustrate the change by supposing that α_1 is repeated μ times, i.e., $\alpha_1 = \alpha_2 = \cdots = \alpha_\mu$. Then, for $T_M(z)$ to be a factor of $\psi_L(z)$, it is necessary and sufficient that $\psi_L(\alpha_1) = \psi_L'(\alpha_1) = \cdots = \psi_L^{(\mu-1)}(\alpha_1) = 0$, plus similar conditions for $z = \alpha_j$, $\mu < j \leqslant M$. One readily establishes that the limits corresponding to (11.116) exist. In particular, we have

$$c_{jk} = \frac{(j-1)!}{2\pi i} \int_{\Gamma_R} \frac{T_{k-1}(v) T_M(v) F(v)}{(v - \alpha_1)^j} \, dv,$$

$$1 \leqslant j \leqslant \mu; \qquad 1 \leqslant k \leqslant M \tag{11.121}$$

Thus again we have in the limit $L \rightarrow \infty$ a lower triangular matrix with nonzero diagonal elements. Consequently, since (11.119) remains valid, so does (11.120).

Now set

$$P_L(z)/Q_M(z) = \psi_L(z)/t_L(z)T_M(z) \qquad (11.122)$$

Since $T_M(z)$ is a factor of $\psi_L(z)$, P_L is a polynomial of degree L and Q_M a polynomial of degree M, by construction. Furthermore, by (11.113), P_L/Q_M is exactly equal to $F(v)$ at the $L + M + 1$ points $\beta_i^{(L+M)}$ [provided $Q_M(\beta_i^{(L+M)}) \neq 0$; however, no additional restriction is required since our proof will allow α_i to be among the β's]. Thus P_L/Q_M is the solution of the fitting problem to $F(v)$ on $\beta_i^{(L+M)}$. The equations in Section 8.B show that the general fitting problem equations can be made identical in structure to the Padé equations and hence the solution is unique by Theorem 1.1.

Since for L large enough $t_L(z)$ is uniformly bounded in the interior of and on Γ_σ where it converges to $T_M(z)$ and where, by (11.113), the comment after (11.115), and (11.122), $P_L(z)$ converges to $T_M(z)F(z)$, and $T_M(z)$ and $T_M(z)F(z)$ do not vanish simultaneously, it follows by the arguments identical to those given at the end of Theorem 11.1 that P_L/Q_M converges uniformly on the sphere to $F(v)$ interior to and on $G(z) = \ln \sigma$. By the arguments which led to (11.116) the numerator and denominator are given of the same structure with regard to convergence as $L \to \infty$, as given by (11.36). ∎

We note that Theorem 11.1 is the special case of Saff's theorem when all $\beta_i^{(j)} = 0$.

12

CONVERGENCE OF GENERAL SEQUENCES

In this chapter we will present theorems on point-by-point convergence of general sequences of Padé approximants. They will generally hold only for a subsequence and will involve assumptions on the Padé approximants themselves rather then purely on the function being approximated, as was the case in the previous chapter.

A. Continuity on the Sphere

In addition to generalizing the notion of convergence as we did in Chapter 11, it is useful to generalize that of continuity as well (Ostrowski, 1925; Hille, 1962, Vol. 2).

Definition. A function $f(z)$ defined on a region R is *continuous on the sphere* at $z = z_0$ if, given any $\epsilon > 0$, there exists a δ depending on ϵ and z_0, such that

$$D(f(z), f(z_0)) < \epsilon, \qquad |z - z_0| < \delta \qquad (12.1)$$

when z and z_0 are in R and $D(x, y)$ is the chord distance given by Eq. (9.21).

In the same way as it follows for ordinary continuity, it also follows for continuity on the sphere that if a function is continuous on the sphere at every point of a closed region R (a region is an open, connected set plus some, all, or none of its boundary points; a closed region contains all its boundary points), it is uniformly continuous on the sphere in R. That is, given any $\epsilon > 0$, there exists a δ, independent of z_0, such that (12.1) holds

166

for all z_0 in R, if R is closed.

We will now show that Padé approximants are continuous on the sphere.

Lemma 12.1. If the $[L/M]$ Padé approximant exists, it is continuous on the sphere at any point z_0, as is any meromorphic function.

Proof: Every Padé approximant is analytic at every point, except for its M poles. At an analytic point it follows from (9.21) that

$$D(f(z), f(z_0)) < 2|f(z) - f(z_0)| \tag{12.2}$$

and so, since an analytic function is automatically continuous, continuity on the sphere follows directly from ordinary continuity. At a pole, $1/f(z)$ is analytic and we can use the same argument as before when we use Eq. (11.2) for $D(x, y)$. The same arguments hold for a meromorphic function. ∎

In addition to continuity, there is the concept of equicontinuity which applies to a sequence of functions.

Definition. A sequence of functions $\{f_n(z)\}$ all defined in a region R is *equicontinuous on the sphere* on R, if, for each z_0 in R and each $\epsilon > 0$, there exists a δ depending on z_0 and ϵ, but independent of n, such that

$$D(f_n(z), f_n(z_0)) < \epsilon, \qquad |z - z_0| < \delta \tag{12.3}$$

where z is in R and $D(x, y)$ is the chord distance given by (9.21).

Again, as before, if R is closed, then one can show that if $\{f_n(z)\}$ is equicontinuous on the sphere for z_0 in R, then it is uniformly equicontinuous on the sphere in R.

We are now in a position to relate the convergence of a sequence of Padé approximants on the sphere to equicontinuity on the sphere.

Theorem 12.1. If $P_n(z)$ is any sequence of meromorphic functions which converges uniformly on the sphere in some closed region R to some limit $f(z)$, then the limit is a meromorphic function in the interior of R and continuous on the sphere in R, and $\{P_n(z)\}$ is uniformly equicontinuous on the sphere in R.

Proof: First we will show that the limit function is continuous on the sphere. By the triangle inequality we have

$$D(f(z), f(z_0)) \leqslant D(f(z), P_n(z)) + D(P_n(z), P_n(z_0)) + D(P_n(z_0), f(z_0))$$

$$\tag{12.4}$$

By assumption we can pick an N such that the first and third terms on the right-hand side of (12.4) are less than ϵ for all z and z_0 in Δ. By Lemma 12.1 we can, after picking a particular $n > N$, select a δ such that the second term is also less than ϵ for all z and z_0 in R satisfying $|z - z_0| < \delta$. Thus $D(f(z), f(z_0)) \leqslant 3\epsilon$ and so f is continuous on the sphere. The uniformity follows from R closed.

We can now show that $\{P_n(z)\}$ is equicontinuous. Again by the triangle inequality

$$D(P_n(z), P_n(z_0)) \leqslant D(P_n(z), f(z)) + D(f(z), f(z_0)) + D(f(z_0), P_n(z_0))$$

$$(12.5)$$

by convergence, by continuity on the sphere of $f(z)$, and given an ϵ, we can find a δ such that for all $n > N$ the right-hand side of (12.5) is less than 3ϵ. Thus, with at most a finite number of exceptions ($n \leqslant N$), $\{P_n(z)\}$ is equicontinuous on the sphere. But the exceptions are, by Lemma 12.1, also continuous on the sphere, so the conclusion about $\{P_n(z)\}$ holds, with uniformity given by R being closed.

Finally, we will show that $f(z)$ is meromorphic. Let z_0 be in the interior of R. First suppose that $f(z_0)$ is finite. Then, by convergence and the equicontinuity of $\{P_n(z)\}$ there exist N, δ, and M such that for all $|z - z_0| < \delta$

$$|P_n(z)| < M, \qquad n > N \tag{12.6}$$

But if $P_n(z)$ is bounded, it is necessarily analytic by virtue of being a meromorphic function. Thus by the standard theorems (Copson, 1948) on a convergent, bounded sequence of analytic functions, $f(z)$ is analytic at $z = z_0$. If, on the other hand, $f(z_0) = \infty$, then by equicontinuity there exist N, δ, and M such that for all $|z - z_0| < \delta$

$$|1/P_n(z)| < M, \qquad n > N \tag{12.7}$$

But if $[P_n(z)]^{-1}$ is bounded, it is again analytic and thus $1/f(z)$ is analytic at $z = z_0$. Thus, since $f(z)$ is at worst the reciprocal of an analytic function, it is at worst meromorphic. ∎

Theorem 12.2. If $P_n(z)$ is any infinite sequence of meromorphic functions which is uniformly equicontinuous on the sphere over a closed region R, then at least a subsequence of the $P_n(z)$ converges uniformly on the sphere to a limit $f(z)$ continuous on the sphere in R and meromorphic in the interior of R.

Proof: We need only prove the existence of a convergent subsequence on the sphere. The rest of the conclusions of the theorem will then follow by Theorem 12.1.

Our first step is to construct a countable set of points which is dense in R. We pick them from the numbers

$$2^{-m}(n + ip) \tag{12.8}$$

and order them first by m. For a fixed m there are only a finite number of n and p in R. We eliminate those numbers that we already have from a smaller value of m. From this sequence, which we label $\{z_k\}$, we can find a member as close as we please to any z in Δ.

Now consider the projection onto the Riemann sphere of $\{P_n(z_1)\}$. It must necessarily be that there is a limit point on the sphere of this sequence. To see this result, divide the sphere in half. In at least one half there must be an infinite number of projections of $\{P_n(z_1)\}$; divide it in half again. Again there will be an infinite number in at least one half. By repeated halving, we can show that there are an infinite number of projections of $\{P_n(z_1)\}$ within as small a distance on the sphere as we please. Therefore we can select a subsequence $\{P_{1,n}(z)\}$ such that

$$\lim_{n \to \infty} P_{1,n}(z_1) = w_1 \qquad \text{(on the sphere)} \tag{12.9}$$

We can now repeat the process to construct a subsequence of $\{P_{1,n}(z)\}$ which converges on the sphere at $z = z_2$, and so on for $z_3, z_4, \ldots$. If we now pick the subsequence

$$f_n(z) = P_{n,n}(z) \tag{12.10}$$

then by construction $\{f_n(z)\}$ converges at every point of $\{z_k\}$. Let us consider the convergence criterion (Definition 11.1). By the triangle inequality we have

$$D(f_n(z), f_m(z)) \leqslant D(f_n(z), f_n(z_k)) + D(f_n(z_k), f_m(z_k))$$

$$+ D(f_m(z_k), f_m(z)) \tag{12.11}$$

Since $\{P_n(z)\}$ is equicontinuous, $\{f_n(z)\}$ is, too. Therefore given an ϵ, we select a δ, and then a z_k, $|z - z_k| < \delta$, which makes the first and third terms on the right-hand side of (12.11) less than ϵ. By the convergence at the z_k we can now pick an $N(\epsilon, z_k)$ such that the second term on the right-hand side of (12.11) is less than ϵ also. Since we can cover R by a finite number of regions $|z - z_k| < \delta$, we can select an $N(\epsilon)$ independent of z and hence have shown that the subsequence $\{f_n(z)\}$ converges uniformly on the sphere in R. The rest of the conclusions follow by Theorem 12.1. ∎

B. Convergence Uniqueness Theorem

We have seen in the two previous theorems that if a sequence of Padé approximants converges, it must necessarily be equincontinuous and if it is equicontinuous, this property is sufficient to ensure the convergence of at least a subsequence. We now show that the limiting function is unique (in a given region) and that, given equicontinuity, the whole sequence converges to it. This theorem is a theorem of the convergence continuation type, where a weaker property over a wide region is used to extend proven meromorphy in a small region or even on only a countable set of points over the wide region. The first theorem of this nature was due to Stieltjes (1894), who showed if a sequence of analytic functions converged in a subregion S contained in a closed region R in which the sequence was uniformly bounded, then the sequence converged to an analytic function in the interior of R. This result was subsequently improved by others, and Vitali (1903) replaced convergence on S by convergence only on a sequence of points $\{z_k\}$ with a limit point interior to R. Montel (1927) extended the result from analytic to meromorphic functions.

Theorem 12.3. If $P_k(z)$ is any sequence of $[L/M]$ Padé approximants such that: (i) $L + M$ tends to infinity with k, and (ii) $\{P_k(z)\}$ is uniformly equicontinuous on the sphere in a closed region R which includes the origin as an interior point, then $\{P_k(z)\}$ converges to a limit on the sphere in R and the limit is that meromorphic function defined by the given power series.

Proof: By Theorem 12.2 there exists at least one convergent subsequence of the $\{P_k(z)\}$ and by Theorem 12.1 any convergent subsequence must converge to a meromorphic limit function. Suppose that there are two limit functions $f(z)$ and $g(z)$. By the defining equations $P_k(0)$ is independent of k, so that we must have $f(0) = g(0) \neq \infty$. Since both $f(z)$ and $g(z)$ are meromorphic, and the origin is an interior point of R, there exist a ρ and an M such that for $|z| \leqslant \rho$ we have $|f(z)| \leqslant M$ and $|g(z)| \leqslant M$. Furthermore, by the convergence of the subsequences $\{f_n(z)\}$ and $\{g_n(z)\}$ of $\{P_k(z)\}$ we also have $|f_n(z)| \leqslant 2M$ and $|g_n(z)| \leqslant 2M$ for $|z| \leqslant \rho$. Now consider $h(z) = f(z) - g(z)$.

By the defining equations for the Padé approximants, we can, by choosing n large enough, make the first N coefficients of $f_n(z)$ and $g_n(z)$ agree. The power series coefficients $f_n(k)$ and $g_n(k)$ of z^k must, by Cauchy's inequalities [see, for instance, Copson (1948, Sec. 4.33)], be bounded by $2M/\rho^k$ independent of n. Thus we obtain

$$|f_n(z) - g_n(z)| \leqslant 4M \sum_{k=N}^{\infty} \left(\frac{z}{\rho}\right)^k = \frac{4M(z/\rho)^N}{1 - (z/\rho)} \tag{12.12}$$

By choosing N, which implies n, large enough we can make (12.12) as small as we like for any z, $|z| < \rho$. Thus we have shown that $f(z) = g(z)$ for any $|z| < \rho$. But this conclusion extends to all of R, for if w is any point in R, we can find a finite sequence of intermediate points $b_0 = 0, b_1, b_2, \ldots, b_m = w$ such that each point b_k is the center of a circle $|z - b_k| \leqslant \rho_k$ which is in R and $|b_k - b_{k-1}| < \rho_{k-1}$. Form the Taylor series to $h(z) = f(z) - g(z)$,

$$\sum_{n=0}^{\infty} h_{n,k}(z - b_k)^n, \qquad k = 0, 1, \ldots, m \tag{12.13}$$

For $k = 0$ all the $h_{m,0} = 0$, as we have shown. Since b_1 is in the circle, all the $h_{m,1}$ are zero. Using the series for $k = 1$, we can next establish that all the $h_{n,2} = 0$, and so on until we finally show

$$h(w) = h_{0,m} = 0 \tag{12.14}$$

Thus $h(z) = 0$ throughout the interior of R. The limit functions $f(z)$ and $g(z)$ are continuous on the sphere in R, by Theorem 12.1, so their identity holds also on the boundary, by continuity on the sphere.

Now let us consider the complete sequence $\{P_k(z)\}$. Since $P_k(0) = f(0) \neq \infty$ by the Padé equations, and since by equicontinuity there exists a δ such that $D(P_k(0), P_k(z)) < \epsilon$ for $|z| < \delta$, it follows easily that $|P_k(z)|$ is bounded in $|z| < \delta$ for $\epsilon(f(0))$ small enough. By using the arguments associated with Eq. (12.12) we can establish that $\{P_k(z)\}$ converges to $f(z)$ in $|z| < \delta$. But this conclusion extends to all of R. We can use the equicontinuity to select a finite sequence of intermediate points, as before, in which $P_k(z) - f(z)$ is bounded, or in the case of a polar singularity of $f(z)$, in which $\{[1/P_k(z)] - [1/f(z)]\}$ is bounded. The argument proceeds as before. Thus we have proved that the sequence $\{P_k(z)\}$ converges to the unique meromorphic function defined by the power series in the interior of R and the result also holds on the boundary, by continuity on the sphere. ∎

This theorem, together with Theorem 12.1, shows that equicontinuity is both necessary and sufficient for the convergence of a sequence of Padé approximants.

An additional theorem is particularly useful for the study of the convergence of other N-point Padé approximants and cases where the origin is not an interior point of the region of convergence for ordinary Padé approximants.

Theorem 12.4. Let $R_k(z)$ be any sequence of meromorphic functions such that (i) $\{R_k(z)\}$ is uniformly equicontinuous on the sphere in a closed region R, (ii) the limit as k tends to infinity of $R_k(z)$ exists for every z in $\{z_n\}$, and (iii) the set $\{z_n\}$ of distinct points has a limit point interior to R. Then the sequence $R_k(z)$ converges in R to a unique function, continuous on the sphere, which is meromorphic in the interior of R.

Proof: By Theorem 12.2 there exists at least a subsequence of the $R_k(z)$ that converges to a meromorphic limit $f(z)$ in the interior of R. If the entire sequence does not converge, there must exist a point w in R such that there is a subsequence of the $R_k(w)$ that does not converge to $f(w)$. Again by Theorem 12.2 we can select a convergent subsequence from that subsequence which converges to a meromorphic function $g(z)$ such that $g(w) \neq f(w)$, but since

$$\lim_{k \to \infty} R_k(z_n) \tag{12.15}$$

exists for $n = 1, 2, 3, \ldots$, we must have

$$f(z_n) = g(z_n), \qquad n = 1, 2, 3, \ldots \tag{12.16}$$

Next consider the subsequence $\{\zeta_n\}$ of $\{z_n\}$, which tends to a limit ζ in the interior of R. Consider the limit

$$\lim_{n \to \infty} f(\zeta_n) = \lim_{n \to \infty} g(\zeta_n) = b \tag{12.17}$$

If $b = \infty$, we will consider the sequence $[R_k(z)]^{-1}$ instead of $R_k(z)$, since it satisfies as well all the conditions of the theorem. Now consider $h(z) = f(z) - g(z)$. Since $f(z)$ and $g(z)$ are meromorphic, so is $h(z)$. By the properties of $h(z)$ there must exist a circle $|z - \zeta| < \rho$ in R where

$$h(z) = \sum_{m=0}^{\infty} a_m (z - \zeta)^m \tag{12.18}$$

but by (12.17), $h(\zeta) = 0$, so that $a_0 = 0$. Next we have

$$a_1 = h'(\zeta) = \lim_{n \to \infty} \frac{h(\zeta_n) - h(\zeta)}{\zeta_n - \zeta} = \lim_{n \to \infty} \frac{h(\zeta_n)}{\zeta_n - \zeta} = 0 \tag{12.19}$$

since $h(\zeta_n) = 0$ and $\zeta_n \neq \zeta$ by (12.16) for all n. We can proceed in this way to show successively that all the $a_m = 0$ and hence $h(z) = 0$ for $|z - \zeta| < \rho$. But we have now reduced the problem to that treated in the proof of Theorem 12.3, i.e., a meromorphic function which vanishes over a disk in R. Thus the conclusions of this theorem now follow by the arguments given there. ■

C. Convergence of General Sequences

There have been a number of theorems proven on the convergence of general sequences of Padé approximants. Quite a few of them follow as corollaries of the following general theorem (Baker, 1975). The proof of this theorem requires two classical lemmas due to Schwarz (1869) and Cartan (1928). We give here a special case of Cartan's lemma. The more general version will be given in Chapter 14.

Lemma 12.2 (Schwarz). If $f(z)$ is analytic in $|z| < R$ and continuous in $|z| \leqslant R$, $|f(z)| \leqslant M$ on $|z| = R$, and $f(0) = f'(0) = \cdots = f^{(n)}(0) = 0$, then

$$|f(z)| \leqslant M(|z|/R)^{n+1} \qquad (12.20)$$

holds for $|z| \leqslant R$.

Proof: First we establish the principle of the maximum modulus. By Cauchy's integral formula if $g(z)$ is analytic in $|z| < R$ and continuous in $|z| \leqslant R$, and $|g(z)| \leqslant M$ for $|z| = R$, then for any a in $|z| < R$

$$\{ g(a) \}^m = \frac{1}{2\pi i} \int_{|z| = R} \{ g(z) \}^m \frac{dz}{z - a} \qquad (12.21)$$

If $\delta = R - |a| > 0$, then (12.21) gives

$$|g(a)|^m \leqslant M^m R / \delta \qquad (12.22)$$

or

$$|g(a)| \leqslant M(R/\delta)^{1/m} \qquad (12.23)$$

Since the left-hand side of (12.23) is independent of m and we can pick m as large as we please, we deduce $|g(a)| \leqslant M$ for $|z| \leqslant R$, which is the principle of the maximum modulus. If we apply this principle to $f(z)/z^{n+1}$, we have

$$\left| \frac{f(z)}{z^{n+1}} \right| \leqslant \max_{|z|=R} \left| \frac{f(z)}{z^{n+1}} \right| = \frac{1}{R^{n+1}} \max_{|z|=R} |f(z)|$$

$$|f(z)| \leqslant M(|z|/R)^{n+1} \qquad (12.24)$$

which establishes Schwarz's lemma. ∎

Lemma 12.3 (Cartan). If $P(z) = \Pi_{\nu=1}^{n}(z - z_\nu)$, then for any $H > 0$ the inequality

$$|P(z)| > (H/e)^n \qquad (12.25)$$

holds outside at most n circles, the sum of whose radii is at most $2H$.

Proof: The idea of the proof is to first classify the given z_ν, then to construct a set of at most n circles, the sum of whose radii is at most $2H$, and finally to prove, for any z outside those circles, that (12.25) holds.

Let λ_1 be the largest integer ($\leqslant n$) for which there is a circle C_1 of radius $\lambda_1 H/n$ containing exactly λ_1 of the z_ν. Suppose there is no such λ_1. In particular this assumption implies that there exists no circle of radius H/n containing exactly one z_ν. Thus any circle containing one z_ν must contain at least two. So, too, must the concentric circle of radius $2H/n$. However, any circle of radius $2H//n$ which contains two z_ν must contain at least three, by assumption. Therefore the concentric circle of radius $3H/n$ contains at least three of the z_ν. By induction we finally show that the concentric circle of radius nH/n contains $n + 1$ of the n z_ν—a contradiction. Therefore λ_1 exists. We put the λ_1 points z_ν that are in circle C_1 in the first class and call them of rank λ_1. We continue in the same way until all the z_ν are grouped into p classes, and $\lambda_1 + \lambda_2 + \cdots + \lambda_p = n$. The sum of the radii of the C_k is thus H.

Let S be any circle of radius $\lambda H/n$. If it contains at least λ of the points z_ν, then at least one of those points is of a rank greater than or equal to λ. If $\lambda > \lambda_2$, then by construction there is no circle of radius $\lambda H/n$ which contains only points z_ν not of class one having λ points z_ν in it. Therefore at least one of the points must be of class one. Similarly, if $\lambda_2 \geqslant \lambda > \lambda_3$, we repeat the argument for points not in the first or second class. We can continue from $\lambda = n$ to $\lambda = 1$ in this manner, thereby establishing the statement.

Now select $\Gamma_1, \Gamma_2, \ldots, \Gamma_p$ to be circles concentric with $C_1, C_2, \ldots, C_p$ but of twice the radius. The Γ_i are the circles referred to at the beginning of the proof. Let z be any point outside the Γ_i. Then a circle S centered on z contains at most $\lambda - 1$ of the z_ν. For, any point of rank λ is inside a C_i of radius at least $\lambda H/n$ and thus inside a Γ_i of radius $2\lambda H/n$. Hence it must be a distance from z of more than $\lambda H/n$. Therefore since S contains no point of rank λ, it can contain at most $\lambda - 1$ of the points z_ν. Next let the z_ν be arranged in order of increasing distance from z; the first one must be at least H/n from z, the second $2H/n$, and so on. Therefore

$$|P(z)| \geqslant (H/n)^n n! \geqslant (H/e)^n \qquad (12.26)$$

which establishes the lemma since the sum of the radii of the Γ_i is less than or equal to $2H$. ■

We can now prove the following theorem (Baker, 1975).

Theorem 12.5. Let there be given a function $f(z)$, with l zeros and m poles (counting multiplicities) meromorphic in an open, simply connected region R of the extended complex plane bounded by a simple closed curve, and containing the origin. Let $\{P_k(z)\}$ be a sequence of $[L/M]$ Padé approximants (assumed to exist) to $f(z)$, with $L + M$ going to infinity with k such that the sum of the number of poles and zeros $n_k(d)$ of $P_k(z)$ in R and more distant on the sphere than any $d > 0$ from the boundary of R satisfies

$$\lim_{k \to \infty} [n_k(d)/(L_k + M_k)] = 0 \qquad (12.27)$$

Further, let T be an arbitrary, closed (on the sphere) set in the interior of R. (We select d so that none of T is closer on the sphere to the boundary of R than d.) Let there be exactly l_1 zeros and m_1 poles (counting multiplicity) of $f(z)$ in the interior of T and an equal number of poles and zeros of $\{P_k(z)\}$ in T and no other limit point of poles or zeros in T. Then the sequence $\{P_k(z)\}$ converges on the sphere uniformly in T to $f(z)$. The conditions on $\{P_k(z)\}$ are also necessary, provided T contains the origin as an interior point and contains no limit point of external poles or zeros.

Proof: First we will prove the necessity. If $\{P_k(z)\}$ converges on the sphere to $f(z)$ in T, there must be at least l_1 zeros and m_1 poles (counting multiplicity) in T since $f(z)$ has them. Suppose there is another pole in T. It must also be that limit $P_k(z) = f(z)$ at its limit point (necessarily in T). However, as $D(\infty, f(z)) \neq 0$, we can uniformly bound $D(\infty, P_k(z)) \geqslant E \geqslant 0$, so $\{P_k(z)\}$ could not be equicontinuous on the sphere in T as the pole moves as close as we like to its limit point. But since this situation contradicts Theorem 12.1, there are no other poles in T. Similarly for zeros. If $L+M$ fails to tend to infinity with k, there must be an infinite number of repetitions of some finite-order Padé approximant in the sequence. Since by the Padé equations these necessarily differ in some derivative at the origin from $f(z)$, they cannot be $f(z)$ and therefore cannot tend to $f(z)$ in the neighborhood of the origin in set T, by Theorems 12.1 and 12.4. Hence it is necessary that $L + M + 1$, the order of the first nonexact derivative, tends to infinity with k. Finally, we give the following counterexample to relaxing condition (12.27). Let $f(x) = (1 - x)^{-1}$ and pick the sequence $\{P_k(x) = [k/0]\}$. We select the sets R and T as shown in Fig. 12.1. The approximants are given by

$$[k/0] = \sum_{j=0}^{k} x^j = (1 - x^{k+1})/(1 - x) \tag{12.28}$$

Thus the zeros are the $(k + 1)$st roots of unity, except for $x = 1$. Therefore, no matter how small a "neck" we choose on set R, a finite fraction of the zeros will lie in that neck. We see directly from Eq. (12.28) that the $[k/0]$ Padé approximants converge to $f(z)$ on that part of T inside the unit circle, but do not in that part outside the unit circle. For a detailed study of the clustering of zeros on the circle of convergence of a Taylor series see Dienes (1957).

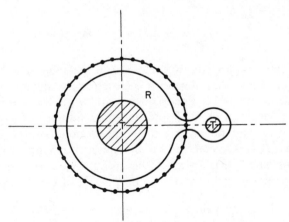

Fig. 12.1 The sets defining the counterexample to relaxing condition (12.27). The dots represent the zeros of the Padé approximants. The sets R and T are those defined in Theorem 12.5.

The idea of the proof of sufficiency is to map R into the unit disk, then to establish the convergence of the logarithmic derivative, and finally to establish the existence of an integration contour linking the origin or a point near it to any point in T. If R is not bounded, we may map it so that it is bounded by means of the transformation $z' = Az/(1 + Bz)$. This mapping preserves the rational fraction character of the Padé approximants by Corollary 9.1. We shall therefore assume, without loss of generality, that R is bounded. The only change in the subsequent arguments is to replace $L + M$ by $2\max(L, M)$, to account for the explicit representation of the pole (zero) at infinity. We note that $(L - M)/(L + M)$ must tend to zero if R includes the point at infinity in order that condition (12.27) be satisfied.

By Riemann's mapping theorem [see, e.g., Hille (1962, Vol. 2)] there exists a function $w = \varphi(z)$ which maps, in a one-to-one fashion, the interior of R onto $|w| < 1$, with $\varphi(0) = 0$, as well as the inverse function $\psi(w) = z$ which maps, in a one-to-one fashion, $|w| < 1$ onto the interior of R, with $\psi(0) = 0$. Under this mapping the Padé approximant

$$P_k(z) = \frac{\Pi_{i=1}^{\lambda}[1 - z/p_i(k)]}{\Pi_{i=1}^{\mu}[1 - z/q_i(k)]} \, h(z) \tag{12.29}$$

where only the zeros and poles in R and more distant on the sphere than d from the boundary of R are displayed explicitly, becomes

$$\mathcal{P}_k(w) = P_k(\psi(w)) = \frac{\Pi_{i=1}^{\lambda}[1 - \psi(w)/p_i(k)]}{\Pi_{i=1}^{\mu}[1 - \psi(w)/q_i(k)]} \, H(w) \tag{12.30}$$

where $H(w)$ is analytic and nonzero for $|w| < 1 - \theta d$. Since the mapping is one-to-one and each factor vanishes for one particular z, we can factor (12.30) as

$$\mathcal{P}_k(w) = \frac{\Pi_{i=1}^{\lambda}[1 - w/u_i(k)]}{\Pi_{i=1}^{\mu}[1 - w/v_i(k)]} \, G(w)H(w) \tag{12.31}$$

where $|u_i| < 1 - \theta d$, $|v_i| < 1 - \theta d$, and $G(w)$ is analytic and nonvanishing for $|w| < 1$. The constant θ depends only on R. Let us define the polynomial

$$\Pi_k(w) = \prod_{j=1}^{\nu} (w - \omega_j) \tag{12.32}$$

where $\nu = \lambda + \mu + l + m$ and the ω_j consist of the u_i, the v_i, and the map of the location of the zeros and poles of $f(z)$ in R. Let $\delta_1 > 0$ be the minimum distance between the set $\varphi(T)$ and the limit points of poles and zeros in $\varphi(R)$ of the $\{P_k(w)\}$ except for the l_1 and m_1 ones assumed in T.

If 2δ is the minimum of θd and δ_1, then, with at most a finite number of exceptions, all the external poles and zeros lie at least δ from any point of T. Thus, since there are only $M + L$ total poles and zeros, we bound, by using the chain rule on (12.29), and inequalities of the type

$$|z - p_i| \geqslant \min_{\text{path}} \{|\psi'(w)|\}|w - u_i|$$

$$\left\| \left[\frac{d \ln f(\psi(w))}{dw} - \frac{d \ln \mathcal{P}_k(w)}{dw} \right] \Pi_k(w) \right\|$$

$$\leqslant \frac{M + L + l + m}{\delta} 2^\nu \frac{\max\limits_{|w| \leqslant 1 - \theta d} \{|\psi'(w)|\}}{\min\limits_{|w| \leqslant 1 - \theta d} \{|\psi'(w)|\}} + K$$

$$= B_k(\delta) \tag{12.33}$$

for w in $\varphi(T)$ and where 2^ν is an upper bound for $\Pi_k(w)$. The constant K comes from the nonsingular part of $d(\ln f(z))/dz$. As $\psi(w)$ is not infinite and $\psi'(w) \neq 0$, the ratio of extrema is a fixed constant depending only on R. By construction the left-hand side of (12.33) is the absolute value of an analytic function in R. For now it suffices to note that if from T we delete small circles of radius ρ about the poles and zeros of $f(\psi(w))$ and $\mathcal{P}_k(w)$ to form a new set T'_k, then $|\Pi_k(w)| \geqslant \rho^\nu$ in T'_k. Thus in T'_k we have, by the Padé equations and Lemma 12.2,

$$\left\| \left[\frac{d \ln f(\psi(w))}{dw} - \frac{d \ln \mathcal{P}_k(w)}{dw} \right] \right\| \leqslant B_k(\hat{\delta})\rho^{-\nu}|w|^{M+L} \tag{12.34}$$

where $\hat{\delta} = \min(\delta, \rho)$. Since $|w| < 1$, and by (12.27) the quantity $\nu/(M + L)$ tends to zero, the right-hand side of (12.34) tends to zero uniformly in T'_k.

Let us now look more closely to see in what part of R the left-hand side of (12.34) need not tend to zero. By Cartan's lemma 12.3, given any $H > 0$, $\Pi_k(w)$ exceeds $(H/e)^\nu$ everywhere except inside at most ν circles of combined radius at most $2H$. Therefore we have, in $|w| < 1 - \theta d$,

$$\left\| \left[\frac{d \ln f(\psi(w))}{dw} - \frac{d \ln \mathcal{P}_k(w)}{dw} \right] \right\| \leqslant B_k(\hat{\delta})\left(\frac{e}{H}\right)^\nu |w|^{M+L} \tag{12.35}$$

except in at most ν circles whose radii satisfy

$$\sum_{i=1}^{\nu} r_i \leqslant 2H \tag{12.36}$$

So again the left-hand side tends to zero except in those certain circles. Within a circle of radius $5H$ of any pole or zero of $f(\psi(w))$ in R one can find a circle around it on which, by (12.35), we have convergence of the logarithmic derivative. Since integration of the logarithmic derivative [the principle of the argument; see, for example, Copson (1948)] around a circle counts the number of zeros less the number of poles inside, we conclude that inside a circle of diameter no more than $5H$ about every pole (or zero) of $f(z)$ in R there is a net of the proper number (counting multiplicity) of poles (or zeros). However, in T this enumeration of poles and zeros exhausts those present by assumption. Thus since this result holds for any $H > 0$, we conclude that $\{(d/dw) \ln \mathcal{P}_k(w)\}$ converges on the sphere to $(d/dw)\ln f(\psi(w))$ on T, and since T is a closed set, uniformly so. If the origin is in T, then in order to prevent the conclusion that the Padé approximants converge from following from what we have proved, a contour of length at least $2\pi\delta$ must be nonconvergent, since every limit point of external poles or zeros is at least δ from a point of T. Otherwise we can find a finite-length contour of integration between any point in T and the origin. However, by Cartan's lemma and (12.35) we can only block a length $2H$, which we can make as small compared to $2\pi\delta$ as we please. Therefore we can integrate to any point in T, over a path on which $(d/dw)\ln \mathcal{P}_k(w)$ is as well converged as we please. This result establishes the convergence of the $\{P_k(z)\}$ on the sphere in T if the origin is in T.

If T does not contain the origin, then an additional argument is required. Let C' be the largest circle of radius $2r$ in R centered at the origin in which $f(z)$ is analytic. In the circle C, $|z| \leqslant r$, $f(z)$ is necessarily bounded. Let

$$f(z) = \sum_{n=0}^{\infty} a_n z^n, \qquad g(|z|) = \sum_{n=0}^{\infty} |a_n z^n| \tag{12.37}$$

Then $g(|z|)$ is also necessarily bounded in C. Define $[t(x)]_n$ to be only those terms in x of degree higher than n. Then the Padé equations (1.16) can be written as

$$Q_M(z)f(z) - P_L(z) = [Q_M(z)f(z)]_{L+M} \tag{12.38}$$

Let us divide the roots of $Q_M(z)$ into two classes. In the first class we put the, necessarily less than or equal to $n_k(d)$, roots in C' and in the second class we put all the others. For z in C we can write

$$|f(z) - \frac{P_L(z)}{Q_M(z)}| \leqslant \frac{|[Q_M(z)f(z)]_{L+M}|}{|Q_M(z)|}$$

$$\leqslant \frac{\{\Pi_{i=1}^M(|z| + |z_i|)\}[g(|z|)]_L}{\Pi_{i=1}^M(|z - z_i|)} \qquad (12.39)$$

For the at most M roots outside of C' and z in C, we have $|z - z_i| \geqslant \frac{1}{2}|z_i|$ and $(|z| + |z_i|) \leqslant 2|z_i|$. For the roots in C', we have $(|z| + |z_i|) \leqslant 3r < 4r$. Thus, if $g(|z|) \leqslant K$ in C, (12.39) becomes, if n_Q is the number of roots of $Q_M(z)$ in C',

$$|f(z) - P_L(z)/Q_M(z)| \leqslant 4^M Kr^{n_Q} / \prod_{i=1}^{n_Q} (|z - z_i|) \qquad (12.40)$$

If we apply Schwarz's lemma 12.2 to (12.40), we obtain

$$|\prod_{i=1}^{n_Q} (z - z_i)[f(z) - P_L(z)/Q_M(z)]| \leqslant 4^M Kr^{n_Q}|z/r|^{L+M} \qquad (12.41)$$

By Cartan's lemma 12.3, given an H, we have

$$|f(z) - P_L(z)/Q_M(z)| \leqslant 4^M K(er/H)^{n_Q}|z/r|^{L+M+1} \qquad (12.42)$$

except on at most n_Q circles of combined radii less than $2H$. For $|z| < 0.2r$, and for whatever $H > 0$ we pick, (12.42) proves the convergence of the Padé approximants to $f(z)$ outside those certain circles for $|z| \leqslant 0.2r$. Thus, since we cannot affect all of these points (r fixed and the H arbitrarily small), we have established the convergence of the Padé approximant at some points. From these points we conclude as before by integration of (12.35) to any point in T that the Padé approximant converges at any point of T. ∎

We can immediately extend Theorem 12.5 by dropping some of the restrictions on R.

Corollary 12.1. Let the assumptions of Theorem 12.5 be satisfied, except that now we only assume R to be a region containing the origin at an interior point. Then the conclusions of Theorem 12.5 hold.

Proof: Consider any point t in T. Since R is connected, we can choose a simply connected subregion of R bounded by a simple, closed curve which contains that point and the origin as interior points. We now apply Theorem 12.5 to this subregion and prove convergence at t. The uniformity follows because T is a closed region on the Riemann sphere. ■

The generality of the theorems given makes it worthwhile to give several special cases, which are perhaps more easily used. The first two special cases are slight generalizations of the theorems of Baker (1965) and Walsh (1967), which involve boundedness assumptions.

Corollary 12.2 (Baker). Let $P_k(z)$ be an infinite sequence of $[L/M]$ Padé approximants to a formal power series, where $L + M$ tends to infinity with k. If $|P_k(z)|$ is uniformly bounded in any region D_1 that contains the origin as an interior point and $|P_k(z)|^{-1}$ is uniformly bounded in any region D_2 that contains the origin as an interior point, then the P_k converge to a meromorphic function $f(x)$ in the interior of the union of D_1 and D_2.

Proof: This corollary follows directly from Theorem 12.3. By the uniform boundedness of the $P_k(z)$ in D_1 it follows that they are analytic, and hence equicontinuous in the ordinary sense in any closed region made up of interior points of D_1. By Eq. (12.2) this implies equicontinuity on the sphere. Likewise we obtain, by considering $[1/P_k(z)]$ and by using form (11.2) of the definition of chordal distance, again equicontinuity on the sphere in any closed region made up of interior points of D_2. Thus the corollary follows by Theorem 12.3. ■

Corollary 12.3 (Walsh). Let Δ be a Jordan region of the extended plane containing the origin, whose boundary is denoted by Γ. Let $w = \varphi(z)$, with $\varphi(0) = 0$, map Δ conformally and one-to-one onto $|w| < 1$ and let Γ_m denote generically the locus $|\varphi(z)| = m$, $0 < m < 1$, in Δ. Let $f(z)$ be analytic at the origin and on Γ, and meromorphic with precisely ν poles in Δ, and suppose the Padé approximants $\{P_k(z)\}$, where $L + M$ goes to infinity with k, are bounded on Γ:

$$|f(z) - P_k(z)| \leqslant M, \qquad z \quad \text{on} \quad \Gamma \qquad (12.43)$$

Suppose $P_k(z)$ has precisely N_k poles in Δ, and $N_k/(L + M) \to 0$. Then we have

$$\lim_{k \to \infty} \sup \left\{ \max[D(f(z), P_k(z))], z \text{ on } T \right\}^{1/(L+M)} \leqslant \max|\varphi(z)|, \qquad z \quad \text{on} \quad T$$

$$(12.44)$$

where T is an arbitrary, closed set in Δ containing only one pole of $P_k(z)$

for each pole of $f(z)$ in the interior of T, and no other limit point of poles.

Proof: This corollary follows from Theorem 12.5 and Eq. (12.34). All that is required is to show that the hypotheses here establish those of Theorem 12.5. Except for (12.27), and that T be free of other limit points of extraneous zeros, these are immediate. We need to bound the number of zeros inside $R = $ interior of Δ, a distance on the sphere $d > 0$ from Γ. Let $\omega(d) = \max[\varphi(z): z$ in Δ, and $D(z, \Gamma) \geqslant d]$. Clearly $\omega(d) < 1$ for $d > 0$. Also let $\psi(w) = z$ be the inverse function to $w = \varphi(z)$ and

$$F_k(w) = G_k(w)P_k(\psi(w)) \tag{12.45}$$

where $G_k(w)$ is a polynomial of degree N_k which vanishes at the N_k poles of $P_k(\psi(w))$ in Δ. By hypothesis $f(z)$ is bounded on Γ, and thus by (12.43) so, too, are the $P_k(z)$, uniformly in k. We can now relate the number of zeros of $F_k(w)$ to this bound. Let $n(r)$ be the number of zeros of $F_k(w)$ for $|w| < r$. Then by the calculus of residues formula (the principle of the argument)

$$n(r) = \frac{1}{2\pi i} \int_{|z|=r} \frac{F_k'(w)}{F_k(w)} \, dw \tag{12.46}$$

By construction $n(r)$ is an increasing function of r. Thus we can write

$$n(\omega)\ln \omega^{-1} = n(\omega) \int_\omega^1 t^{-1} \, dt \leqslant \int_\omega^1 t^{-1} n(t) dt \leqslant \int_0^1 t^{-1} n(t) dt$$

$$\leqslant (2\pi)^{-1} \int_0^{2\pi} \ln|F_k(e^{i\theta})| \, d\theta - \ln|F_k(0)|$$

$$\leqslant (2\pi)^{-1} \int_0^{2\pi} \ln|P_k(\psi(e^{i\theta}))| \, d\theta - \ln|P_k(0)| + N_k \tag{12.47}$$

Thus we can conclude from (12.47), for fixed $\omega(d) < 1$, that

$$n(\omega) \leqslant (C + N_k)/\ln \omega^{-1} \tag{12.48}$$

Hence the sum of the number of poles and zeros satisfies (12.27) by the assumptions of the theorem. In the set T_k', which excludes certain small circles about the poles and zeros of $f(z)$ and $P_k(z)$, the error estimate (12.44) follows directly from integration of (12.34) by taking the $(L + M)$th root and maximizing over T_k', where (12.2) is used. By use of (12.46), where the integration contour is selected as in T_k' but surrounding a pole or zero of $f(z)$, we can show that there is also a (net) pole or zero of $P_k(z)$, provided k is large enough. This identification accounts, by assumption, for all the poles of $P_k(z)$. There cannot be any surplus zeros of $P_k(z)$ over those of $f(z)$, by the convergence of $P_k(z)$ in T'_k and (12.46). Thus the poles and zeros converge to those of $f(z)$. If a boundary point τ of T is a possible limit point of zeros, we can consider $U = T \cup D$, where D is a disk centered at τ of radius $\frac{1}{2}\delta$. By applying what we have shown so far to U instead of T, we conclude such is the case only if $f(z)$ vanishes there. Thus, T is free of extraneous limit points of zeros as the limit points of poles are assumed to be at least δ from T. By integrating (12.34), as explained in the proof of Theorem 12.5, we obtain a bound on the error which yields (12.44) on taking the $(L + M)$th root. ∎

Another application of our results is to the problem of convergence in the disk, when one assumes only that there are no poles of the Padé approximant in the disk beyond those needed to approximate the poles of the function itself. Experience with numerical examples indicated that the only defects in convergence to be expected are pairs of close poles and zeros intruding into the region of convergence. If one assumed that there were no such extra poles, it was hoped one could then prove convergence. Weaker results were, however, all that were obtained initially. Baker (1965) showed that if the poles were assumed absent from the disk $|z| < R$, then the diagonal $[L/M]$ Padé approximants converged inside $|z| < 0.3R$. Chisholm (1966) improved the result to $|z| < (\sqrt{2} - 1)R$, and more recently Zinn-Justin (1971b) showed the result for $|z| < R/\sqrt{3}$. We can now give a corollary to Theorem 12.5 which is valid for $|z| < R$ (Baker, 1975).

Corollary 12.4. Let $f(z)$ be a function which is regular and nonzero at $z = 0$, and meromorphic in the region $|z| < S$ with precisely l zeros and m poles. Let $\{P_k(z)\}$ be any sequence of Padé approximants to $f(z)$ such that $L + M$ tends to infinity with k and such that there are exactly l zeros and m poles (counting according to their multiplicity) of $\{P_k(z)\}$ in $|z| < S$; then $\{P_k(z)\}$ converges to $f(z)$ on the sphere in $|z| < S$.

Proof: As with the previous corollary, the proof consists in showing that the conditions given establish the assumptions of Theorem 12.5. We can select $|z| < S$ as the region R of that theorem and if we are concerned with point z_0, $|z| \leqslant \frac{1}{2}(S + |z_0|)$, for the set T. Since the number of poles and zeros of $\{P_k(z)\}$ is asymptotically finite, condition (12.27) holds and thus this corollary follows immediately. ∎

Clearly we also have the extension of Corollary 12.4.

Corollary 12.5. Let the assumptions of Corollary 12.4 hold except that the region $|z| < S$ is replaced by any open region containing the origin as an interior point. Then the conclusions continue to hold.

We conclude the chapter with a theorem due to Beardon (1968a) which is of a slightly different character. It makes no limiting assumption on the number of poles inside the region R, but instead treats either only a more limited sequence of Padé approximants or a more limited class of functions. Beardon's theorem depends on an estimate for the error of the Padé approximants. We use an estimate similar to Eq. (12.39). Suppose that the function $f(z)$ we are Padé-approximating is of the form $S(z)/T(z)$, where $S(z)$ is an infinite series with a radius of convergence greater than or equal to R and $T(z)$ is a polynomial of degree m all of whose roots lie in the disk $|z| \leqslant R$. Then the Padé equation (1.16) can be written as

$$S(z)Q_M(z) - T(z)P_L(z) = [Q_M(z)S(z)]_{L+M} \qquad (12.49)$$

provided $M \geqslant m$, where the notation is that of (12.38). If we further define

$$S(z) = \sum_{j=0}^{\infty} a_j z^j, \qquad \mathcal{S}(|z|) = \sum_{j=0}^{\infty} |a_j z^j| \qquad (12.50)$$

then we can write, if z_i are the roots of $Q_M(z)$,

$$\left| \frac{S(z)}{T(z)} - \frac{P_L(z)}{Q_M(z)} \right| = \frac{|[Q_M(z)S(z)]_{L+M}|}{|Q_M(z)T(z)|}$$

$$\leqslant \prod_{i=1}^{M} \left(\frac{|z| + |z_i|}{|z - z_i|} \right) \frac{\mathcal{S}(|z|)}{|T(z)|} \qquad (12.51)$$

By the use of Schwarz's inequality (Lemma 12.2), we can derive from (12.51), by noting that $SQ - PT$ is analytic in $|z| < R$ and by using the Padé equations,

$$\left| \frac{S(z)}{T(z)} - \frac{P_L(z)}{Q_M(z)} \right| \leqslant \left(\frac{z}{R} \right)^{L+M+1} \prod_{i=1}^{M} \left(\frac{|z| + |z_i|}{|z - z_i|} \right) \frac{\mathcal{S}(|z|)}{|T(z)|} \qquad (12.52)$$

With this inequality we can prove Beardon's theorem (1968a).

Theorem 12.6 (Beardon).   Let $f(z)$ be analytic in $|z| < R$ except for m nonzero poles there, and let E be any closed set of $|z| < R$ on which $f(z)$ is analytic. For each $\delta > 0$ there exists a number k ($k \geqslant 1$ and depends only on f, E, and δ) such that if $[L/M]$ is any sequence of Padé approximants to $f(z)$ whose poles are at least δ from E and for which $L \geqslant kM \geqslant M \geqslant m$, then $[L/M]$ converges uniformly to $f(z)$ on E as $L + M \to \infty$. In particular this result is applicable when $L = kM$ or when M is constant.

Proof: By hypothesis $S(|z|)/|T(z)|$ takes on a fixed, finite maximum for z in E. Also, by hypothesis, since $|z - z_i| > \delta$ for z in E we can bound

$$\frac{|z| + |z_i|}{|z - z_i|} \leqslant \frac{2|z| + |z - z_i|}{|z - z_i|} \leqslant \frac{2|E|}{\delta} + 1 \tag{12.53}$$

where $|E|$ is defined as the maximum of $|z|$ for z in E. Thus (12.52) becomes (using $L = kM$)

$$\left| \frac{S(z)}{T(z)} - \frac{P_L(z)}{Q_M(z)} \right| \leqslant \left\{ \max_{z \in E} \left[\left| \frac{z}{R} \right| \frac{S(|z|)}{|T(z)|} \right] \right\} \left[\left(1 + \frac{2|E|}{\delta} \right) \left(\frac{|E|}{R} \right)^{k+1} \right]^M \tag{12.54}$$

Since $|E|/R < 1$, by hypothesis, we can select a k so that the quantity raised to the Mth power in (12.54) is less than unity. Thus we can make the right-hand side as small as we please as M tends to infinity. Clearly the same conclusion holds, $M \geqslant m$ fixed, as $L \to \infty$. ∎

We can give a generalization of Beardon's subsidiary result for meromorphic functions.

Corollary 12.6 (Beardon). Let $f(z)$ be analytic at $z = 0$, meromorphic in the complex plane, and $\{P_k(z)\}$ be any sequence of Padé approximants for which M tends to infinity with k. If E is any closed set such that there are no poles of $f(z)$ and no limit points of poles of the $\{P_k(z)\}$ in E, then $\{P_k(z)\}$ converges uniformly to $f(z)$ on E as $k \to \infty$.

Proof: Since E is free of poles and closed, the poles must lie (with a finite number of exceptions) at least a distance $\delta > 0$ from E. If we select R so that

$$\left(1 + \frac{2|E|}{\delta} \right) \frac{|E|}{R} < 1 \tag{12.55}$$

then convergence follows from (12.54). ∎

If L does not also tend to infinity, then the size of the set E on which the hypotheses can be satisfied is limited by Theorem 11.1 and Corollary 11.3.

Beardon's theorem and its corollary can easily be extended to allow a pole of the approximant for each one of $f(z)$ in circles of meromorphy of $f(z)$. Then convergence on the sphere results.

13

THE DISTRIBUTION OF POLES AND ZEROS

In Chapter 11 we showed for a variety of functions that the vertical and horizontal sequences of Padé approximants reproduce the function $f(z)$ being approximated in circles (centered on the origin) of meromorphy. Each pole and zero of $f(z)$ is approximated, in order of its distance from the origin, to the extent possible with the particular sequence selected. In the simple case of a meromorphic function and a vertical sequence of Padé approximants, $[L/m]$ with m fixed, the first m poles of $f(z)$ are approximated by the Padé approximant, and almost all of the zeros lie on a circle whose radius is the distance from the origin to the $(m + 1)$th pole. [See Dienes (1957) for a discussion of the clustering of zeros of sections of a Taylor series on the circle of convergence.] By the method of Chapter 11, Section F, we can construct sequences of Padé approximants in which both L and M go to infinity together. However, these are still basically either vertical or horizontal sequences. Although they converge in the entire complex plane when the function being approximated is meromorphic, their mode of convergence is still centered on the origin. This feature of this type of approximant is apparent when we consider their behavior in the presence of a "smooth," nonpolar singularity. In this case convergence is confined to the largest circle (center at the origin) of meromorphy. For vertical (or near vertical) sequences the extra poles cluster at the nonpolar singularity, and almost all the extra zeros lie on the circle of convergence. The behavior of vertical or horizontal sequences of Padé approximants is quite well characterized, or characterizable, by rigorous theorems for functions whose closest nonpolar singularity is "smooth." For more general functions, as Perron's example showed, the situation is less well defined. The appearance of poles with very close zeros in the region of convergence has made general results difficult to obtain. In Perron's example it turned out that, in spite of every point of the complex plane being a possible limit point of poles, there existed a subsequence of Padé approximants which converges at every point of the complex plane. This result was extended, for certain vertical sequences, to certain types of entire functions.

In Chapter 12 we turned our attention to characterizing necessary and sufficient conditions for convergence and to examining the consequences of convergence when and where it occurs. We found that where the Padé approximant converges, it converges to the meromorphic function defined by the Taylor series. A most useful way to consider the problem is in terms of Riemann's spherical representation of the complex numbers. In these terms we found that the convergence of a sequence of Padé approximants is equivalent to that sequence being equicontinuous on the sphere. That is, a single δ can be selected for the whole sequence so that if $|z - z_0| < \delta$, then the distance on the sphere between any Padé approximant at z and at z_0 is less than a preassigned ϵ for all z and z_0 in some relevant closed region R. This result reduces a more difficult property, convergence, to a simpler one, equicontinuity. The property of equicontinuity necessarily excludes the appearance of the arbitrarily close poles and zeros which we found in Perron's example. We showed convergence of general sequences of Padé approximants on the assumption that extraneous poles and zeros are excluded from a particular region. Various assumptions about the distribution of extraneous poles and zeros and the character of the function outside this region lead to conclusions of varying strengths.

At this point there are two approaches to the treatment of convergence. The first, which we consider in this chapter, concerns itself with the existence of a subsequence of Padé approximants which converge in the usual or point-by-point sense. The second approach seeks to define the region of convergence in some weaker sense, such as convergence in measure, convergence almost everywhere, or convergence in the mean on the sphere. Examples will be given which show that this type of convergence is the strongest that can be obtained for the entire sequence of, say $[M/M]$ Padé approximants. Once the distribution of poles and zeros is known the results of the previous and next chapters should suffice to illuminate the convergence properties of the Padé approximants. Unfortunately, we do not in general know these distributions. Therefore we will present the Padé conjecture of Baker *et al.* (1961), for which there is currently neither a proof nor a counter example and which summarizes a great deal of what is surmised about the important diagonal sequence of Padé approximants.

A. The Baker – Gammel – Wills Conjecture

The following conjecture was made by Baker *et al.* (1961) concerning the existence of convergent subsequences of Padé approximants. We first give it as originally stated and then restate it, and generalize it somewhat in terms of the concepts of Chapter 12.

Conjecture 1. If $P(z)$ is a power series representing a function which is regular for $|z| \leq 1$, except for m poles within this circle and except for $z = +1$, at which point the function is assumed continuous when only points $|z| \leq 1$ are considered, then at least a subsequence of the $[M/M]$ Padé approximants converges uniformly to the function (as M tends to infinity) in the domain formed by removing the interiors of small circles with centers at these poles.

We remark that the study of any sequence of $[M + k/M]$ Padé approximants can be reduced to the study of the $[M/M]$ Padé approximants by simply considering Padé approximants to

$$g(z) = f(z)z^{-k} \tag{13.1}$$

if $k \leq 0$ or

$$g(z) = z^{-k}\left[f(z) - \sum_{j=0}^{k-1} f_j z^j \right] \tag{13.2}$$

if $k > 0$, where f_j are the coefficients of the power series expansions of $f(z)$.

In line with the concepts of Chapter 12, we can rephrase and generalize the Conjecture 1 as follows.

Conjecture 2. If $P(z)$ is a power series representing a function which is meromorphic in $|z| < 1$ and continuous on the sphere in $|z| \leq 1$, then at least a subsequence of the $[M/M]$ Padé approximants is equicontinuous on the sphere in $|z| \leq 1$.

By application of Theorem 12.3 the original conjecture follows from Conjecture 2. Conjecture 2 is broader than the original conjecture because it relaxes the assumptions on the function being approximated. When applied to the same class of functions the equivalence of the conjectures follows directly from Theorem 12.3.

B. Extensions of the Conjectures

We can use Theorem 9.1 to extend the scope of the conjecture. Theorem 9.1 showed that the process of Padé approximation is, for the $[M/M]$, invariant under linear fractional trnsformations. For example, we have the following result.

Theorem 13.1. If Conjecture 2 holds and $f(z)$ is regular, except for a finite number of poles in a closed half-plane containing the origin, and continuous in that half-plane at infinity, then at least a subsequence of the $[M/M]$ Padé approximants converges to it uniformly on the sphere in that half-plane.

Proof: By Theorem 9.1 we may study equivalently $g(w) = f(A^{-1}w/(1-w))$. If the bounding line for the closed half-plane passes thru the point $-1/(2A)$ and is perpendicular to the ray from the origin at that point, then the transformation $w = Az/(1 + Az)$ maps the closed half-plane onto the unit circle. The point $-1/(2A)$ goes into -1, and the point at infinity goes into $+1$. The origin is left unchanged. Thus $g(w)$ is regular in the unit circle except for a finite number of poles in $|w| < 1$ and is continuous at $w = +1$. Hence, by Conjecture 2, the conclusion of this theorem follows. ∎

This theorem can be immediately extended to yield the following result.

Theorem 13.2. If Conjecture 2 holds and $f(z)$ is regular except for a finite number of poles in a circle which contains the origin as an interior point and continuous on the sphere at the boundary of the circle, then at least a subsequence of $[M/M]$ converges to $f(z)$ in and on the circle.

Proof: By Theorem 9.1 it suffices to show that any such circle can be mapped into the unit circle by a linear fractional transformation. For a circle with center at C and of radius R the transformation

$$w = z / [R(1 - |C/R|^2) + C^*z/R] \tag{13.3}$$

accomplishes the job, where C^* is the complex conjugate of C. Since the origin is an interior point, $|C/R| < 1$, so the transformation is one-to-one. The theorem now follows from Conjecture 2 as did Theorem 13.1. ∎

Corollary 13.1. If Conjecture 2 holds and $f(z)$ is regular in the union of any number of circles and half-planes, except for a finite number of poles in each, and is continuous on the sphere on the boundary of this region, then there exists at least a subsequence of $[M/M]$ which converges at any point of this region to $f(z)$.

Proof: As any point of the region considered is in some one circle or half-plane, the conclusion follows at once from Theorem 13.1 or 13.2 respectively. ∎

To illustrate these results, we will look again at the functions $t(x)$, $w(x)$, $u(x)$, $v(x)$, $s(x)$, $c(x)$, and $d(x)$ introduced in Chapter 10, Eqs. (10.1), (10.3) and (10.16). The limiting figure will always be found, by Corollary 13.1, by drawing circles through the origin and the nonpolar singularities. In Fig. 13.1 we illustrate this construction for the function $w(x)$. The region outside circle B includes the origin and excludes the singularities at $x = (12 \pm 5i)/13$, while the inside of circle A includes the origin and excludes the nonpolar singularities. The entire complex plane, except for the shaded area, is thus included in the region of convergence. The only points which cannot be so included are those on the arc of a circle through the origin and the singular points. As we observed in Chapter 10, all the poles and extraneous zeros do in fact lie on the arc of the circle, as predicted by Corollary 13.1. In this case the location of the poles and zeros is completely specified by the conjecture and the subsequence is the whole sequence.

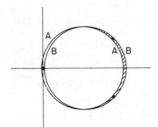

Fig. 13.1 Construction of the location of cuts from Corollary 13.1 for the function $w(x)$.

The cuts shown in Fig. 10.1 for the function $u(x)$ are precisely given by the conjecture as the limit of very large circles passing to the right of $+i$, to the left of the origin, and to the right of $-i$, and vice versa. Figure 10.1 also shows the areas which must, by the conjecture, enclose the extraneous poles and zeros for $t(x)$ and $d(x)$. For all these cases the conjecture is borne out by the numerical examples. In fact, as pointed out in Chapter 10, the limiting behavior of the poles and zeros seems to be a cut which lies within the prescribed area. We are of course assured convergence by Theorem 12.5 in any such area that is free from extraneous poles and zeros.

As a further illustration of the implications of Corollary 13.1, Fig. 13.2 shows the regions implied by the nonpolar singularities of the functions $v(x)$, $s(x)$, and $c(x)$. Again the evidence of the numerical calculations confirms the conjecture.

The general picture as seen at the present time is as follows. For a particular sequence of Padé approximants the vast majority of the poles and zeros, beyond those required to represent the poles and zeros of the function, cluster along a curve or curves which cut the complex plane in such a way as to leave the function single valued in that part of the plane connected to the origin. For horizontal or vertical sequences the boundary curve is expected to be a circle centered on the origin [there may be

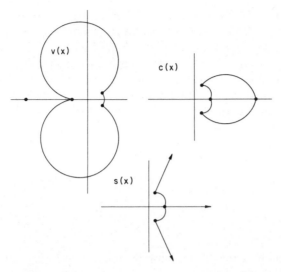

Fig. 13.2 The curves in this figure divide the complex plane into two regions for each function. Corollary 13.1 implies convergence in the region containing the origin.

subsequences which do better than this; see Dienes (1957) for a discussion of overconvergence]. In the case of the N-point Padé approximant the circles are distorted into the level contours of $G(z)$ (Section 11.H). Beyond these boundaries convergence of those sequences to $f(z)$ generally fails. For diagonal $[M/M]$ sequences of Padé approximants the locations of the curves where the extraneous poles and zeros cluster seem to be very much determined by the invariance properties of the Padé approximants with respect to the linear fraction group. In the case of meromorphic functions there are an infinite number of poles and zeros which are relevant for those of the Padé approximants to converge to. Essential singularities, as can be illustrated by the formula

$$\exp\left(\frac{1}{1+x}\right) = \sum_{n=0}^{\infty} \frac{1}{n!}\left(\frac{1}{1+x}\right)^n \tag{13.4}$$

look very much, if one is not too close, like a high-order pole. The Padé approximants treat them in just this way and cluster poles and zeros at the singular point, rather than on a curve.

In the case of a function with a natural barrier, i.e., a curve of singularities which separates the complex plane into two or more disconnected pieces, poles and zeros are clustered along the natural barrier. Convergence may or may not extend beyond, according to how solid the barrier is (Gammel, 1974).

Recently Nuttall (1973) has found how the Padé approximants define their cuts for a certain class of functions which have $2l$ square-root-type branch points in arbitrary locations. These functions can be made single valued by connecting, in any order, the branch points by cuts. The Padé approximants make the smallest possible cuts in a very special sense. If we look at the cuts in the $1/z$ plane, then the correct set is that set with the smallest transfinite diameter. By the transfinite diameter of a set E in the w plane we mean

$$\rho(E) = \lim_{n \to \infty} \max_{w_j \in E} \left| \det \begin{vmatrix} 1 & w_1 & \cdots & w_1^{n-1} \\ 1 & w_2 & \cdots & w_2^{n-1} \\ \vdots & \vdots & \ddots & \vdots \\ 1 & w_n & \cdots & w_n^{n-1} \end{vmatrix} \right|^{2/n(n-1)} \quad (13.5)$$

The determinant is the Vandermonde determinant and is the product of the $\frac{1}{2}[n(n-1)]$ differences $w_i - w_j$. The transfinite diameter of a straight line segment is one-fourth its length, that of a circle (or solid disk also) is its radius. The concept of transfinite diameter is related to Tschebycheff polynomials and logarithmic capacity (Hille, 1962). Nuttall finds a subsequence of the diagonal Padé approximants which converges point by point outside these cuts.

14

CONVERGENCE IN HAUSDORFF MEASURE

In this chapter we will discuss a weaker type of convergence than we did in the previous chapters. Instead of making assumptions on the location of the poles of the Padé approximants, we will prove that if they disrupt the convergence, the size of the region over which this disruption occurs is in some sense small. The first theorem of this type was due to Nuttall (1970), although Chisholm (1966) gave a transitional theorem where under mild assumptions the area of the disruption was proven to tend to zero. Later refinements have made by Wallin (1972), Zinn-Justin (1971b), and Pommerenke (1973). Our presentation in this chapter is based significantly on all of these results. We give examples due to Gammel (1970) and Wallin (1972) which show that these results are the best possible for the whole sequence of Padé approximants. The question of point-by-point convergence of a subsequence is not yet settled.

A. Hausdorff Measure

In order to make precise the size of the regions in which the convergence is not controlled, we define the following measure of the size of a bounded set E in the complex plane:

$$\Lambda_\alpha(E) = \min\left\{ \sum_{i=1}^{\infty} [\delta(D_i)]^\alpha \right\} \qquad (14.1)$$

where the D_i are a denumerable family of circular disks which cover every point of the set E, and $\delta(D_i)$ is the diameter of the circular disk D_i. As an example, for $\alpha = 2$, Eq. (14.1) differs from the area of the covering disks by a factor of $\frac{1}{4}\pi$. It is easy to work out that if E is a sufficiently simple region of area A, by considering a covering of disks of the same small radius centered on the sites of a triangular lattice (see Fig. 14.1), then

$$8A/3\sqrt{3} \geqslant \Lambda_2(E) \geqslant 4A/\pi \tag{14.2}$$

Thus $\Lambda_2(E)$ is very much like the ordinary area. The quantity $\Lambda_1(E)$ is in a sense like the diameter of a set. If E is a simply connected region, then $\Lambda_1(E)$ is the diameter of the smallest circumscribed circle. Thus $\Lambda_1(E)$ is related to the length of a set. If the set E is a finite-length curve, it will have $\Lambda_2(E) = 0$, since it can be covered by disks of arbitrarily small area, but $\Lambda_1(E) \neq 0$. Since any finite set can be covered by of the order of N^2 disks of diameter $1/N$, we can bound $\Lambda_\alpha(E)$ by $N^{2-\alpha}$. Thus if $\alpha > 2$, $\Lambda_\alpha(E) = 0$, Our definition of $\Lambda_\alpha(E)$ is not quite the same as the usual one for Hausdorff measure, but has been adapted for our purposes. However, sets with $\Lambda_\alpha(E) = 0$ also have Hausdorff measure zero and vice versa. The relation between α and other concepts of the dimension of the set $[\Lambda_\alpha(E) = 0$ if α is greater than the "dimension" of the set] is discussed in detail by Hurewicz and Wallman (1941).

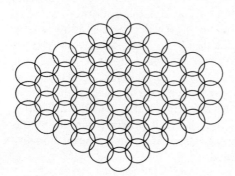

Fig. 14.1 A covering of the plane by disks of uniform radius. Their centers form a triangular lattice.

In the proof of Theorem 12.5 the notion of uncontrolled regions of size $\Lambda_1 \to 0$ played a key role. The essential tool in bounding Λ_1 was Cartan's lemma 12.3. It is easily extended (Cartan, 1928) to give:

Lemma 14.1 (Cartan).   If $P(z) = \prod_{\nu=1}^{n}(z - z_\nu)$, then for any $H > 0$ the inequality

$$|P(z)| > (H/e^{1/\alpha})^n \tag{14.3}$$

holds outside at most n circles, the radii of which satisfy

$$\sum_{i=1}^{n} r_i^\alpha \leqslant (2H)^\alpha \tag{14.4}$$

Proof: This proof proceeds just as did that for Lemma 12.3, except that where the radii of circles C_i were selected to be $\lambda_i H/n$ and $\lambda H/n$, we now select $(\lambda_i/n)^{1/\alpha}H$ and $(\lambda/n)^{1/\alpha}H$ instead. The conclusion then follows as before. ∎

B. Error Formula

We next need to derive a good general bound on the difference between the Padé approximant and the function being approximated. We make no assumptions on the properties of the Padé approximants themselves, beyond their existence. However, the existence of an infinite sequence is assured by Theorem 2.4 and direct extensions thereof.

Lemma 14.2. Suppose $f(z)$ is regular in $|z| \leqslant R$, except for a finite number of poles and a finite number μ of essential singularities, and $|f(z)|$ is bounded on $|z| = R$; then there exists an M_0 such that for $M > M_0$, $\mu L < \lambda M$, $\infty > \lambda > 0$,

$$|f(z) - [L/M]| \leqslant \left|\frac{z}{R}\right|^{L+M+1} \frac{K \max_{|z|=R}\{|Q_M(z)R_M(z)|\}}{(1 - |z|/R)|Q_M(z)R_M(z)|\Delta^p} \quad (14.5)$$

everywhere in $|z| \leqslant R$ except for z further than some distance ω from any essential singularity, where $R_M(z)$ is a polynomial satisfying:

(a) $R_M(z)f(z)$ is bounded at poles of $f(z)$.
(b) $R_M(z)$ vanishes at essential singularities of $f(z)$ like $(z - w_k)^p$, where $p/(M + L) \geqslant \rho$, as $M + L$ tends to infinity.
(c) The degree of $R_M(z)$ is $\leqslant M$ and the coefficient of the highest power of z is one.

Proof: The proof of this lemma is based on Hermite's formula for an interpolating polynomial (11.106). The function $f(t) = t^{L+M+1}$ for the $[L/M]$ Padé approximant. Thus, rewriting (11.106), we obtain by Cauchy's theorem

$$R_M(z)Q_M(z)f(z) - R_M(z)P_L(z) = \frac{z^{L+M+1}}{2\pi i} \oint \frac{R_M(t)Q_M(t)f(t)\,dt}{(t - z)t^{L+M+1}}$$

$$(14.6)$$

where $U(t)$ is taken to be $R_M(z)Q_M(z)f(z)$. That $R_M(z)P_L(z)$ is the correct interpolating polynomial $\psi(z)$ follows because by definition ψ is of degree less than or equal to $L + M$ and the coefficients of z^k, $k = L + 1, \ldots, L + M$, in $Q_M(z)f(z)$ vanish by the Padé equations.

The contour of integration in Eq. (14.6) is a circle about the origin which includes no poles or singularities of $f(t)$. Since we may choose M as large as we need, we will assign zeros of $R_M(z)$ as follows. First, we assign a zero at every pole of $f(z)$, according to multiplicity—that is, a double zero of $R_M(z)$ at a double pole of $f(z)$, etc. We next assign p zeros to each of the μ (μ may be zero) essential singularities of $f(z)$ inside $|z| < R$. Finally, we assign the remainder (up to a maximum of M zeros) of the zeros in any way that is convenient. In case there are not enough zeros selected in this way, $R_M(z)$ will be of degree less than M.

We now distort, by Cauchy's theorem, the contour of integration to lie on the circle $|z| = R$. Since the integrand is now regular at the poles of $f(t)$, we may pass over them without difficulty. The essential singularities may not be so crossed and we must adjoin small circles (traversed clockwise) surrounding these μ points w_k. If we denote the radius of these circles by δ_k, then we can bound their contributions by

$$R^{-L-M-1} \sum_{k=1}^{\mu} \max_{|t-w_k|=\delta_k} \left\{ \frac{|Q_M(t)f(t)R_M(t)|\delta_k{}^p}{|t-z|\,|t-w_k|^p|w_k/R|^{M+L+1}} \right\} \quad (14.7)$$

Here $|f(t)|$ is a fixed constant, independent of L and M and dependent only on δ_k. The maxima of Q_M and $R_M(t)/(t-w_k)^p$, by the principle of the maximum modulus [(12.21)–(12.23)], are bounded by their maxima on $|t| = R$. By selecting

$$\delta_k|R/w_k|^{(M+L+1)/(p-1)} < 1 \quad (14.8)$$

we can, by letting p go to infinity and restricting t to lie outside small circles around the essential singularities, bound (14.6) by

$$R^{-L-M-1}(K/\Delta^p) \max_{|t|=R} \{|Q_M(t)R_M(t)|\} \quad (14.9)$$

where Δ is the minimum of $(R - |w_k|) > 0$, and k is a constant independent of M or L. In order to do this, we must select δ_k independent of L and M. Thus we choose $M/\mu > p \geqslant \rho(L + M + 1)$, where $\rho > 0$. Hence there exists an M_0 such that (14.8) is valid for $M > M_0$ and this choice of p. If $\mu = 0$, of course, these terms do not arise, and we may set $\Delta = 1$ in that case. The bound (14.9) is clearly still valid if we replace those $\Delta > 1$ by $\Delta = 1$. Since $f(z)$ is bounded, by assumption, on $|z| = R$, the integral on this contour is again bounded by a term of the form (14.9) with $\Delta = 1$; thus the conclusion (14.5) follows directly from (14.6). ∎

The combination of this error formula with Cartan's lemma will lead to the convergence theorems of this chapter.

C. Convergence Theorems

There are two principal types of theorems available. First, we consider diagonal-type sequences ($\infty > \lambda^{-1} \geqslant L/M \geqslant \lambda > 0$) of Padé approximants. The proof here requires explicit assumptions on the analytic properties of the function being approximated throughout the entire complex plane. Then we consider nondiagonal sequences of Padé approximants, i.e., L/M tends to zero or infinity. Here we need only make assumptions on the analytic properties inside a circle. Finally, we show how these results can be related to some other forms of convergence.

Theorem 14.1. Let $f(z)$ be regular in $|z| \leqslant R$ for all $0 \leqslant R < \infty$, except for a finite number (counting multiplicity) of poles $\nu(R)$ and essential singularities $\mu(R)$. Then, given any $\epsilon, \delta > 0$ and $1 \geqslant \lambda > 0$, there exists an M_0 such that for all $M > M_0$, $\lambda^{-1}M > L > \lambda M$,

$$|f(z) - [L/M]| < \epsilon \tag{14.10}$$

for all z in a given closed, bounded region $\mathcal{R}$ of the complex plane except for a set of points $E(L/M)$, where $\Lambda_\alpha(E(L/M)) < \delta$, for any fixed $\alpha > 0$.

Proof: Without loss of generality we may select $\mathcal{R}$ as the unit disk because any allowed $\mathcal{R}$ can be enclosed in a finite circle and by a scale change we can transform it into the unit circle. First we select

$$R = \tfrac{1}{4}\left[24(2e/\delta)^{1/\alpha}\right]^{1+2/\lambda} \tag{14.11}$$

as the radius of the circle we will use to apply Lemma 14.2. This choice of R will determine a Δ, the distance of the closest essential singularity in $|z| < R$ to $|z| = R$. Next we choose the ratio $p/M > 0$ to be less than the minimum of $-(\ln \Delta)/\ln 2$ and $\tfrac{1}{2}\lambda/\mu(R)$. We then choose $p \to \infty$ with M according to this ratio. This choice ensures that the condition on p in Lemma 14.2 is satisfied, that the degree of $R_m(z)$ can be made $\leqslant M$, and that $\Delta^{p/M} > \tfrac{1}{2}$. We choose not to have any additional roots of $R_M(z)$ beyond the $\nu(R) + p\mu(R)$ required to use Lemma 14.2. The maximum value of $R_M(z)$ on $|z| = R$ is then bounded by

$$|R_M(z)| \leqslant (2R)^{\nu(R) + p\mu(R)} \tag{14.12}$$

since each factor can be at most $2R$. Also, we choose the circles about the essential singularities to be of radius less than $(\tfrac{1}{2}\delta/\mu)^{1/\alpha}$ and $\tfrac{1}{2}|w_k/R|^{2\mu(1+\lambda^{-1})}$. Thus with these choices we know that there exists an M_0 such that we can apply Lemma 14.2 and hence obtain

$$|f(z) - [L/M]| \leqslant |\frac{z}{R}|^{L+M+1} \frac{K(2R)^{\nu(R)+p\mu(R)} \max_{|z|=R}\{|Q_M(z)|\}}{(1-|z|/R)|Q_M(z)R_M(z)|\Delta^p}$$

(14.13)

Let us now divide the roots of $Q_M(z)$ into two groups so that

$$|z_i| \leqslant 2R, \quad i \leqslant m; \qquad |z_i| > 2R, \quad i > m \qquad (14.14)$$

For $|z| \leqslant R$ we can write the bounds

$$|z - z_i| \leqslant 3R, \quad i \leqslant m; \qquad \tfrac{1}{2}|z_i| \leqslant |z - z_i| \leqslant \tfrac{3}{2}|z_i|, \quad i > m$$

Thus for $|z| \leqslant R$

$$\frac{\max_{|z|=R}\{|Q_M(z)|\}}{|Q_M(z)|} \leqslant \frac{(3R)^m}{\prod_{i=1}^m |z - z_i|} \prod_{i=m+1}^{M} \frac{\tfrac{3}{2}|z_i|}{|z - z_i|} \leqslant \frac{3^M R^M}{\prod_{i=1}^m |z - z_i|} \qquad (14.15)$$

Since $R > 1$. If we now apply Lemma 14.1 to the polynomial $R_M(z)\prod_{i=1}^m(z - z_i)$, we have, except for z in at most $m + \nu(R) + p\mu(R)$ circles which satisfy (14.4),

$$|f(z) - [L/M]| < |\frac{z}{R}|^{L+M+1} \frac{K(3R)^{M+\nu(R)+p\mu(R)}}{\Delta^p(1-|z|/R)(H/e^{1/\alpha})^{M+\nu(R)+p\mu(R)}}$$

$$< K'\left[\left(\frac{3Re^{1/\alpha}}{H}\right)^{1+[\nu(R)+p\mu(R)]/M} \Delta^{-p/M}R^{-1-\lambda}\right]^M$$

(14.16)

Since $|z| \leqslant 1$. By choosing $H = \tfrac{1}{2}(\tfrac{1}{2}\delta)^{1/\alpha}$, so that $\Lambda_\alpha < \tfrac{1}{2}\delta$, and using (14.11) for R, and with $\Delta^{-p/M} < 2$, Eq. (14.16) becomes

$$|f(z) - [L/M]| < 2^{-M}K' \qquad (14.17)$$

for M sufficiently large. This relation holds except for the $\mu(R)$ small circles about the essenual singularities, and the at most $M + \nu(R) + p\mu(R)$ other small circles. If E is the set of all these disks, then $\Lambda_\alpha(E) < \tfrac{1}{2}\delta + \tfrac{1}{2}\delta = \delta$. Thus the theorem is established. ∎

Next we treat nondiagonal sequences. The nature of these results, insofar as the nature of the region of convergence is concerned, is not different from that found in Wilson's theorem 11.2, Wallin's corollary 11.2, and the duality extension corollary 11.3. In those and the following results the natural region of convergence is the largest circle in which the function being approximated is meromorphic. The immediate cause of the failure in this case is that we cannot meet the conditions of our error formula, Lemma 14.2, and have $L/M \to \infty$ unless the number of essential singularities μ is zero. Generally speaking, we feel that this feature is likely to be a true reflection of the nature of the convergence properties of the Padé approximants.

Theorem 14.2. Let $f(z)$ be regular in $|z| \leqslant R$ except for a finite number (counting multiplicity) of poles ν. Then given any ϵ, $\delta > 0$ and a sequence $\{P_k(z)\}$ of Padé approximants for which $L/(M + 1) \to \infty$, and evenutally $M \geqslant \nu$, there exists a k_0 such that for all $k > k_0$

$$|f(z) - P_k(z)| < \epsilon \qquad (14.18)$$

for all $|z| < R$ except for a set of points E_k where $\Lambda_\alpha(E_k) < \delta$ for any fixed $\alpha > 0$.

Proof: The proof again relies on the error formula, Lemma 14.2, and Cartan's lemma 14.1. Since there are no essential singularities, the application is much simpler than in the previous theorem. We set $\Delta = 1$, so the choice of p is irrelevant. The polynomial $R_M(z)$ is chosen to vanish at the poles of $f(z)$ according to their multiplicity and is thus of degree ν, which by assumption is less than M, eventually. We can use Eq. (14.15) to bound the terms involving $Q_M(z)$. Thus we have

$$|f(z) - P_k(z)| \leqslant \left| \frac{z}{R} \right|^{L+M+1} \frac{K \cdot 3^M R^M (2R)^\nu}{(1 - |z|/R) \Pi_{i=1}^n |z - z_i| \Pi_{j=1}^\nu |z - w_j|}$$

$$(14.19)$$

where w_j are the poles of $f(z)$. By Lemma 14.1 applied to the polynomial of degree at most $M + \nu$ in the denominator of Eq. (14.19) we can, by selecting $H = \frac{1}{2} \delta^{1/\alpha}$, bound Eq. (14.19), except in at most $M + \nu$ circles whose radius satisfies (14.4), by

$$|f(z) - P_k(z)| \leqslant \left| \frac{z}{R} \right|^L \left\{ \left| \frac{z}{R} \right|^{M+1} \frac{K \cdot 3^M 2^\nu R^{M+\nu}}{(1 - |z|/R) \left[\frac{1}{2} (\delta/e)^{1/\alpha} \right]^{M+\nu}} \right\}$$

$$(14.20)$$

Now since $|z/R| < 1$ and $L/M \to \infty$, we can make (14.20) as small as we please, in particular, less than $\frac{1}{2}\epsilon$, by simply picking k large enough. ■

Corollary 14.1. Let $f(z)$ be regular in $|z| \leqslant R$ except for a finite number (counting multiplicity) of zeros η. Then, given any $\epsilon, \delta > 0$ and a sequence $\{P_k(z)\}$ of Padé approximants for which $M/(L+1) \to \infty$ and eventually $L \geqslant \eta$, there exists a k_0 such that for all $k > k_0$

$$|f(z) - P_k(z)| < \epsilon$$

for all $|z| < R$ except for a set of points E_k where $\Lambda_\alpha(E_k) < \delta$.

Proof: By hypothesis, $1/f(z)$ satisfies the conditions of Theorem 14.2; therefore, the conclusions of that theorem hold for Padé approximants to $1/f(z)$. We can, by the duality theorem 9.2 on the reciprocal of Padé approximants, conclude at once

$$\left| [f(z)]^{-1} - [P_k(z)]^{-1} \right| < \epsilon' \tag{14.21}$$

for all $|z| < |R|$ except for a set with $\Lambda_\alpha < \delta'$. In this set there are only a finite number of poles of $f(z)$ and a finite number of poles of $P_k(z)$. Thus we can factor

$$f(z)P_k(z) = G(z)/F(z) \tag{14.22}$$

where $G(z)$ is bounded in this set and $F(z)$ is a polynomial of finite degree s. Thus by Lemma 14.1 $|F(z)| > [\frac{1}{2}(\frac{1}{2}\delta/e)^{1/\alpha}]^s$ outside a set with $\Lambda_\alpha \leqslant \frac{1}{2}\delta$. Thus if we choose

$$\delta' = \frac{1}{2}\delta, \qquad \epsilon' = \epsilon \left[\frac{1}{2}(\frac{1}{2}\delta/e)^{1/\alpha} \right]^s / \max\{G(z)\} \tag{14.23}$$

the conclusion follows directly by multiplying Eq. (14.21) by $|f(z)P_k(z)|$. ■

Corollary 14.2. Let $f(z)$ be meromorphic with only a finite number of poles and zeros (counting multiplicity) in any closed, bounded region $\mathcal{R}$ of the complex plane. Then, given any $\epsilon, \delta > 0$ and any sequence $\{P_k(z)\}$ for which $L, M \to \infty$ with k, there exists a k_0 such that for all $k > k_0$

$$|f(z) - P_k(z)| < \epsilon$$

for all z in $\mathcal{R}$ except for a set of points E_k where $\Lambda_\alpha(E_k) < \delta$ for any fixed $\alpha > 0$.

Proof: We select from the original sequence two subsequences. In the first we take all those Padé approximants for which $L \geqslant M$. In the second we take those for which $L < M$. Together they comprise the whole series. Consider the first subsequence. By direct consideration of Eq. (14.20) we see, if the radius of the circle of meromorphy satisfies the condition

$$R > \left[\frac{6z^2}{(\delta/e)^{1/\alpha}} \right] \qquad (14.24)$$

that we can make our error arbitrarily small by selecting M large enough. But we can take R as large as necessary because $f(z)$ is assumed meromorphic in the whole plane. The result for the second sequence follows by a duality argument as in Corollary 14.1. ■

Corollary 14.3. Let the conditions of Theorem 14.1, Theorem 14.2, Corollary 14.1, or Corollary 14.2 hold; then there exists a contour C', which is nowhere further than any given $\delta > 0$ from any given contour C not through a pole or zero of $f(z)$ in the domain of convergence of the relevant theorem or corollary, such that the number of poles less the number of zeros of $P_k(z)$ inside C' is exactly equal to that for the function $f(z)$ being approximated, provided k is sufficiently large.

Proof: In every cited case we have proved that $\Lambda_1(E_k)$ tends to zero. Thus if we select the parameter δ of the relevant theorem or corollary to be less than, say, $\frac{1}{2}\delta$ and consider a family of contours lying just inside C at distance ρ, $0 < \rho < \delta$, then since the maximum width of the set of uncontrolled points is only $\frac{1}{2}\delta$, at least one of these contours, call it C_k', is completely in the region of convergence. It follows by the calculus of residues that

$$n_p - n_z = \frac{1}{2\pi i} \oint_{C'} \frac{P_k'(z)}{P_k(z)} \, dz \qquad (14.25)$$

where n_p is the number of poles and n_z the number of zeros. Since the error formula (14.5) can also be established by the same line of proof for $P_k'(z)$ as for $P_k(z)$, and by assumption $f(z)$ does not go to zero or infinity on C and thus, also, not on C', we can make the difference

$$\left| \frac{f'(z)}{f(z)} - \frac{P_k'(z)}{P_k(z)} \right| \qquad (14.26)$$

as small as we like at every point of C'. Since C' is assumed to be of finite length and n_p and n_z are necessarily integers, we thus find $n_p - n_z$ to be exactly the same as for $f(z)$, provided only k is large enough. ■

A consequence of this last corollary is that the extra poles and zeros, which may be present when we have convergence in Hausdorff measure, "pair off" so that wherever there is an extra pole, very close by there is also an extra zero. This result is in agreement with observations in numerical experiments.

One can show that, if a sequence of functions differs from its limiting value by more than any preassigned small number only over a set E_k having $\Lambda_\alpha(E_k)$ arbitrarily small, there exists at least a subsequence that converges everywhere except on an exceptional set E which has $\Lambda_\alpha(E)$ less than any $\delta > 0$ we please. This property is called convergence *almost everywhere*.

The proof depends on the following property of $\Lambda_\alpha(E)$:

$$\Lambda_\alpha\left(\bigcup_{n=1}^\infty E_n\right) \leqslant \sum_{n=1}^\infty \Lambda_\alpha(E_n) \tag{14.27}$$

where $\bigcup_{n=1}^\infty$ stands for union, i.e., every point in any E_n. To show this property from the definition, Eq. (14.1), we remark that by slightly relaxing the minimum we can exhibit a set of disks $D_{n,i}$ such that $D_{n,i}$ cover E_n and

$$\sum_{i=1}^\infty \left[\delta(D_{n,i})\right]^\alpha \leqslant \Lambda_\alpha(E_n) + 2^{-n}\epsilon \tag{14.28}$$

where $\epsilon > 0$. Now since the $D_{n,i}$ cover all the E_n, if we take the series of sets $D_{1,1}, D_{2,1}, D_{1,2}, D_{3,1}, D_{2,2}, D_{1,3}, D_{4,1}, \ldots$, we will cover $\bigcup_{n=1}^\infty$ and if we use these disks in (14.1), we have

$$\Lambda_\alpha\left(\bigcup_{n=1}^\infty E_n\right) \leqslant \sum_{n=1}^\infty \Lambda_\alpha(E_n) + \epsilon \tag{14.29}$$

But since ϵ is arbitrary, we can conclude that (14.27) holds. This property of Λ_α is referred to as being countably subadditive.

We are now in a position to select a sequence. Suppose we are given $\epsilon', \delta' > 0$ and we require an error of no more than ϵ' outside a set E having $\Lambda_\alpha(E) \leqslant \delta'$. We simply do this selection by choosing the first member of our sequence satisfying (14.18) except for a set with $\Lambda_\alpha(E_k) = \frac{1}{2}\delta'$. The next will have $\Lambda_\alpha(E_k) = \frac{1}{4}\delta'$, and so on. By (14.27) every member of this sequence will satisfy (14.18) in a set less than $(\frac{1}{2} + \frac{1}{4} + \cdots)\delta' = \delta'$. Hence we have show the following result.

Corollary 14.4. If the conditions of Theorem 14.1, Theorem 14.2, Corollary 14.1, or Corollary 14.2 hold, then there exists a subsequence of the $P_k(z)$ which converges almost everywhere in the sense of Λ_α, $\alpha > 0$.

Wallin (1972) has given sufficient conditions on $f(z)$ for the entire sequence to converge almost everywhere in the sense Λ_α, $\alpha > 0$. He starts from error formula (12.39). He shows, in essence, that if the coefficients of z^n in an entire function decay faster than $(\ln n)^{-n\theta}$ and $\alpha\theta > 1$, then the entire sequence converges except at most on a set with $\Lambda_\alpha = 0$. (His criterion is a more precise one than our simplified version.) When we remember that $\Lambda_\alpha = 0$ for any region of the complex plane for $\alpha > 2$, we see that this result is useful only for functions bounded by $\theta > \frac{1}{2}$. This class is, however, very extensive, relative to Theorem 11.6, for example.

D. Convergence in the Mean on the Riemann Sphere

In addition to the results given on convergence except on a small set of points, it is interesting to see what we can deduce about convergence on the Riemann sphere. Since, by Eq. (9.21), we have for the chordal metric

$$D(f, g) \leqslant 2|f - g| \qquad (14.30)$$

all the results of the previous section can be taken over directly to convergence on the Riemann sphere. Since also $D(f, g) \leqslant 2$ for all f and g, it will turn out that integrals of D will converge for Padé approximants. This type of convergence is familiar from the theory of Fourier series, where one finds convergence in the mean (Franklin, 1940)

$$\lim_{n\to\infty} \int_0^{2\pi} |f(x) - f_n(x)|^2 \, dx = 0 \qquad (14.31)$$

or, in other words, the mean-squared error tends to zero. Because of "Gibbs phenomena," the maximum error does not necessarily tend to zero for Fourier series, as can be shown by actual examples, but the affected region instead shrinks to zero. As a preliminary to stating our results, we point out that the error formula (14.5) applies equally well, as do all the conclusions deduced from it, for

$$|f(z) - [L/M]|/|z|^j \qquad (14.32)$$

where $L + M > j$, and j is a fixed (i.e., independent of L and M) integer. With this observation we are in a position to prove the following extension of a result due to Baker (1973).

Theorem 14.3. Let $P_k(z)$ be a sequence of Padé approximants, and let the conditions of Theorem 14.1, Theorem 14.2, Corollary 14.1, or Corollary 14.2 hold. Let $\mathcal{R}$ be a closed, bounded region and let $\mathcal{C}$ be an arc of finite length; then

$$\lim_{k\to\infty} \int_{\mathcal{R}} dx\, dy\, D^p(f(z), P_k(z))/z^j = 0 \qquad (14.33)$$

$$\lim_{k\to\infty} \int_{\mathcal{C}} |dz|\, D^p(f(z), P_k(z))/z^j = 0 \qquad (14.34)$$

where $p > 0$, and j is arbitrary and real.

Proof: Let μ be the area of $\mathcal{R}$. Let us select an $\epsilon = \frac{1}{2}(\epsilon'/2\mu)^{1/p}$ and a $\delta = \epsilon'/2^{p+1}$. By the relevant theorem or corollary for $\alpha = 2$, (14.30), and (14.32), we can for k large enough divide (14.33) into two integrals, one over a set E_k having $\Lambda_2(E_k) < \delta$ and the other having

$$D^p(f(z), P_k(z))/|z|^j \leqslant 2^p |f(z) - P_k(z)|^p/|z|^j \leqslant \epsilon'/2\mu \quad (14.35)$$

Thus, since the area of the second region is at most μ,

$$\int dx\, dy\, D^p(f(z), P_k(zz))/z^j \leqslant 2^p \epsilon'/2^{p+1} + (\epsilon'/2\mu)\mu = \epsilon' \quad (14.36)$$

and since ϵ' is as small as we please, Eq. (14.33) follows at once. Equation (14.34) follows in exactly the same way, except that we use the result for $\alpha = 1$, which implies that the length of arc which is uncontrolled shrinks to zero. ∎

E. Examples

The purpose of these examples is twofold. First to illustrate the theorems already proved, and second to serve as counterexamples to their extension. These examples are due to Gammel (1970) and Wallin (1972). They show how an entire function can be constructed for which many diagonal Padé approximants have poles in arbitrarily prescribed places. For ease of presentation we stick mainly to a discussion of Gammel's example. It constructs a subsequence of $[M/M]$ Padé approximants whose denominators are of real degree unity. Wallin's example generalizes this to construct a sequence $[n_v/n_v]$ which has denominators of prescribed real degree $m_v = n_v - 2n_{v-1}$. His results are the corresponding generalizations of Gammel's. Gammel defines his function as

$$f(x) = \sum_{j=0}^{\infty} f_j x^j = 1 + \sum_{v=1}^{\infty} \alpha_v \sum_{n=n_v}^{2n_v} r_v^{-n} x^n \qquad (14.37)$$

where

$$n_{v+1} = 2n_v + 1, \qquad n_0 = 0, \qquad \alpha_v = \left[(2n_v)! \; \max(|r_v|^{-n_v}, |r_v|^{-2n_v})\right]^{-1}$$

(14.38)

and $r_v \neq 0$ is any sequence of complex numbers having any desired set of points in the complex plane as limit points. By comparison with the exponential series it is clear that (14.37) is an entire function. It is also clear from the Padé equations that

$$[n_v/n_v] = \sum_{j=0}^{n_v-1} f_j x^j + \alpha_v x^{n_v} (1 - x/r_v)^{-1}$$

(14.39)

This approximant is of course also equal to the $[n_v/1]$, so that an $(n_v - 1) \times (n_v - 1)$ block exists in the Padé table (see Chapter 2). Clearly (14.39) has a single pole at $x = r_v$, as required. The residue at this pole is $-\alpha_v r_v$. Thus the area where the pole term is greater than ϵ is given by $\Lambda_\alpha (E) = (2\alpha_v r_v/\epsilon)^\alpha$, which tends to zero like $(2n)^{-2\alpha n}$ for large enough v and r_v near a given point $z_0 \neq 0$. When we remember that $n_v = 2^v - 1$ we see that the uncontrolled area decays extremely rapidly. More elaborate examples can be constructed on the same principle (Wallin, 1972). These involve Padé denominators $(1 - x/r_v)^{\beta n_v}$ with β near unity, and a complicated bounding comparison series behaving generally like $w_\theta(z) = \Sigma[z/(\ln n)^\theta]^n$, which is entire, $\theta > 0$. These results show that Wallin's criterion is the best possible for the entire series of Padé approximants to convergence almost everywhere. It is still true of these examples, of course, that a subsequence can always be found that converges almost everywhere in the sense Λ_α. In the case of Gammel's and Wallin's examples, Baker (1973) has shown that a subsequence exists where point-by-point convergence occurs.

PART III

SERIES OF STIELTJES AND PÓLYA

15

SERIES OF STIELTJES, INEQUALITIES

In this part we discuss two important classes of functions whose Padé approximants can be characterized in great detail. In these cases analytic properties of the functions can be related to simple properties of the determinants of the Padé equations; that is, to properties of the C table (Chapter 2). From these determinantal properties it is possible to develop a complete characterization of the location of the poles and zeros of the Padé approximants, to prove certain monotonicity properties, and, finally, to establish the convergence properties.

A. Series of Stieltjes

This particular class of functions turns out to be particularly interesting not only in its own right, but also because a large number of problems in physics, chemistry, mathematics, and engineering are expressible in terms of them. We will discuss some of these applications in the last part of this book.

By a series of Stieltjes [a great many of the fundamental results are due to Stieltjes (1894), although we do not always follow his methods] we mean that

$$f(z) = \sum_{j=0}^{\infty} f_j(-z)^j \tag{15.1}$$

is a series of Stieltjes if, and only if, there is a bounded, nondecreasing function $\varphi(u)$, taking on infinitely many values in the interval $0 \leqslant u < \infty$, such that

$$f_j = \int_0^{\infty} u^j \, d\varphi(u) \tag{15.2}$$

where we mean a Stieltjes integral, i.e., φ may have a finite step. It is to be noted that we do not necessarily assume that the series is convergent. Let us introduce the determinants

$$
D(m, n) = \det \begin{vmatrix} f_m & f_{m+1} & \cdots & f_{m+n} \\ f_{m+1} & f_{m+2} & \cdots & f_{m+n+1} \\ \vdots & \vdots & \ddots & \vdots \\ f_{m+n} & f_{m+n+1} & \cdots & f_{m+2n} \end{vmatrix}
$$

$$
= (-1)^{(m+n)(n+1)} C(m+n/\ \ n+1) \tag{15.3}
$$

in terms of the C table of Chapter 2.

This definition of a series of Stieltjes turns out to be equivalent to the conditions

$$
D(0, n) > 0, \qquad D(1, n) > 0, \qquad n = 0, 1, 2, \ldots \tag{15.4}
$$

It is not hard to see that (15.4) follows from (15.2). In case the function $\varphi(u)$ takes on only a finite number of different values, that is, $\varphi(u)$ is a step function with only a finite number of jumps; then we find that (15.2) gives

$$
f_j = A\delta_{j0} + \sum_{k=1}^{p} \alpha_k u_k^{\ j}
$$

From the theory of the block structure of the Padé table (Sections 2.B and 11.B) we see that for $n \geqslant p$ we have $D(m, n) = 0$ and $[p/p] = f(z)$. In this case questions of convergence are trivial. For ease of presentation we will usually limit our discussion to nonrational series of Stieltjes. It is to be understood that the same results hold for rational series of Stieltjes, except that, past a certain point, all the Padé approximants are the same.

It is much more difficult to see that (15.2) follows from (15.4). We will return to this question in the next chapter.

Lemma 15.1. If the f_j are given by Eq. (15.2), then the corresponding $D(m, n)$, $n, m \geqslant 0$, given by (15.3) are all positive.

Proof: Consider the quadratic form

$$
\sum_{p,\,q=0}^{n} f_{p+q+m} x_p x_q = \int_0^\infty u^m (x_0 + x_1 u + \cdots + x_n u^n)^2 \, d\varphi(u) > 0 \tag{15.5}
$$

since the integrand is $\geqslant 0$. By assumption $d\varphi(u) \neq 0$ for an infinite number of values of u. Since the squared quantity can vanish only a finite number of times, the result (15.5) follows. By the variational principle the minimum of (15.5), subject to the normalization condition $\Sigma x_j^2 = 1$, is the smallest eigenvalue of the matrix of $D(m, n)$. Since this quadratic form is real symmetric, all the eigenvalues are real and hence positive definite. Since a determinant is the product of the eigenvalues of its matrix, it is necessarily positive. ∎

The theory of series of Stieltjes is closely related to the "moment problem." [See Shohat and Tamarkin (1963) for a further discussion.] This problem asks how we can tell, given a series of numbers f_j, if they are the moments of some distribution. Clearly, Lemma 15.1 gives a necessary condition, if $x \geqslant 0$, for the Stieltjes moment problem. If x is not so restricted, we have the Hamburger moment problem. That is,

$$g_j = \int_{-\infty}^{+\infty} u^j \, d\psi(u) \tag{15.6}$$

where $d\psi \geqslant 0$ and ψ takes on infinitely many values. By the same proof that gave Lemma 15.1 we can prove, for even m, the following lemma.

Lemma 15.2.   If the g_j are given by Eq. (15.6), then the corresponding $D(2m, n)$, $n, m \geqslant 0$, given by (15.3) are all positive.

We can recognize, in a simple way, the reciprocal of a series of Stieltjes by determinantal conditions. Using Hadamard's formula (4.100) and the definition (15.3), we have at once

$$\overline{D}(m, n) = (-1)^{\frac{1}{2}m(m+1)+n} D(2 - m, m + n - 1)/f_0^{m+2n+1} \tag{15.7}$$

where $\overline{D}$ are the D's formed from the series coefficients in the expansion of $[1/f(z)]$. Clearly, then, by (15.7) the determinantal condition (15.4) is equivalent to

$$(-1)^{n+1}\overline{D}(1, n) > 0, \qquad (-1)^{n+1}\overline{D}(2, n) > 0 \tag{15.8}$$

We are now in a position to prove the following useful property of series of Stieltjes.

Lemma 15.3.   If the determinantal condition (15.4) holds for $f(z)$, then it also holds for $F(z)$, where $f(z) = f_0/[1 + zF(z)]$.

Proof: From Eq. (15.7) we have the determinantal conditions implied for $[1 + zF(z)]/f_0$. By the sign convention (15.1) every element of the determinant for $F(z)$ is multiplied by -1 relative to $[1 + zF(z)]/f_0$, since

$f_0 > 0$. Thus the determinant is multiplied by $(-1)^{n+1}$ and the value of m is decreased by one. Hence (15.8) becomes (15.4) for $F(z)$ directly. ∎

We get a similar result for the Hamburger moment problem. Here

$$D(0, n) > 0, \qquad n = 0, 1, 2, \ldots \qquad (15.9)$$

plays the role of condition (15.4) for the Stieltjes case.

Lemma 15.4. If the determinantal condition (15.9) holds for $f(z)$, then it also holds for $G(z)$, where $f(z) = f_0/[1 + (f_1/f_0)z - z^2 G(z)]$.

Proof: From Eq. (15.7) we have the determinantal conditions implied for $[1 + (f_1/f_0)z - z^2 G(z)]$. By the sign convention (15.1) and the minus sign in the definition of $G(z)$ every element of the determinant for $G(z)$ is multiplied by -1 relative to that for (15.7). Since the matrix is $(n + 1) \times (n + 1)$ and since the definition of a determinant for G reduces the apparent value of m by two, we get

$$D_G(m - 2, n) = (-1)^{\frac{1}{2}m(m+1)+1} D(2 - m, m + n - 1)/f_0^{m+2n+1} \qquad (15.10)$$

so that (15.9) for $f(z)$ implies the same condition for $G(z)$. ∎

In order to illustrate the foregoing results, let us consider the following example:

$$f_j = \int_0^\infty x^j e^{-x}\, dx = j! \qquad (15.11)$$

which corresponds to the formal series [see also Eq. (10.13)]

$$f(z) = 1 - z + 2z^2 - 6z^3 + 24z^4 - 120z^5 + 720z^6$$
$$- 5040z^7 + 40{,}320z^8 - \cdots \qquad (15.12)$$

This series is, of course, just $_2F_0(1, 1; -z)$, a special confluence of Gauss's hypergeometric function, which we discussed in Section 5.E. For this case if we define

$$\Omega_{2j} = D(0, j), \qquad \Omega_{2j+1} = D(1, j), \qquad j = 0, 1, 2, \ldots \qquad (15.13)$$

then we have, by Eqs. (4.84), (4.85), (5.49), and (15.3), the recursion relation

$$\Omega_k = \left[\frac{k + 1}{2}\right] \frac{\Omega_{k-1}\Omega_{k-2}}{\Omega_{k-3}}, \qquad k \geq 3 \qquad (15.14)$$

where $[x]$ means the greatest integer $\leq x$. Clearly, by (15.14), since we can verify $\Omega_0 = \Omega_1 = \Omega_2 = 1$ by direct calculation, all the other determinants are positive, as expected from Lemma 15.1. The first few are easily calculated directly and agree with (15.14):

$$\Omega_0 = D(0, 0) = 1, \qquad \Omega_1 = D(1, 0) = 1, \qquad \Omega_2 = D(0, 1) = 1,$$
$$\Omega_3 = D(1, 1) = 2, \qquad \Omega_4 = D(0, 2) = 4, \qquad \Omega_5 = D(1, 2) = 24,$$
$$\Omega_6 = D(0, 3) = 144, \qquad \Omega_7 = D(1, 3) = 3456$$

$$(15.15)$$

If we compute the reciprocal series, we get

$$[f(z)]^{-1} = 1 + z - z^2 + 3z^3 - 13z^4 + 71z^5 - 461z^6$$
$$+ 3447z^7 - 29{,}093z^8 + \cdots \tag{15.16}$$

By inspection, the sign pattern is right for $F(z)$ and $G(z)$ of Lemmas 15.3 and 15.4 to be series of Stieltjes. Direct calculation of the determinants verifies Eq. (15.7), and thereby Lemma 15.3.

B. Inequalities

The importance of the determinantal inequalities (15.4) is clearly seen from the solution (1.27) of the Padé equation. $D(L - M + 1, M - 1)$ is the determinant of the Padé equations and determines its singularity or nonsingularity. Thus Lemma 15.1 implies that at least the lower left half $(J = L - M \geqslant -1)$ of the Padé table is normal (Section 2.B). We shall now show that these determinantal inequalities also determine the location of all the poles of the Padé approximants $[L/M]$ for $J \geqslant -1$.

Theorem 15.1. If $\sum f_j(-z)^j$ is a series of Stieltjes, then the poles of the $[M + J/M], J \geqslant -1$, Padé approximants are on the negative real axis. Furthermore, the poles of successive approximants interlace, and all the residues are positive. The roots of the numerator also interlace those of the denominator.

Proof: We start from expression (3.44) for the Padé approximant. Taking account of the sign convention (15.1), and canceling an overall sign factor against one from the numerator, we can write the denominator of $[M + J/M]$ as

$$\Delta_M^{(J)}(x) = \det \begin{vmatrix} f_{1+J} + xf_{2+J} & f_{2+J} + xf_{3+J} & \cdots & f_{M+J} + xf_{M+J+1} \\ f_{2+J} + xf_{3+J} & f_{3+J} + xf_{4+J} & \cdots & f_{M+J+1} + f_{M+J+2} \\ \vdots & \vdots & \ddots & \vdots \\ f_{M+J} + xf_{M+J+1} & f_{M+J+1} + xf_{M+J+2} & \cdots & f_{2M+J-1} + xf_{2M+J} \end{vmatrix}$$

$$(15.17)$$

We note that Δ_{M-r} is the coaxial minor formed by striking off the last r rows and columns and also the denominator of $[M + J - r/M - r]$. We now assert that $\Delta_M, \Delta_{M-1}, \ldots, \Delta_1, \Delta_0 = 1$ is a Sturm sequence; that is, if $\Delta_k = 0$, then Δ_{k-1} and Δ_{k+1} have opposite signs. To see this result, we use Sylvester's determinant identity (2.10):

$$A A_{M-1, M; M-1, M} = A_{M-1, M-1} A_{M, M} - A_{M-1, M} A_{M, M-1} \qquad (15.18)$$

But since A is symmetric, $A_{M-1, M} = A_{M, M-1}$. Thus if $A_{M, M} \equiv \Delta_{M-1} = 0$, it must be that $\Delta = A$ and $\Delta_{M-2} = A_{M-1, M; M-1, M}$ have opposite signs by Eq. (15.18).

To locate the roots, we first note that $\Delta_M(0) = D(1 + J, M - 1) > 0$ for all M. Now

$$\Delta_1(x) = f_{1+J} + x f_{2+J} \qquad (15.19)$$

and since all the $f_k = D(k, 0) > 0$, the root of $\Delta_1(x)$ is real and negative. Since $\Delta_2(0)$ is positive, and by (15.18) negative at the real, negative root of $\Delta_1(x)$, it must vanish between zero and that root. The coefficient of x^2 in $\Delta_2(0)$ is $D(j + 2, 1)$, by (15.17). Therefore as $x \to -\infty$, $\Delta_2(x) \to +\infty$. Since $\Delta_2(x)$ is a polynomial of degree two, we see that its two roots must be both real and negative. In general, we see that the first M roots of $\Delta_{M+1}(x)$ lie in the M intervals formed by zero and the M roots of $\Delta_M(x)$. Since the coefficient of x^{M+1} is $D(J + 2, M)$, the single remaining root must also be real and negative.

To demonstrate the positiveness of the residues, we use the identity (3.5), which becomes in our present notation

$$[M + J + 1/\ \ M + 1] - [M + J/M] = \frac{(-x)^{2M+J+1}[D(1 + J, M)]^2}{\Delta_{M+1}(x)\Delta_M(x)}$$

$$(15.20)$$

By the interlacing property the residue at the first (counting in the negative direction from the origin) root of $\Delta_{M+1}(x)$ must be positive on the right and so, too, on the left of (15.20). We now pass the first and only the first root of $\Delta_M(x)$ before we come to the second root of $\Delta_{M+1}(x)$. Since both $\Delta_M(x)$ and $\Delta_{M+1}(x)$ are now negative, the residue of the second root is also positive. We can repeat this process until we have shown that all the residues are positive. This result, together with the sign oscillations of the denominator, forces the roots of the numerator to interlace those of the denominator. ∎

The following theorem is fundamental to the study of series of Stieltjes, for it provides upper and lower bounds for the exact sum of the series, whether convergent or not.

Theorem 15.2. The Padé approximants for series of Stieltjes obey the following inequalities, where $f(z)$ stands for the limit as M goes to infinity of $[M/M]$ or $[M - 1/M]$, and z is real and nonnegative:

$$(-1)^{1+J}\{[M + 1 + J/\quad M + 1] - [M + J/M]\} \geqslant 0 \quad (15.21)$$

$$(-1)^{1+J}\{[M + J/M] - [M + J + 1/\quad M - 1]\} \geqslant 0 \quad (15.22)$$

$$[M/M] \geqslant f(z) \geqslant [M - 1/M] \quad (15.23)$$

$$[M/M]' \geqslant f'(z) \geqslant [M - 1/M]' \quad (15.24)$$

where $J \geqslant -1$. These inequalities have the consequence that the $[M/M]$ and $[M - 1/M]$ form the best upper and lower bounds obtainable using only the given number of coefficients, and that the use of additional coefficients (higher M) improves the bounds. Equations (15.21) and (15.22) are valid when differentiated once, provided $J \geqslant 0$ for (15.21).

Proof: Inequality (15.21) follows directly from Theorem 15.1 and Eq. (15.20), since the Δ's are positive, and so $(-1)^{1+J}$ times the right-hand side of (15.20) is positive or zero in this range. To prove (15.22), we start with identity (3.9), which becomes in our present notation

$$[M + J/M] - [M + J + 1/\quad M - 1]$$

$$= \frac{[D(2 + J/\quad M - 1)]^2(-x)^{2M+J+1}}{\Delta_M^{(J)}(x)\Delta_{M-1}^{(J+2)}(x)} \quad (15.25)$$

Again the denominator of the right-hand side is nonnegative, so that (15.22) follows directly as before. To derive (15.23), we use identity (3.7) with an identical proof for the outside part of the inequality. That identity becomes

$$[M/M] - [M - 1/M] = \frac{x^{2M}D(0, M)D(1, M - 1)}{\Delta_M^{(0)}(x)\Delta_M^{(-1)}(x)} \quad (15.26)$$

By (15.21) both the $[M/M]$ and $[M - 1/M]$ sequences are monotonic for any z real and nonnegative. Therefore $f(z)$ is correctly placed. Since the denominator of Eq. (15.26) is of degree $2M$, with all positive coefficients, if we divide numerator and denominator by x^{2M} and then differentiate, we get the outside part of (15.24) directly. The derivative inequalities for (15.21) and (15.22) follow the same way by differentiating (15.20) and

(15.25), since the degree of the numerator is greater than or equal to that of the denominator. ■

By way of illustration of these results let us compute a few Padé approximants to series (15.12). They are

$$[1/0] = 1 - x, \qquad [0/1] = \frac{1}{1+x}, \qquad [2/0] = 1 - x + 2x^2,$$

$$[1/1] = \frac{1+x}{1+2x}, \qquad [0/2] = \frac{1}{1+x-x^2},$$

$$[3/0] = 1 - x + 2x^2 - 6x^3, \qquad [2/1] = \frac{1 + 2x - x^2}{1 + 3x}$$

$$[1/2] = \frac{1 + 3x}{1 + 4x + 2x^2}, \qquad [0/3] = (1 + x - x^2 + 3x^3)^{-1},$$

$$[2/2] = \frac{1 + 5x + 2x^2}{1 + 6x + 6x^2} \tag{15.27}$$

In Table 15.1 we list the values of this part of the Padé table at $x = 1$. Note the lower left-hand part of this table, i.e., for $L - M \geqslant -1$. It will be observed that the even-J diagonals decrease in value and the odd ones increase, as expected from Eq. (15.21). If we look at the counter diagonals, we see an alternating increase and decrease of value as we proceed from lower left to upper right. We also notice that the best values listed are the [2/2] and the [1/2] since (Section 10.C) $0.6154 > 0.5963 > 0.5715$. In Table 15.2 we have listed some of the poles of these Padé approximants. Note that, as expected from Theorem 15.1, the poles interlace when the degree of both numerator and denominator increases by one. Other monotonicity properties, such as shown by the roots of the [1/2] and the [2/2], can be proven from other two-term identities [Eq. (3.7) in this case]. The residues (numerator over the derivative of the denominator evaluated at the poles) are all positive, as expected (Theorem 15.1).

TABLE 15.1 Padé Values to Euler's Series

L M	0	1	2	3
0	1	1/2	1	4
1	0	2/3	4/7	
2	2	1/2	8/13	
3	−4			

TABLE 15.2 Poles and Residues of Padé Approximants to Euler's Series

Padé	Pole	Residue
[0/1]	−1.000	1.000
[1/2]	−0.293, −1.707	0.043, 1.457
[1/1]	−0.500	0.250
[2/2]	−0.211, −0.789	0.010, 0.490

By changing the sign and reversing the inequalities in (15.24) and dividing by (15.23), we have the following result.

Corollary 15.1. For x real and nonnegative and $f(x)$ a series of Stieltjes we have

$$\frac{-d\ln[M/M]}{dx} \leqslant \frac{-d\ln f(x)}{dx} \leqslant \frac{-d\ln[M-1/M]}{dx} \qquad (15.28)$$

The $f(x)$ which appears both in Theorem 15.1 and Corollary 15.1 can be extended to be any series of Stieltjes satisfying Eq. (15.2), as we will show at the end of Section 17.A.

16

SERIES OF STIELTJES, CONVERGENCE

A. Existence of a Limit

The first step in our study is to show that any particular diagonal sequence of Padé approximants tends to a limit. Then we investigate the conditions under which the limit is unique. Generally speaking, they will agree with the condition that a function be uniquely defined by a formal power series.

Theorem 16.1.   Any sequence of $[M + J/M]$ Padé approximants to a series of Stieltjes converges to an analytic function in the cut complex plane $-\infty \leqslant z \leqslant 0$ as M tends to infinity, $J \geqslant -1$.

Proof: The proof will proceed by means of Theorem 12.4. We first need to establish boundedness of the sequence of Padé approximants and second to establish convergence at an infinite number of points.

From Theorem 15.1 we can write

$$[M + J/M] = \sum_{p=0}^{J} f_p(-z)^p + (-z)^{1+J} \sum_{p=1}^{M} \frac{\beta_p}{1 + \gamma_p z} ,$$

$$\beta_p > 0, \quad \gamma_p > 0 \tag{16.1}$$

since all the poles are real and negative with positive residues. Now we can bound

$$|[M + J/M]| \leqslant \left| \sum_{p=0}^{J} f_p(-z)^p \right| + |z|^{1+J} \sum_{p=1}^{M} \frac{\beta_p}{|1 + \gamma_p z|}$$

$$\leqslant \left| \sum_{p=0}^{J} f_p(-z)^p \right| + |z|^{1+J} \sum_{p=1}^{M} \beta_p, \qquad \mathrm{Re}\,(z) \geqslant 0$$

$$\leqslant \left| \sum_{p=0}^{J} f_p(-z)^p \right| + \frac{|z|^{1+J}}{|\mathrm{Im}\,(z)|} \sum_{p=1}^{M} \frac{\beta_p}{\gamma_p} , \qquad \mathrm{Re}\,(z) < 0 \tag{16.2}$$

However, by (16.1) and the Padé equations

$$\sum_{p=1}^{M} \beta_p = f_{1+J} \tag{16.3}$$

and thus

$$\|[M + J/M]\| \leqslant \left| \sum_{p=0}^{J} f_p(-z)^p \right| + |z|^{1+J}f_{1+J} \tag{16.4}$$

for Re $(z) > 0$. By the determinantal solution (1.27), in the limit $z \to \infty$ we identify in (16.1)

$$f_J - \sum_{p=1}^{M} (\beta_p/\gamma_p) = D(J, M)/D(1 + J, M - 1) > 0 \tag{16.5}$$

and thus, if Re $(z) < 0$,

$$\|[M + J/M]\| \leqslant \left| \sum_{p=0}^{J} f_p(-z)^p \right| + f_J|z|^{1+J}/|\mathrm{Im}\ z| \tag{16.6}$$

Therefore if $\mathcal{R}$ is any closed, bounded region in the cut complex plane $-\infty \leqslant z \leqslant 0$, every point of $\mathcal{R}$ must be further than some finite distance $\delta > 0$ from the cut. Hence in $\mathcal{R}$, Eqs. (16.4) and (16.6) imply the existence of a uniform upper bound for $\|[M + J/M]\|$. Thus (see Chapter 12) $[M + J/M]$ is equicontinuous as M goes to infinity and *a fortiori*, is equicontinuous on the sphere.

Now by Theorem 15.2 every sequence $[M + J/M]$ is monotonic for z real and positive [Eq. (15.21)]. Further, for J even all such sequences are decreasing and bounded from below by $[0/1]$, and, for J odd all such sequences are increasing and bounded from above by $[0/0]$ [Eqs. (15.22) and (15.23)]. Thus $[M + J/M]$ must converge at any real, positive point and in particular at a sequence $\{z_n\}$ with a limit point, say $z = 1$, of them. Therefore by Theorem 12.4 the $[M + J/M]$ converge to a meromorphic (and in this case analytic by the uniform bound) function in $\mathcal{R}$. When we remember that all Padé approximants equal f_0 at $z = 0$ by definition, we have established this theorem. ∎

B. Uniqueness of the Limit

By means of the following lemma we can establish the convergence results for any sequence of Padé approximants when the series of Stieltjes has a nonzero radius of convergence.

Lemma 16.1.   If $f(z) = \sum_{p=0}^{\infty} f_p(-z)^p$ is a series of Stieltjes ($L \geqslant M - 1$), then

$$0 \leqslant (-1)^p [L/M]^{(p)}/(p!) \leqslant f_p \tag{16.7}$$

where the superscript (p) denotes the pth derivative at zero.

Proof: If $p \leqslant L + M$, the right-hand inequality of (16.7) holds as an equality. Otherwise let us write

$$f_p - (-1)^p [L/M]^{(p)}/(p!)$$

$$= (-1)^p \left\{ [L + 1/\ M + 1]^{(p)}/(p!) - [L/M]^{(p)}/(p!) \right\}$$

$$+ \cdots + (-1)^p \left\{ (-1)^p f_p - [M_\alpha + J/M_\alpha] \right\} \tag{16.8}$$

where M_α is large enough so that the last term in braces vanishes, by the Padé equations. The terms in braces are nothing but single terms in $(-1)^p$ times the right-hand side of Eq. (15.20). We can write the right-hand side of Eq. (15.20) as

$$(-z)^{2M+J+1} K \prod_{k=1}^{2M+1} (1 + \alpha_k z)^{-1}$$

$$= K \sum_{\nu = 2M+1+J}^{\infty} \left[\sum_{\substack{\nu_1 + \nu_2 + \cdots + \nu_{2M+1} \\ = \nu - 2M - 1 - J}} \prod_{k=1}^{2M+1} \alpha_k^{\nu_k} \right] (-z)^\nu \tag{16.9}$$

where K and α_k are positive, by Theorem 15.1. The sign of z^p in (16.9) is thus $(-1)^p$ and hence the right-hand side of (16.8) is positive. Since the sign of $[L/M]^{(p)}$ is $(-1)^p$, by an argument similar to (16.9), we obtain (16.7).   ∎

Theorem 16.2.   Let $f(z)$ be a series of Stieltjes with radius of convergence $R > 0$; then if $\{P_k(z)\}$ is any sequence of Padé approximants such that $M - 1 \leqslant L \leqslant \lambda M$, $1 \leqslant \lambda \leqslant \infty$, that sequence converges geometrically like ψ^{L+M} everywhere in

$$\psi = \left| \frac{(1 + z/R)^{1/2} - 1}{(1 + z/R)^{1/2} + 1} \left(\frac{z}{R} \right)^{(\lambda - 1)/(\lambda + 1)} \right| < 1 \tag{16.10}$$

to the analytic function $f(z)$ defined by the power series. For any particular z in the cut, complex plane, $-\infty \leqslant z \leqslant -R$ there exists a $\lambda > 1$ so that z is in the region of convergence. Any sequence $L \geqslant M - 1$ converges in $|z| < R$.

Proof: First we will show that all Padé approximants, $L \geqslant M - 1$, have all their poles on the line $-\infty \leqslant z \leqslant -R$. First by Theorem 15.1 all the poles lie on $-\infty \leqslant z \leqslant 0$. Suppose for some $[L/M]$ a pole lay in the range $-R < z < 0$. If there is more than one, we will consider the one closest to the origin. Call it γ_1. By Lemma 16.1 and Eq. (16.1) and the formula for the radius of convergence of a power series,

$$R^{-1} = \lim_{p \to \infty} \sup(f_p)^{1/p} \geqslant \lim_{p \to \infty} \sup\{[L/M]^{(p)}/(p)!\}^{1/p} = \gamma_1^{-1} > R^{-1}$$

$$(16.11)$$

by assumption, which is a contradiction. Thus all the poles must lie on $-\infty \leqslant z \leqslant -R$. Hence for this case the inequalities (16.4) and (16.6) can be improved to

$$|[M + J/M]| \leqslant \left| \sum_{p=0}^{J} f_p(-z)^p \right| + (|z|^{1+J}f_J/\rho)$$

$$(16.12)$$

where ρ is the distance between z and the cut from $-R$ to $-\infty$. If $J \geqslant -1$ and $|z| < R$, then inequality (16.12) provides a uniform bound with respect to M, and hence we can conclude convergence by Theorem 12.3, since uniform boundedness implies equicontinuity.

By (16.12), for $|z| > R$, and by the assumption that $f(z)$ has a radius of convergence R we can select a constant K such that

$$|[M + J/M]| \leqslant (K/\rho)|z/R|^{1+J}$$

$$(16.13)$$

where use has been made of Cauchy's inequalities for the derivatives of an analytic function. Now consider the special mapping

$$z = \frac{4Rw}{(1-w)^2}, \qquad w = \frac{(1+z/R)^{1/2} - 1}{(1+z/R)^{1/2} + 1}$$

$$(16.14)$$

which one can verify maps the cut z plane into the unit circle in the w plane.

If we now think of $[M + J/M] - f(z)$ as a function of w, it is bounded on the circle $|w| = \omega < 1$ by the maximum of (16.13) plus the maximum of $f(z)$ on that circle. Also the first $2M + J$ derivatives vanish at the origin.

Thus by Schwarz's lemma 12.2 applied in the w plane we have

$$|[M + J/M] - f(z)| \leq \frac{K'}{\rho} \left\{ \left[\frac{4\omega}{(1 - \omega)^2} \right]^{(1+J)/(2M+J+1)} \omega^{2M+J+1} \right\}$$

(16.15)

for

$$\left| \frac{(1 + z/R)^{1/2} - 1}{(1 + z/R)^{1/2} + 1} \right| < \omega$$

(16.16)

For M and L very large (16.15) implies that if

$$\omega \left(\frac{4\omega}{(1 - \omega)^2} \right)^{(\lambda - 1)/(\lambda + 1)} < 1$$

(16.17)

then by inequality (16.15), we have convergence. By transformation (16.14) this result is equivalent to (16.10) and proves that part of the theorem.

Since the whole, cut, complex plane is mapped into $|w| < 1$, we can find an ω such that any particular z is included in (16.16). Since any number not zero or infinity to the zero power is one, we can satisfy (16.17) by picking $\lambda > 1$ small enough. Finally, since $|z| < R$, the left-hand side of (16.10) is the product of two factors, the first less than one, and the second less than or equal to, one. For any $\lambda \geq 1$ inequality (16.10) is satisfied and the last statement of the theorem follows. The uniqueness of the limit function follows by Theorem 12.3 ∎

By way of an illustration of this theorem, and as a counterexample to substantial extension, let us consider the example

$$f(x) = \int_0^1 \frac{du}{1 + ux} = \frac{1}{x} \ln(1 + x) = 1 - \frac{1}{2}x + \frac{1}{3}x^2 - \frac{1}{4}x^3 + \cdots$$

(16.18)

In this case many of the calculations can be carried through directly. In particular we can inquire to what extent inequality (16.12) can be improved. We will look at x real and positive. For large x, $L \geq M$, we find by Eqs. (1.27) and (7.67)

$$[L/M] \approx x^{L-M} \frac{C(L + 1/M + 1)}{C(L/M)} \approx 4^{-L-M-1} x^{L-M}$$

(16.19)

Thus, the best we can expect in this example is, for x real and positive, to get convergence for

$$\frac{(1+x)^{1/2}-1}{(1+x)^{1/2}+1} x^{(\lambda-1)/(\lambda+1)} < 4 \qquad (16.20)$$

instead of one as in (16.10). As an illustration, we have computed the $[4M/M]$ Padé approximants. Equation (16.20) definitely implies divergence for x outside this range, i.e., $x \gtrsim 20$. Equation (16.10) implies convergence for $x \lesssim 4.5$. In Table 16.1 we give several values of the $[4M/M]$ Padé approximants. It will be observed from the table that the values for $x = 1.0$ and 4.0 are converging to the desired values and those for $x = 6.0$ and 9.0 are diverging by oscillation. The correct region of convergence in this case is fairly close to that predicted by the theorem and much smaller than the region given by (16.20), outside of which we can show divergence.

TABLE 16.1 $[4M/M]$ Padé Approximants to $x^{-1} \ln(1+x)$

x \ M	1	2	3	4	$x^{-1} \ln(1+x)$
1.0	0.69242424	0.69314873	0.69314718	0.69314718	0.69314718
6.0	−0.80000000	2.1535393	−2.6374816	5.1116289	0.32431836
9.0	−4.3735294	28.911655	−176.68475	1091.7095	0.25584279

It is interesting to see how, in accordance with the remarks at the end of Chapter 13, the extra zeros not associated with poles lying along the cut, $-\infty \leqslant z \leqslant -1$ mark out the boundary of the region of convergence. We have plotted them in Fig. 16.1. We see that the zeros border a generally heart-shaped region of the general nature of that described by Theorem 16.2. Since a series of Stieltjes does not vanish anywhere in the cut plane, as we shall see in the next chapter, the curve of zeros certainly limits the region of convergence.

On the basis of this example and Theorem 16.2 we expect only those Padé approximants for which $\lim_{M\to\infty} (L/M) = 1$ to converge in general in the whole, cut, complex plane. Clearly, they do not even converge in Hausdorff measure outside this heart-shaped region. In the case of essential singularities at least convergence in Hausdorff measure can be proven for $[\lambda M/M]$, $0 < \lambda < \infty$, sequences of Padé approximants (Theorem 14.1). When a branch point occurs, it is necessary to employ an essentially diagonal sequence to obtain convergence in the whole, cut, complex plane.

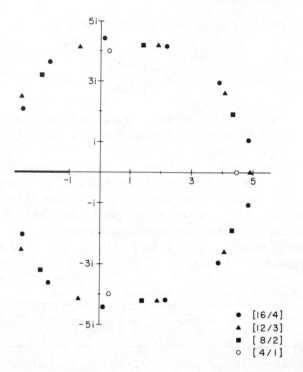

Fig. 16.1   The zeros (not on the negative real axis) of the $[4M/M]$ Padé approximants. These zeros define the edge of the region of convergence of this sequence to $x^{-1} \ln(1 + x)$.

We now turn our attention to diagonal sequences of Padé approximants whether formed from a series with a nonzero radius of convergence or not.

Theorem 16.3.   Let $f(z) = \sum_{p=0}^{\infty} f_p(-z)^p$ be a series of Stieltjes. If $\sum_{p=1}^{\infty}(f_p)^{-1/(2p+1)}$ diverges, then all $[M + J/M]$ sequences tend to a common analytic limit in the cut $(-\infty \leqslant z \leqslant 0)$ complex plane as $M \to \infty$, J fixed and greater than or equal to -1.

Proof: By Theorem 16.1 every $[M + J/M]$ sequence tends to an analytic limit in the cut $(-\infty \leqslant z \leqslant 0)$ complex plane. We now need to establish the uniqueness of this limit, even for certain types of divergent series. [Note that if $f_p = (2p)!$, then

$$\sum_{p=1}^{\infty} (f_p)^{-1/(2p+1)} \sim \sum (1/p) = \infty$$

so that this theorem would apply to such a series.] The condition in this theorem is, by a theorem of Carleman (1926), necessary and sufficient that there be at most one analytic function which is regular in some region which permits the approach to the origin on all rays and which has the derivatives implied by the power series as we approach along any ray. These conditions are clearly satisfied, except for the negative real axis. To apply Carleman's theorem, it suffices to consider a circular arc with center on the imaginary axis. Now suppose there is a pole of the approximant at a small distance d from the origin on the negative real axis. This pole will be a distance of order d^2 from the circular arc; however, by considering the third derivative and Lemma 16.1, the residue will be less than $d^3 f_3$; hence the contribution to the value on the circular arc will be of order d and hence we can uniformly bound the value on the circular arc and obtain convergence there, and therefore apply Carleman's theorem. Thus under the conditions given in the theorem a series of Stieltjes corresponds by Carleman's theorem to a unique analytic function. Our present theorem shows that when there is such a unique function all the diagonal $[M + J/M]$ Padé approximant sequences converge to it. To establish the proof of uniqueness, we will show only that the $\lim\{[M/M] - [M - 1/M]\} = 0$. The proof for other sequences is similar. By Theorem 15.2 the $[M/M]$ and $[M - 1/M]$ approximants provide the best bounds.

The proof rests on some of the identities we obtained in Chapter 3. It is convenient to rewrite them in the following notation:

$$\Omega_{2M-1} = \Delta_M^{(-1)}(z) / [D(1, M - 1)z^{M-(1/2)}],$$

$$\Omega_{2M} = \Delta_M^{(0)}(z) / [D(0, M)z^M]$$

$$\omega_{2M} = [D(0, M)]^2 / [z^{1/2}D(1, M - 1)D(1, M)]$$

$$\omega_{2M+1} = [D(1, M)]^2 / [z^{1/2}D(0, M)D(0, M + 1)] \qquad (16.21)$$

where $\Delta_M^{(J)}$ is given by (15.17). Then the two-term identities (3.7) and (3.8) become

$$[M/M] - [M - 1/M] = z^{1/2}/\Omega_{2M}\Omega_{2M-1},$$

$$[M/M + 1] - [M/M] = -z^{1/2}/\Omega_{2M+1}\Omega_{2M} \qquad (16.22)$$

and the three-term identities (3.23) and (3.24) lead to

$$\Omega_{p+1} = \Omega_{p-1} + \omega_p\Omega_p \qquad (16.23)$$

We now seek to show for z real and positive that $\Omega_p \to \infty$, and thus by (16.22) the difference between the $[M/M]$ and the $[M - 1/M]$ tends to zero. Equation (16.23) can be used to give, or Eq. (3.28) can be re-expressed as,

$$\Omega_{p+3} = [(\omega_p + \omega_{p+2} + \omega_p \omega_{p+1} \omega_{p+2})\Omega_{p+1}/\omega_p] - (\omega_{p+2}\Omega_{p-1}/\omega_p) \quad (16.24)$$

Thus, since all the ω_2 are positive, by discarding the term $\omega_p \omega_{p+1} \omega_{p+2}$ and taking absolute values, we obtain

$$|\Omega_{p+3}| - |\Omega_{p+1}| \geqslant |(\omega_{p+2}/\omega_p)|(|\Omega_{p+1}| - |\Omega_{p-1}|) \quad (16.25)$$

If we compute Ω_0, Ω_1, Ω_2, and Ω_3 from (16.21) directly and pick $k = \min[f_0^{-1}, z^{1/2}, f_2/(f_0 f_2 - f_1^2)]$, then we have

$$|\Omega_2| - |\Omega_0| \geqslant k\omega_1, \qquad |\Omega_3| - |\Omega_1| \geqslant k\omega_2,$$
$$|\Omega_0| \geqslant k, \qquad\qquad\quad |\Omega_1| \geqslant k \quad (16.26)$$

Applying (16.25) to (16.26) recursively, we obtain

$$|\Omega_{2p}| \geqslant k(1 + \omega_1 + \omega_3 + \cdots + \omega_{2p-1}),$$

$$|\Omega_{2p+1}| \geqslant k(1 + \omega_2 + \omega_4 + \cdots + \omega_{2p}) \quad (16.27)$$

It is clearly sufficient for at least either the odd or the even series to diverge for $\sum_{i=1}^{\infty} \omega_i$ to diverge. Now by Carleman's inequality (Lemma 16.2)

$$e \sum_{i=1}^{\infty} \omega_i \geqslant \sum_{p=1}^{\infty} (\pi_p)^{1/p}, \qquad \pi_p = \prod_{i=1}^{p} \omega_i \quad (16.28)$$

where e is the base of the natural logarithms. From (16.21) we find

$$\pi_{2p+1} = \frac{f_1 D(1, p)}{f_0 D(0, p+1)} z^{-p-(1/2)}, \qquad \pi_{2p} = \frac{f_1 D(0, p)}{f_0 D(1, p)} z^{-p} \quad (16.29)$$

Now the numerator of π_{2p+1} is the cofactor of f_{p+1} in the denominator. We can make all other elements in the first row zero besides f_{p+1} by multiplying the other columns by the coefficients of the denominator of the $[p/p + 1]$ Padé approximant, where the signs of the coefficients alternate and we normalize the coefficient z^{p+1} to be $+1$. If $(-1)^{p+1}A$ is the multiplier of the last column, then for p odd

$$\pi_{2p+1}$$

$$= (f_1/f_0)\left\{z^{p+(1/2)}\left[f_{p+1} - A\left(f_{p+2} - \frac{[p/p+1]^{(2p+2)}}{(2p+2)!}\right)\right]\right\}^{-1}$$

$$\geqslant (f_1/f_0)\left\{z^{p+1}f_{p+1}\right\}^{-1} \qquad (16.30)$$

by Lemma 16.1. Thus, omitting all but the odd-p terms, we have

$$\sum_{i=1}^{\infty} \omega_i \geqslant e^{-1} \sum_{r=1}^{\infty}\left(\frac{f_1}{f_0 f_{2r} z^{2r-(1/2)}}\right)^{1/(4r-1)} \qquad (16.31)$$

Now since any f_r is given by some $[M - 1/M]$ Padé approximant with the properties implied by (16.1), we can apply the Hölder inequality [see Beckenbach and Bellman (1965)] to show $(f_0/f_s)^{1/s} \leqslant (f_0/f_r)^{1/r}$ for $r \leqslant s$. Thus the divergence of (16.31) is equivalent to the divergence of

$$\sum_{p=1}^{\infty} (f_0/f_p)^{1/(2p-1)} \qquad (16.32)$$

But this divergence implies, by (16.22), (16.27), and (16.28), the coincidence of the $[M/M]$ and $[M - 1/M]$ limits, which together with a similar proof for the other sequences completes the proof of this theorem. ∎

In the proof of the preceding theorem we used the following inequality due to Carleman (1926).

Lemma 16.2 (Carleman). Let u_j, $j = 1, 2, \ldots$, be a given sequence of nonnegative numbers. Then

$$\sum_{j=1}^{n} (u_1 u_2 \cdots u_j)^{1/j} < e \sum_{j=1}^{n} u_j \qquad (16.33)$$

where e is the base of the natural logarithms.

Proof: The proof rests on the geometric-mean–arithmetic-mean inequality

$$\left(\prod_{i=1}^{m} w_i\right)^{1/m} \leqslant \frac{1}{m} \sum_{i=1}^{m} w_i \qquad (16.34)$$

provided $w_i \geqslant 0$. Beckenbach and Bellman (1965) give 12 proofs of this inequality. Perhaps the simplest is based on the concavity of $\ln x$ ($d \ln x / dx > 0$, $d^2 \ln x / dx^2 < 0$), so that

$$\ln\left(\frac{w_1 + w_2 + \cdots + w_m}{m} \right) \geqslant \frac{1}{m} \left(\ln w_1 + \ln w_2 + \cdots + \ln w_m \right)$$

(16.35)

which gives (16.34) by exponentiation.

Let us select the auxiliary sequence $v_p = p[1 + (1/p)]^p$. This sequence has the property

$$(v_1 v_2 \cdots v_p)^{1/p} = \left[\prod_{k=1}^{p} \frac{(1 + k)^k}{k^{k-1}} \right]^{1/p} = 1 + p \qquad (16.36)$$

Then we can write

$$\sum_{j=1}^{n} (u_1 u_2 \cdots u_j)^{1/j} = \sum_{j=1}^{n} (u_1 v_1 \cdot u_2 v_2 \cdots u_j v_j / v_1 v_2 \cdots v_j)^{1/j}$$

$$\leqslant \sum_{j=1}^{n} (v_1 v_2 \cdots v_j)^{-1/j} (u_1 v_1 + u_2 v_2 + \cdots u_j v_j)/j$$

$$= \sum_{k=1}^{n} u_k v_k \sum_{j=k}^{n} (v_1 v_2 \cdots v_j)^{-1/j}/j \qquad (16.37)$$

by (16.34). Now, by the definition of the v_p and by (16.36) we can rewrite the right-hand side of (16.37) as

$$\sum_{k=1}^{n} u_k \left(1 + \frac{1}{k} \right)^k k \sum_{j=k}^{n} \frac{1}{j(j+1)} \qquad (16.38)$$

since

$$\sum_{j=k}^{n} \frac{1}{j(j+1)} = \frac{1}{k} - \frac{1}{n} < \frac{1}{k} \qquad (16.39)$$

and since $[1 + (1/k)]^k < e$, we conclude (16.33) from (16.38) directly. ∎

C. Stieltjes Integral Representation

We remarked in Chapter 15 that the determinantal conditions (15.4) were necessary and sufficient for a function $f(z)$ to have all its coefficients given by the integral formula (15.2). We then proved (Lemma 15.1) that the integral formula implied the determinantal conditions. We are now in a position to prove the converse. In fact, we will establish the formal sum of conditions (15.2) as

$$f(z) = \int_0^\infty \frac{d\varphi(u)}{1 + uz} \tag{16.40}$$

To this end, we will concentrate our attention on the $[M - 1/M]$ Padé approximants. First, for this sequence we see by inspection that Theorem 15.1 involves only $D(0, n)$ and $D(1, n)$, and so holds assuming (15.4) only. The same is true of result (15.21), i.e., the monotonic increase of the $[M - 1/M]$ sequence with M. Theorem 15.1 also implies, as in Lemma 16.1, that $[M - 1/M]$ decreases monotonically with z real and positive. Thus, by assuming only (15.4), we see, since $[M - 1/M]$ is bounded uniformly from above by f_0 for z real and positive, and is monotonically increasing in M, that we can conclude Theorem 16.1 under these weaker assumptions for the $[M - 1/M]$ sequence.

By considering form (16.1) for the $[M - 1/M]$ Padé approximant

$$[M - 1/M] = \sum_{p=1}^M \frac{\beta_p}{1 + \gamma_p z}, \qquad \beta_p, \gamma_p > 0 \tag{16.41}$$

we can show directly that

$$\text{Im}\{[M - 1/M](z)\} \gtrless 0 \qquad \text{as} \qquad \text{Im}(z) \gtrless 0 \tag{16.42}$$

Let us now apply Cauchy's theorem to the $[M - 1/M]$. This gives

$$[M - 1/M](z) = \frac{1}{2\pi} \oint_C \frac{\text{Im}\{[M - 1/M](w)\}\,dw}{w - z} \tag{16.43}$$

where the contour C is as shown in Fig. 16.2. This representation is valid for every finite M because $[M - 1/M] \propto z^{-1}$ at $z = \infty$ and all the poles are on the negative real axis. The reason that only the imaginary part appears in (16.43) is that the real part cancels because the direction of the contour is opposite at complex conjugate points, and by representation (16.41) the real part is the same. Now by Theorem 16.1, $[M - 1/M](w)$ tends to an analytic function $f(z)$ at every point of the contour C. Since the

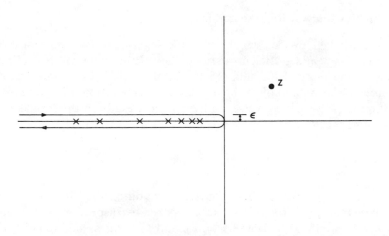

Fig. 16.2 The contour of integration C for integral (16.43).

imaginary part is negative for $\operatorname{Im} z > 0$, and uniformly for every M, $[M - 1/M](0) = f_0$, we can write

$$\lim_{M \to \infty} [M - 1/M](z) = \frac{1}{2\pi} \oint_C \frac{\operatorname{Im}[f(w)]\, dw/w}{1 - z/w} \qquad (16.44)$$

Thus if we define at points of continuity

$$\varphi(u) - \varphi(w) = \frac{1}{\pi} \lim_{\epsilon \to 0} \int_w^u \operatorname{Im}\left\{ f\left[-(v - i\epsilon)^{-1} \right] \right\} \frac{dv}{v} \qquad (16.45)$$

where $\varphi(\infty) - \varphi(0) = f_0$, we have established the existence of a φ such that (16.40) holds. Conditions (15.2) follow in the same way by equating derivatives initially. We have thus established the representation theorem.

Theorem 16.4. The determinantal conditions (15.4) imply the existence of an integral representation (16.40) and the validity of the representation of the series coefficients (15.2).

D. The Hamburger Problem

The results for this case, conditions (15.6), are similar, but weaker, than those for the Stieltjes case. Here we have only half the determinantal conditions [Eq. (15.9)] and expect a representation

$$g(z) = \int_{-\infty}^{+\infty} \frac{d\psi(u)}{1 + uz} \qquad (16.46)$$

Let us begin with condition (15.9). Referring to Eq. (15.17) for the Padé denominators, we see that this condition implies $\Delta_M^{(-1)}(0) > 0$. Thus for $M = 1$ we have a nonzero real root since the g_j [Eq. (15.6)] are all real. Since formula (15.18) still holds, i.e., if $\Delta_M(z)$ vanishes, $\Delta_{M+1}(z)$ and $\Delta_{M-1}(z)$ have opposite signs, and $\Delta_0(z)$ is a positive constant, we conclude that $\Delta_2(z)$ has a real root between the origin and that of $\Delta_1(z)$. Since $\Delta_2(z)$ is a polynomial of degree two, its other root must also be real. This argument can be carried on to any finite M since the $z^{-M}\Delta_M$ form a Sturm sequence. The roots of successive Δ_M interlace each other. The residues are all positive, by the same argument [Eq. (15.20)] as for series of Stieltjes. The root $z = \infty$ is not excluded and corresponds to a Δ_M not of true degree M. Thus we have established the following result.

Theorem 16.5. If the g_j satisfy the determinantal conditions $D_g(0, n) > 0$ for all $n = 0, 1, 2, \ldots$, then the $[M - 1/M]$ Padé approximants all have the form

$$[M - 1/M](z) = \sum_{p=1}^{M} \frac{\beta_p}{1 + \gamma_p z}, \qquad \beta_p > 0 \qquad (16.47)$$

and γ_p real.

We immediately have the following result.

Theorem 16.6. If the g_j satisfy the determinantal conditions $D_g(0, n) > 0$ for all $n = 0, 1, 2, \ldots$, then there exists at least a subsequence of the $[M - 1/M]$ which converges to a function $g(z)$ which is analytic in the half-plane Im $(z) \gtrless 0$. This function is asymptotically equal to the formal power series $\sum_{j=0}^{\infty} g_j(z)^j$ at $z = 0$ approached from the upper or lower half-plane. This function can be represented by form (16.46).

Proof: Since $\Sigma \beta_p = g_0$, we have

$$|[M - 1/M](z)| \leqslant g_0 |z| / |\text{Im } z| \qquad (16.48)$$

by minimizing Eq. (16.47) with respect to the γ_p. Thus for any closed region $\mathfrak{R}$ of the half-plane Im $(z) > 0$ or the half-plane Im $(z) < 0$ we can apply Theorem 12.2 to prove the existence of a convergent subsequence in $\mathfrak{R}$. Next, to prove that the $[M - 1/M]$ are asymptotically equal to the formal defining series, we note that, since by (16.47)

$$\sum_{p=1}^{M} \beta_p \gamma_p^{2j} = g_{2j} \qquad (16.49)$$

it must be that if there is a pole at distance d from the origin,

$$\beta_p \leqslant g_{2j}d^{2j} \tag{16.50}$$

for $M > j$. Thus let us consider a circle passing through the origin with center on the imaginary axis. Near the origin if there is a pole at distance d from the origin, it is a distance of the order of d^2 from the circle. Thus

$$z^{-2k}\left|[M - 1/M](z) - \sum_{j=0}^{2k} g_j(-z)^j\right| \leqslant d^{-2k}g_{2k+4}d^{2k+4-2} \leqslant g_{2k+4}d^2 \tag{16.51}$$

using (16.50) for $j = k + 2$. Since (16.51) holds uniformly in M for $|z| \leqslant d$ and z in that circle just described, we must have

$$\lim_{z \to 0} \lim_{M \to \infty} z^{-2k}\left\{[M - 1/M](z) - \sum_{j=0}^{2k} g_j(-z)^j\right\} = 0 \tag{16.52}$$

where the limit $z \to 0$ is on any path in the circle just described. Thus we have proved everything necessary to conclude this theorem. ■

It now follows that we can define a $\psi(u)$ from the $g(z)$ defined by Theorem 16.6. From the $\psi(u)$ we obtain the representation (16.46), and from Lemma 15.2 we can extend the determinantal condition (15.9) to $D(2m, m) > 0$, $n, m = 0, 1, 2, \ldots$. By the use of these additional determinants we can show that

$$\Delta_M^{(J)}(0) > 0, \qquad J \text{ odd}$$

$$\lim_{z \to \infty} z^{-M}\Delta^{(J)}(z) > 0, \qquad J \text{ even} \tag{16.53}$$

By an argument similar to that for Theorem 16.5 we can show that all the poles lie on the real axis for all $M > 0, J \geqslant -1$, and thus at least a subsequence of every $[M + J/M]$ sequence converges.

According to Carleman's theorem (1926), there can be at most one function which is asymptotically equal to $\sum g_j(-z)^j$ under the conditions we have established if

$$\sum_{j=1}^{\infty} (g_{2j})^{-1/2j} \text{ diverges} \tag{16.54}$$

Thus if this condition holds, for $2M + J$ large enough all $[M + J/M]$ are asymptotically equal [(16.52)] to $\Sigma\, g_j(-z)^j$. Given condition (15.9), they all converge to the same limiting function.

We remark that as an extension of the results on series of Stieltjes, Nuttall (1972) has recently extended the allowed form of (16.40) to include

$$F(z) = \int_a^b \frac{\omega(x)\,dx}{1 - zx} \qquad (16.55)$$

where a, b, and $\omega(x)$ can be complex, but $\omega(x) \neq 0$ on the path of integration, plus some other mild restrictions. Then, generalizing the methods of Chapter 7, he shows convergence of the $[M - 1/M]$ Padé approximants in the cut plane.

We can always find a transformation $z = \alpha w/(1 + \beta w)$ so that $\beta - \alpha a$ and $\beta - \alpha b$ are real. Under this transformation Eq. (16.55) becomes

$$G(w) = (1 + \beta w)^{-1} F\!\left(\frac{\alpha w}{1 + \beta w} \right) = \int_a^b \frac{\omega(x)\,dx}{1 + (\beta - \alpha x)w} \qquad (16.56)$$

The determinants are related by

$$D_G(0, n) = \alpha^{n(n-1)} D_F(0, n) \qquad (16.57)$$

as can be seen by row-and-column operations on D_G. Thus, if $\omega(x) \geqslant 0$, we can recognize this property directly from the determinants if there is an α such that $\alpha^{m(m-1)} D_f(0, m)$ is real and positive. Convergence of the $[M - 1/M]$ follows directly from Corollary 9.1 when we apply the inverse transform to $G(x)$, since it is invariant under this transformation.

SERIES OF STIELTJES, INCLUSION REGIONS

The main consideration in this chapter will be the range of values that the Padé approximants can take on, given certain restrictions, i.e., that the underlying series is a series of Stieltjes (or of Hamburger) and that a certain number of terms are given. Historically this problem seems to have originated with Tschebycheff (1874), who was interested in the "reduced moment problem," that is: To what extent does a given sequence of moments

$$\int_{-\infty}^{+\infty} p(x)x^n \, dx = \mu_n \qquad (17.1)$$

of a probability distribution determine the distribution? As we have seen in the previous chapters, this problem was completely solved by Stieltjes (1894) for the range zero to infinity, and by Hamburger (1920, 1921) for the full range $-\infty$ to $+\infty$. Tschebycheff's (1874) inequalities for the cumulative distribution function were proved almost simultaneously by Markoff (1884) and Stieltjes (1884). Hamburger (1920, 1921) seems to have been the first to give inclusion regions, circles for his problem, for the values for all higher-order Padé approximants. Recently a number of workers have contributed to this subject, including Baker (1969), Common (1968), Gragg (1968, 1970), Gordon (1968), Gargantini and Henrici (1967), and Jones and Thron (1971). Results have been extended from the Hamburger problem to the Stieltjes problem with a nonzero radius of convergence.

A. Nonzero Radius of Convergence

The first step in defining an inclusion region is to use Lemma 15.3 and Theorem 16.4. To summarize, if $f(z)$ is a series of Stieltjes with a radius of convergence R, then

$$f(z) = \int_0^{1/R} \frac{d\varphi(u)}{1 + uz} \tag{17.2}$$

and $g(z)$, defined by

$$f(z) = f_0 / [1 + zg(z)] \tag{17.3}$$

is also a series of Stieltjes. From the representation (17.2) it is clear that $f(z)$ is regular in the cut $(-\infty \leqslant z \leqslant -R)$ complex plane. Except for possible polar singularities, $g(z)$ must be also, by (17.3). However, if $g(z)$ had a pole, then $f(z)$ would vanish, but $f(z) \neq 0$ in the cut plane. Thus $g(z)$ has a radius of convergence of at least R. To see $f(z) \neq 0$, we first see directly from (17.2) that for z real and greater than $-R$, $f(z) > 0$. Now, for z complex

$$\operatorname{Im} f(z) = -\int_0^{1/R} \frac{[\operatorname{Im}(z)]u \, d\varphi(u)}{|1 + uz|^2} \neq 0 \tag{17.4}$$

since $\operatorname{Im}(z) \neq 0$, which shows $f(z) \neq 0$ in the cut plane.

Now since in (17.3) $g(z)$ is in the same class as $f(z)$, namely a series of Stieltjes of radius of convergence at least R, and we are free to iterate that form. When we do we obtain

$$f(z) = \cfrac{f_0}{1 + \cfrac{za_1}{1 + \cfrac{za_2}{1 + \cdots \cfrac{}{ + \cfrac{za_p}{1 + zh_p(z)}}}}} \tag{17.5}$$

where $h_p(z)$ is again a series of Stieltjes with radius of convergence at least R. By using the formulas we developed in Chapter 4 for continued fractions we can re-express (17.5) as

$$f(z) = \frac{A_p(z) + zh_p(z)A_{p-1}(z)C_p}{B_p(z) + zh_p(z)B_{p-1}(z)C_p} \tag{17.6}$$

where the A's and B's are polynomials in z and the C's are constants we have introduced for future convenience. The fractions $A_p(z)/B_p(z)$ fill the stair-step sequence in the Padé table, $[0/0], [0/1], [1/1], [1/2], \ldots$, by Theorem 4.2.

Returning to Eq. (17.2), we note one further restriction that $g(z)$ must satisfy; namely, since $g(z)$ is monotonic in $-R \leqslant z \leqslant 0$, we must have

$$\lim_{z \to -R} [-zg(z)] \leqslant 1 \qquad (17.7)$$

in order for $f(z)$ to be free of singularities in this range. Therefore

$$g(-R) \leqslant 1/R \qquad (17.8)$$

As we iterate (17.2) the $h_p(z)$ obtained will be similarly restricted, although the bound is now a function of all the constants in (17.5), which are determined by the first p Taylor series coefficients of $f(z)$, through Eq. (4.84).

We choose to redefine $h_p(z)$ so that

$$h_p(-R) \leqslant 1/R \qquad (17.9)$$

and to absorb the change in normalization in C_p. We can now easily solve for C_p from (17.6), using the critical equation which makes $z = -R$ a pole of $f(z)$. Thus

$$C_p = B_p(-R)/B_{p-1}(-R) \qquad (17.10)$$

Having established (17.6), subject to (17.9)–(17.10), we see, given the first p coefficients, and that $f(z)$ is a series of Stieltjes with radius of convergence at least R, that the range of $f(z)$ is just the linear fractional transformation of the range of $h_p(z)$.

We now turn our attention to the problem of computing the range of $h_p(z)$, subject to (17.9). Now we can represent

$$h_p(z) = \int_0^{1/R} \frac{d\psi_p(u)}{1 + uz} \qquad (17.11)$$

which evaluated at $z = -R$ is

$$h_p(-R) = \int_0^{1/R} \frac{d\psi_p(u)}{1 - Ru} \leqslant \frac{1}{R} \qquad (17.12)$$

By this equation

$$d\omega_p(u) = R \, d\psi_p(u)/(1 - Ru) \qquad (17.13)$$

is also an allowable measure in a Stieltjes integral with

$$\int_0^{1/R} d\omega_p(u) \leqslant 1 \qquad (17.14)$$

Thus we can write

$$h_p(z) = \frac{1}{R} \int_0^{1/R} \frac{(1 - Ru)\, d\omega_p(u)}{1 + zu} \tag{17.15}$$

with $d\omega_p$ an arbitrary, nonnegative-definite, normalized measure. It follows at once from (17.15) that if H_1 and H_2 are possible values of $h_p(z)$, then $\alpha H_1 + (1 - \alpha)H_2$, $0 \leqslant \alpha \leqslant 1$, are, too, for if ω and ω' are allowed measures in (17.15), then so, too, is $\alpha\omega + (1 - \alpha)\omega'$. Consequently, the range of $h_p(z)$ is a convex region. The integrand of (17.15) is a weighted sum of

$$\frac{1}{R}\left(\frac{1 - Ru}{1 + uz}\right), \qquad 0 \leqslant u \leqslant \frac{1}{R} \tag{17.16}$$

which is a linear fractional transformation of the segment of the u axis. It follows from our discussion (Section 9.C) of the Riemann sphere that linear fractional transformations map the family of circles and straight lines into itself. We can also see this result directly. The equation of a circle is

$$0 = a(x - x_0)^2 + a(y - y_0)^2 - \rho^2 = azz^* + bz^* + b^*z + c \tag{17.17}$$

where $z = x + iy$ and z^* is the complex conjugate of z. Transforming by

$$z = (\alpha + \beta w)/(\gamma + \delta w) \tag{17.18}$$

yields

$$a'ww^* + b'w^* + b'^*w + c' = 0 \tag{17.19}$$

where a' and c' are necessarily real and b' may be complex. Thus the curve in the w plane is a circle, too, or possibly a straight line if $a' = 0$.

Hence the range of $h_p(z)$ is the convex hull of this arc. That is, every point on any line between any two points on the arc in the range. We have illustrated in Fig. 17.1 the sample case, $z = -2R + iR$. The circular arc is tangent, at the origin, to a line through $R + z^*$. The vertical height of the region for any z and R can be computed easily to be

$$0 \leqslant -\mathrm{Im}\left\{h_p(z)\right\} \leqslant \frac{y}{2R\left\{\left[(R + x)^2 + y^2\right]^{1/2} + (R + x)\right\}} \tag{17.20}$$

where $z = x + iy$. The range is the complete, convex, lens-shaped region described previously since any point on the circular arc can be obtained for $d\omega_p(u) = \delta(u - u_0)\, du$, where u_0 is appropriately selected. Any point

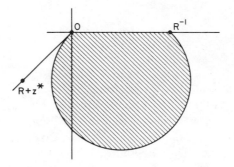

Fig. 17.1   The shaded region is the allowed range of values of $h_p(z)$ for the case $z = -2R + iR$. Note that the circle is tangent to the line $\overline{0, R + z^*}$.

on the real axis portion of the boundary can be obtained by using $d\omega(u) = [\alpha\delta(u) + (1 - \alpha)\delta(u - 1/R)]\,du$. All points in the interior can be obtained in an infinite number of ways as linear combinations of boundary points. These are, of course, not necessarily the only such ways to obtain them. The range of $f(z)$ is therefore the map of this lens-shaped region under (17.6) The resulting region $F_p(z)$ will again be a lens-shaped region, by (17.17)–(17.19). It must remain convex because if $\varphi(u)$ is a measure which, through Eq. (17.2), reproduces the first p moments, and $\hat{\varphi}(u)$ is another, then so, too, does $\alpha\varphi + (1 - \alpha)\hat{\varphi}$. Since that lens-shaped region contains *all* such allowed functions, by construction, it is thus necessarily convex. By the method of constructing regions satisfying more and more restrictions, we have

$$f(z) \in F_p(z) \subset F_{p-1}(z) \subset \cdots \subset F_1(z) \tag{17.21}$$

where $\subset$ means contained in, i.e., every element of $F_p(z)$ also belongs to $F_{p-1}(z)$.

By means of the identity

$$\frac{1}{1 + x} = 1 - x + x^2 - \cdots + (-x)^{n-1} + \frac{(-x)^n}{1 + x} \tag{17.22}$$

we can write

$$f(z) = f_0 - f_1 z + f_2 z^2 + \cdots + f_{n-1}(-z)^{n-1} + \int_0^{1/R} \frac{(-zu)^n \, d\varphi(u)}{1 + uz}$$

$$= f_0 - f_1 z + f_2 z^2 + \cdots + f_{n-1}(-z)^{n-1} + \frac{f_n(-z)^n}{1 + zg(z)} \tag{17.23}$$

where by Lemma 15.3, $g(z)$ is a series of Stieltjes. We can construct, by iterating (17.23) as we did in (17.6), a new sequence of inclusion regions. In terms of the notation of Chapter 3

$$[M + J/M] = P^{(J)}_{M+J}(z)/Q^{(J)}_M(z)$$

we can write out explicitly the relation between $f(z)$ and $h(z)$. For even values of p it is

$$f(z) = \frac{P^{(n)}_{m+n}(z)Q^{(n-1)}_m(-R) + zh_{2m,n}(z)P^{(n-1)}_{m+n-1}(z)Q^{(n)}_m(-R)}{Q^{(n)}_m(z)Q^{(n-1)}_m(-R) + zh_{2m,n}(z)Q^{(n-1)}_m(z)Q^{(n)}_m(-R)} \qquad (17.24)$$

and for p odd it is

$$f(z) = \frac{P^{(n-1)}_{m+n}(z)Q^{(n)}_m(-R) + zh_{2m+1,n}(z)P^{(n)}_{m+n}(z)Q^{(n-1)}_{m+1}(-R)}{Q^{(n-1)}_{m+1}(z)Q^{(n)}_m(-R) + zh_{2m+1,n}(z)Q^{(n)}_m(z)Q^{(n-1)}_{m+1}(-R)} \qquad (17.25)$$

We can use (17.24) and (17.25) to define the inclusion regions $F_{p,n}(z)$ as

$$F_{p,n}(z) = \left\{ f(z) | h_{p,n}(z) = v(1 - Ru)/(1 + uz), \quad 0 \leqslant u, v \leqslant 1/R \right\}$$

$$(17.26)$$

which depends explicitly via (17.24) and (17.25) on $R, f_0, f_1, \ldots f_{p+n}$. Equation (17.26) means $F_{p,n}(z)$ consists of all $f(z)$ given as u and v vary over the prescribed range. The case $n = 0$ is just the $F_p(z)$ of Eq. (17.21). That equation becomes directly

$$f(z) \in F_{p,n}(z) \subset F_{p-1,n}(z) \subset \cdots \subset F_{1,n}(z) \qquad (17.27)$$

We observe from (17.24) and (17.25) that for every nearest-neighbor pair of approximants in the Padé table there is an inclusion region given in terms of that pair of approximants alone. The best inequalities (Theorem 15.2) were obtained in terms of the diagonal approximants. We are now in a position to find the best inclusion regions available for a fixed number of coefficients.

Theorem 17.1. Let $f(z) = \sum_{j=0}^{\infty} f(-z)^j$ be a series of Stieltjes with radius of convergence R. Then $f(z)$ is contained in $F_{p,n}(z)$ $(p = 1, 2, \ldots;$ $n = 0, 1, 2, \ldots)$ defined by (17.24)–(17.26). The inclusion regions satisfy the properties $(p \geqslant 1; n \geqslant 0)$

$$F_{p,n}(z) \subset F_{p-1,n}(z)$$

$$F_{p,n}(z) \subset F_{p-1,n+1}(z) \qquad (17.28)$$

which imply that the best inclusion region obtainable from a fixed number of coefficients is $F_{p,0}(z)$. This region is the best possible one in terms of the assumed information. The interior angles in the $F_{p,n}(z)$ are $|\arg(R + z)|$. The diameter of the inclusion region goes to zero geometrically in the cut plane like $\{[(1 + z/R)^{1/2} - 1])/[(1 + z/R)^{1/2} + 1]\}^p$ as p goes to infinity.

Proof: We have already demonstrated by construction that $f(z)$ lies in the inclusion regions $F_{p,n}(z)$ for all z in the cut plane, and that the first inclusion relation holds. That the interior angles of the $F_{p,n}(z)$ are $|\arg(R + z)|$ follows (see Fig. 17.1) from a direct calculation of the angles for $h(z)$ and the fact that the transformations (17.24)–(17.25) are conformal, i.e., angle preserving. The only requirement for this result to hold is that the derivative $df/dh \neq 0$. This derivative can be easily evaluated by means of the two-term identities (3.7) or (3.8) and found not to vanish ($z \neq 0$), by the determinantal conditions of Lemma 15.1 and the result $Q(-R) \neq 0$ from Theorems 15.1 and 16.2. The rate of convergence follows from Theorem 16.2 directly using the special case $\lambda = 1$, which is applicable to every fixed-n sequence, since $n > 0$ only contributes to K' in Eq. (16.15).

To establish relation (17.28), we first consider the possible relation

$$F_{2m+1,n}(z) \subset F_{2m,n+1}(z) \tag{17.29}$$

To this end, we define

$$\tau = zh_{2m+1,n}(z)Q_{m+1}^{(n-1)}(-R)/Q_m^{(n)}(-R)$$

$$\sigma = zh_{2m,n+1}(z)Q_m^{(n+1)}(-R)/Q_m^{(n)}(-R) \tag{17.30}$$

If we now solve for τ in terms of σ from Eq. (17.24)–(17.25), we get

$$\tau = \{[P_{m+n}^{(n-1)}(z)Q_m^{(n+1)}(z) - P_{m+n+1}^{(n+1)}(z)Q_{m+1}^{(n-1)}(z)]$$

$$+ [P_{m+n}^{(n-1)}(z)Q_m^{(n)}(z) - P_{m+n}^{(n)}(z)Q_{m+1}^{(n-1)}(z)]\}$$

$$\times \{P_{m+n+1}^{(n+1)}(z)Q_m^{(n)}(z) - Q_m^{(n+1)}(z)P_{m+n}^{(n)}(z)\}^{-1} \tag{17.31}$$

By use of the two-term identities (3.7)–(3.9) we can simplify (17.31) to give

$$\frac{\tau}{z} = \frac{[D(1 + n, 1 + m)]^2}{D(2 + n, m - 1)D(1 + n, m)} + \frac{D(n, m)}{D(2 + n, m - 1)}\frac{\sigma}{z} \tag{17.32}$$

which is a *linear* relation with constant coefficients between the points of

$F_{2m, n+1}$ and $F_{2m+1, n}$. Since (17.32) is a linear transformation, it is automatically a conformal one. Since the constants are real, the real-axis portion of the boundary to the range of σ/z maps into a portion of the real axis of the (τ/z) plane. The origin in the (σ/z) plane maps into a real, negative point in the (τ/z) plane. Since Eq. (17.31) is equivalent to (17.32), we can obtain the value of τ which corresponds to the vaue of σ when $h_{2m, n+1}(z)$ $= 1/R$ by substituting into (17.31) for the particular value $z = -R$. This yields

$$\tau = -Q_m^{(n)}(-R)/Q_{m+1}^{(n-1)}(-R) \qquad (17.33)$$

which implies the mapped value

$$h_{2m+1, n}(-R) = 1/R \qquad (17.34)$$

which is the same maximum value that we had derived previously. Since the map of the (σ/z) plane, straight-line portion of the boundary of the ranges includes all of the straight-line portion of the boundary in the (τ/z) plane, and since the transformation is a linear and shape preserving, we see that the curved boundary lies outside the boundary for the region in the (τ/z) plane (see Fig. 17.2). Thus we have proven (17.29). The similar relation for $F_{2m, n}$ can be proven in an exactly analogous manner, except that the map of the (σ/z) plane range has the curved side in common with the (τ/z) plane range instead of the straight side. Thus inclusion relation (17.28) also holds. The result that $F_{p, 0}(z)$ is the best inclusion region then follows by using (17.28) to systemmatically reduce n to zero without increasing the number of coefficients. That $F_{p, 0}(z)$ is the best possible inclusion region follows because we have shown explicitly how every point may be attained. We have not similarly demonstrated that every point of $F_{p, n}(z)$, $n \geqslant 1$, can be attained, because we have not shown that powers of z can be represented by form (17.2). ∎

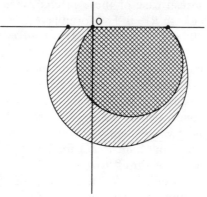

Fig. 17.2 The shaded area is the map of the allowed region in the (σ/z) plane. The cross-hatched area is the region in the (τ/z) plane.

As an illustration of how the inclusion regions $F_{p,0}$ look, we have sketched a typical case in Fig. 17.3. This example uses $z/R = -2 + i$. The successive inclusion regions lie inside one another and the new, smaller region touches the old one at just two points.

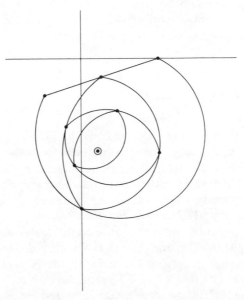

Fig. 17.3   Sketch for the Stieltjes problem of the sequence of inclusion regions $F_{p,0}$ for $z/R = -2 + i$. The successive regions lie inside one another and the new, smaller region touches the old one at just two points. The shape of each region is that of a convex lens.

The best inclusion regions in the cut plane ($-\infty < z \leqslant 0$) for a series of Stieltjes with no (i.e., $R = 0$) assumption on the radius of convergence can be obtained as a special case of these results, since Theorem 17.1 is still valid for the limiting case $R \to 0$. The formulas (17.24)–(17.26) remain valid. By definition $Q_m^{(n)}(0) = D(1 + n, m - 1)$ for all m, n. The range of $h_{p,n}$ follows directly as the limit as $R \to 0$ of Eq. (17.16). It is the complete angular wedge bounded by the real axis and the ray through $R + z^*$. Since a wedge is merely a lens-shaped region with one vertex at infinity, the ranges $F_{p,n}$ for the value of $f(z)$ are again lens-shaped regions. The rate of convergence is no longer geometric. As we saw in the examples of Section 10.C, a power of m^{-1} is frequently encountered in the case of a unique limit. In the case where the limit is not unique the radius of the inclusion region, of course, does not shrink to zero.

We remark as a corollary that the results of this theorem are also available for the Hamburger problem with a finite radius of convergence.

This problem is defined by

$$g(z) = \int_{-1/S}^{+1/R} \frac{d\psi(u)}{1 + uz} \tag{17.35}$$

where $RS \neq 0$. If we make a linear fractional transformation on (17.35), we find

$$\frac{R + S}{S + R(1 + \xi)} g\left(\frac{RS\xi}{S + R(1 + \xi)} \right) = \int_0^1 \frac{d\psi \left[(R^{-1} + S^{-1})v - S^{-1} \right]}{1 + v\xi}$$

$$\frac{R + S}{R + S(1 + \xi)} g\left(\frac{-RS\xi}{R + S(1 + \xi)} \right) = \int_0^1 \frac{d\psi [R^{-1} - (R^{-1} + S^{-1})v]}{1 + v\xi}$$

$$\tag{17.36}$$

which are both series of Stieltjes with radius of convergence unity. Thus we can obtain best possible inclusion regions for $g(z)$ through the use of these modified functions.

The inclusion regions given by Theorem 17.1 assume more than the corresponding inequalities given in Theorem 15.2, since an assumption is made on the radius of convergence. This additional information allows us to give tighter bounds on the values for real arguments than the inequalities given there. In particular, since for z real we have directly from (17.15)

$$0 \leqslant h(z) \leqslant 1/R \tag{17.37}$$

we have the following corollary.

Corollary 17.1.   If $f(z)$ is a series of Stieltjes with radius of convergence R, then for $z \geqslant 0$

$$\frac{P_m^{(0)}(z)Q_m^{(-1)}(-R) + (z/R)P_{m-1}^{(-1)}(z)Q_m^{(0)}(-R)}{Q_m^{(0)}(z)Q_m^{(-1)}(-R) + (z/R)Q_m^{(-1)}(z)Q_m^{(0)}(-R)} \leqslant f(z) \leqslant [m/m]$$

$$\tag{17.38}$$

or

$$[m/m + 1] \leqslant f(z) \leqslant \frac{P_m^{(-1)}(z)Q_{m+1}^{(-1)}(-R)}{Q_{m+1}^{(-1)}(z)Q_m^{(0)}(-R) + (z/R)Q_m^{(0)}(z)Q_{m+1}^{(-1)}(-R)}$$

$$\tag{17.39}$$

When $-R < z < 0$ the sense of the inequality signs in (17.38) reverses, and Eq. (17.39) remains valid.

Proof: These inequalities follow directly by substituting the range (17.37) in Eqs. (17.24) and (17.25). The values 0 and $1/R$ give the end points of the range. All that remains is to identify which one is which. This identification is easily made by using Eq. (15.23) and the observation that $[m/m]$ and $[m/m + 1]$ are lower bounds to $f(z)$ for $-R < z < 0$. ∎

We note that when we make no assumption on R, i.e., $R = 0$, both (17.38) and (17.39) reduce to our previous result (15.23), as they should. This observation extends by Corollary 17.1 the definition of $f(z)$ used in Theorem 15.2 to be *any* series of Stieltjes, rather than just the limit of the $[M/M]$ or $[M - 1/M]$ sequences.

B. The N-Point Problem

With only a few additional considerations we can extend the results of the previous section to the general N-point Padé approximant. To do this, we need the following simple lemma.

Lemma 17.1. If $f(z)$ is a series of Stieltjes with a radius of convergence at least R, then $f(z + w)$ is a series of Stieltjes with a radius of convergence at least $w + R$, $w > -R$.

Proof: Beginning with representation (17.2) and making the change of variables $v = u/(1 + wu)$, we have

$$f(z + w) = \int_0^{1/R} \frac{d\varphi(u)}{1 + u(z + w)}$$

$$= \int_0^{1/(R+w)} \frac{(1 - vw)d\varphi[v/(1 - vw)]}{1 + zv} \tag{17.40}$$

which is of the required form. ∎

With this lemma we can proceed as we did in Section 8.C on Thiele's reciprocal difference method and show, using Lemmas 17.1 and 15.3, successively, that if $\{z_j, j = 0, 1, \ldots, p\}$ are points of the real line $-R < z < +\infty$, then we can write

$$f(z) = \cfrac{a_0}{1 + \cfrac{(z - z_0)a_1}{1 + \cfrac{(z - z_1)a_2}{1 + \cdots \cfrac{(z - z_{p-1})a_p}{1 + (z - z_p)g_{p+1}(z)}}}} \tag{17.41}$$

which defines $g_{p+1}(z)$ as a series of Stieltjes, provided that the a_p are selected so as to fit $f(z)$ at $z = z_0, z_1, \ldots, z_p$. Formulas for doing this are given in Section 8.C. Since Eq. (17.41) can be re-expressed [by (8.24)] as

$$f(z) = \frac{A_p(z) + (z - z_p)g_{p+1}(z)A_{p-1}(z)C_p}{B_p(z) + (z - z_p)g_{p+1}(z)B_{p-1}(z)C_p} \qquad (17.42)$$

where the A's and B's are the polynomial numerators and denominators of successive convergents to (17.41). Let us, in analogy to the procedures of the previous section, renormalize $g_{p+1}(z)$ and absorb this change into C_p. We select the normalization

$$(R + z_p)g_{p+1}(-R) \leqslant 1 \qquad (17.43)$$

That $g_{p+1}(-R)$ is bounded follows as it did for $h_p(-R)$. If we consider $g_{p+1}(w + z_p) = k_{p+1}(w)$, then by Lemma 17.1, we have shown $k_{p+1}(w)$ to be a series of Stieltjes with a radius of convergence of at least $z_p + R$. Hence, as before, the bounds on its range are the convex hull of Eq. (17.16) when one replaces R by $R + z_p$. We can evaluate C_p as before, and we obtain

$$f(z) = \frac{B_{p-1}(-R)A_p(z) + (z - z_p)g_{p+1}(z)B_p(-R)A_{p-1}(z)}{B_{p-1}(-R)B_p(z) + (z - z_p)g_{p+1}(z)B_p(-R)B_{p-1}(z)} \qquad (17.44)$$

This formula defines a lens-shaped inclusion region $F_p(z)$ at any point in the cut, complex plane which is given by

$$F_p(z) = \left\{ f(z) | g_{p+1}(z) = v[1 - (R + z_p)u]/(1 + uz), \right.$$

$$\left. 0 \leqslant u, v \leqslant (R + z_p)^{-1} \right\} \qquad (17.45)$$

This lens-shaped region reduces to an interval for z real. This bound on the value of $f(z)$ is the best possible, for by arguments directly analogous to those of the previous section we can construct a series of Stieltjes to take on any value in the region (17.45).

Theorem 17.2. If the set $\{z_j, -R < z_j < \infty\}$ possesses a limit point on the real line, $-R < z < +\infty$, then the necessary and sufficient conditions that the set of values $\{f(z_j)\}$ are values of a series of Stieltjes with radius of convergence R are that

$$f(z_{p+1}) \in F_p(z_{p+1}), \qquad p = 0, 1, \ldots \qquad (17.46)$$

where $f_p(z)$ is given by (17.45).

Proof: If $f(z)$ is a series of Stieltjes with radius of convergence R, then (17.46) must hold. Thus (17.46) is necessary.

On the other hand, if (17.46) holds, we can construct a sequence of series of Stieltjes $[S_p(z) = A_p(z)/B_p(z)]$ which agree with $f(z)$ at the points $(z_0, z_1, \ldots, z_p)$. Since by arguments analogous to those of Section 17.A we must have

$$f(z) \in F_p(z) \subset F_{p-1}(z) \subset \cdots \subset F_1(z) \subset F_0(z) \tag{17.47}$$

since $F_0(z)$ is uniformly bounded over any closed bounded region in the interior of the cut $(-\infty \leqslant z \leqslant -R)$ complex plane, so every $S_p(z)$ is uniformly bounded on p. By Theorem 12.2 there exists at least a subsequence of the $S_p(z)$ which converges to an analytic function in any closed region in the interior of the cut complex plane, since if a sequence of analytic functions converges uniformly to a limit which is analytic, it follows by Cauchy's residue formula that all the derivatives at a point converge as well. Hence the determinantal conditions

$$D_p(m, n) \geqslant 0 \tag{17.48}$$

satisfied by the $S_p(z)$ also converge and hold in the limit as p tends to infinity. If $R = 0$, so the origin is not an interior point, then we can still conclude (17.48) by computing the determinants at any real $\epsilon > 0$, where, by Lemma 17.1, they are still positive. Result (17.48) then follows in the limit as $\epsilon \to 0$. We can then conclude that (17.46) implies that there exists a series of Stieltjes which has the values $f(z_j)$, by Theorem 16.4. ■

C. The Hamburger Problem

Since we have already treated the case of a nonzero radius of convergence [Eqs. (17.35)–(17.36)], we will confine our attention here to the general case where there is no assumption on the radius of convergence. The method of solution for possible values of Hamburger functions is generally the same as it was for series of Stieltjes. In this case we start from Lemma 15.4. Then, for $g(z)$ represented by Eq. (16.46), we can show that

$$g(z) = \cfrac{g_0}{1 + h_1 z - \cfrac{g_1 z^2}{1 + h_2 z - \cfrac{g_2 z^2}{1 + \cdots \cfrac{-g_p z^2}{1 + h_{p+1} z - z^2 k_{p+1}(z)}}}} \tag{17.49}$$

defines $k_{p+1}(z)$, which is also a Hamburger function of form (16.46). By direct calculation, if $g(z)$ is of form (16.46),

$$zg(z) = \int_{-\infty}^{+\infty} \frac{z\,d\psi(u)}{1 + uz} = \int_{-\infty}^{+\infty} \frac{(z + u|z|^2)\,d\psi(u)}{|1 + uz|^2} \qquad (17.50)$$

so that $\text{Im}[zg(z)] > 0$ if $\text{Im}\,(z) > 0$. By Section 4.D we see that (17.49) is just the associated or Jacobi-type continued fraction expansion of $g(z)$. Thus we must have, by the fundamental recursion relations (4.5)

$$g(z) = \frac{(1 + h_{p+1}z)A_p(z) - z^2 k_{p+1}(z)A_{p-1}(z)}{(1 + h_{p+1}z)B_p(z) - z^2 k_{p+1}(z)B_{p-1}(z)} \qquad (17.51)$$

Thus $g(z)$ is a linear fractional transformation of $zk_{p+1}(z)$. As we saw in Eqs. (17.17)–(17.19), a linear fractional transformation maps the family of circles and straight lines into itself. Therefore we expect that the allowed values of $zg(z)$ will be a circle in the same half-plane as z. Taking account of the degrees of the numerator and the denominator of the convergents to (17.49) and the number of coefficients which match those in the formal series expansion of $g(z)$, we identify

$$A_p(z)/B_p(z) = [p/p + 1] \qquad (17.52)$$

so that the h_j and g_j can be also identified from the Frobenius identity (3.41). Clearly, since the region defined at the pth stage contains all possible solutions, it contains the region defined at the $(p + 1)$st stage, which is subject to additional restrictions. In point of fact the region defined at the $(p + 1)$st stage is a circle which is tangent to the circle defined at the pth stage from the inside at exactly one point. That it touches at one and only one point follows from the observation that the boundary values of the $zk_p(z)$ are all real. The relation is

$$zk_p(z) = g_p z / \left[1 + h_{p+1}z - z^2 k_{p+1}(z)\right] \qquad (17.53)$$

with h_{p+1} and g_p both real, by (3.41). If $\alpha = zk_{p+1}(z)$ is real, then

$$\text{Im}[zk_p(z)] = g_p(\text{Im}\,z)/|1 + (h_{p+1} - \alpha)z|^2 \qquad (17.54)$$

The only solution to (17.54) is $\alpha = \infty$. Thus the circles touch at one and only one point. Consequently, the circles are nested and monotonically decrease in size. If the solution to the Hamburger moment problem is unique, their diameters shrink to zero. Otherwise there will be a nonzero minimum diameter. One can compute (Shohat and Tamarkin, 1963) that the radius of the pth circle is given by

$$\frac{1}{r_p(z)} = 2y \sum_{k=1}^{p} \frac{|Q_k^{(-1)}(z)|^2}{D(0, k + 1)D(0, k)} \qquad (17.55)$$

where the $Q_M^{(J)}(z)$ are normalized as in Chapter 3. Alternatively, this direct calculation shows that the size of the radius shrinks. The determinate case holds if and only if the sum in (17.55) diverges.

We show in Fig. 17.4 an example of the inclusion regions.

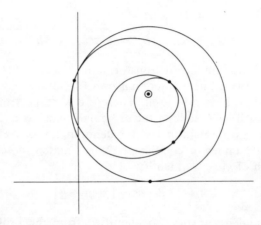

Fig. 17.4 Sketch for the Hamburger problem of the sequence of inclusion circles. The new, smaller circle touches the old one at just one point.

D. Tschebycheff's Inequalities

The question we seek to answer in this section is, given that $g(z)$ is of the form (16.46) and given the first $2M + 2$ coefficients of its formal power series expansion, what bounds can be put on the function $\psi(u)$? We will normalize $\psi(-\infty) = 0$.

It is evident by inspection of Eqs. (16.46) and (16.47) that the $[M/M + 1]$ Padé approximant provides the approximation

$$d\psi(u) = \sum_{p=1}^{M+1} \beta_p \, \delta(u - \gamma_p)du \qquad (17.56)$$

to $d\psi(u)$, where the β_p are positive and γ_p real and $\delta(x)$ is Dirac's delta function [the function $\delta(x) = 0$, $x \neq 0$, and $\int_{-\epsilon}^{+\epsilon}\delta(x)dx = 1$]. In like manner $[M/M]$ provides the approximation

$$d\hat{\psi}(u) \approx \hat{\beta}_0 \, \delta(u)du + \sum_{p=1}^{M} \hat{\beta}_p \, \delta(u - \hat{\gamma}_p)du \qquad (17.57)$$

where again the $\hat{\beta}_p$ are positive and the $\hat{\gamma}_p$ real. Thus Padé approximation is equivalent to a step function approximation to $\psi(u)$. Define the polynomial

$$\mathcal{Q}_{M+1}(z) = zQ_M^{(0)}(z)Q_{M+1}^{(-1)}(z_0) - z_0Q_M^{(0)}(z_0)Q_{M+1}^{(-1)}(z) \qquad (17.58)$$

which vanishes at $z = z_0$. It is of degree $M + 1$ in z and, by the Padé equations, has the property

$$\mathcal{Q}_{M+1}(z)f(z) - zP_M^{(0)}(z)Q_{M+1}^{(-1)}(z_0) + z_0Q_M^{(0)}(z_0)P_M^{(-1)}(z) = O(z^{2M+2}) \qquad (17.59)$$

Thus

$$f(z) = [\mathcal{P}_{M+1}(z)/\mathcal{Q}_{M+1}(z)] + O(z^{2M+2}) \qquad (17.60)$$

where $\mathcal{P}_{M+1}(z)$ is a polynomial of degree at most $M + 1$. Now, by use of Eq. (15.26) and arguments similar to those of Theorem 15.1, we see that the roots of $Q_{M+1}^{(-1)}(z)$ and $Q_M^{(0)}(z)$ interlace. Thus, $\mathcal{Q}_{M+1}(0) < 0$, and if z_j are the ordered roots of $Q_M^{(0)}(z)$, then $\mathcal{Q}_{M+1}(z_1) > 0$, directly from (17.58). Thus there is a real root of $\mathcal{Q}_{M+1}(z)$ between zero and z_1. By repeating this argument we establish that all the roots of $\mathcal{Q}_{M+1}(z)$ are real. Similarly it follows that all the residues of $\mathcal{P}_{M+1}(z)/\mathcal{Q}_{M+1}(z)$ are positive. Let the $M + 1$ roots of $\mathcal{Q}_{M+1}(z)$ be ξ_j^{-1}; by construction one of them is $\xi_k^{-1} = z_0$. We next define a polynomial $S(t)$ of degree $\leqslant 2M + 1$ by the properties

$$S(\xi_j^{-1}) = 1, \quad 1 \leqslant j \leqslant k, \qquad S(\xi_j^{-1}) = 0, \quad k < j \leqslant M + 1,$$

$$S'(\xi_j^{-1}) = 0, \quad j = 1, 2, \ldots, M + 1 \qquad (17.61)$$

In the region $-\infty < t \leqslant z_0$ we must have $S(t) \geqslant 1$, and the inequality is strict if $t \neq \xi_j, j \leqslant k$. In the region $z_0 \leqslant t < \infty$ we must have $S(t) \geqslant 0$ and again the inequality is strict if $t \neq \xi_j, j > k$. If we use the representation

$$\frac{\mathcal{P}_{M+1}(z)}{\mathcal{Q}_{M+1}(z)} = \rho_0 + \sum_{j=1}^{M+1} \frac{\rho_j}{1 + z\xi_j}, \qquad d\Psi(t) = \sum_{j=0}^{M+1} \rho_j \, \delta(t - \xi_j) \, dt \qquad (17.62)$$

where $\rho_j > 0$ and $\xi_0 = 0$, then we have

$$\int_{-\infty}^{+\infty} S(t) \, d\Psi(t) = \sum_{\{\xi_j < z_0^{-1}\}} \rho_j = \Psi(z_0^{-1} + 0) \qquad (17.63)$$

On the other hand, since $S(t)$ is a polynomial of degree $\leqslant 2M + 1$, we have

$$\int_{-\infty}^{+\infty} S(t) \, d\Psi(t) = \int_{-\infty}^{+\infty} S(t) \, d\Psi(t) \qquad (17.64)$$

Therefore

$$\Psi(z_0^{-1} + 0) = \int_{-\infty}^{+\infty} S(t) \, d\Psi(t) = \int_{-\infty}^{+\infty} S(t) \, d\psi(t)$$

$$= \int_{-\infty}^{z_0^{-1}+0} S(t) \, d\psi(t) + \int_{z_0^{-1}+0}^{\infty} S(t) \, d\psi(t) > \int_{-\infty}^{z_0^{-1}+0} d\psi(t)$$

$$= \psi(z_0^{-1} + 0) \qquad (17.65)$$

Thus this construction gives us directly, in terms of the first $2M + 2$ coefficients, an upper bound on $\psi(z_0^{-1} + 0)$ at any real point z_0. By the use of a different polynomial $S(t)$ which vanishes at ξ_j, $1 \leqslant j < k$, and equals unity at ξ_j, $k \leqslant j \leqslant M + 1$, we can show a lower bound so that

$$\Psi(z_0^{-1} - 0) < \psi(z_0^{-1} - 0) \leqslant \psi(z_0^{-1}) \leqslant \psi(z_0^{-1} + 0) < \Psi(z_0^{-1} + 0)$$

$$(17.66)$$

Thus by this procedure we can construct bounds in terms of the coefficients $g_0, g_1, \ldots, g_{2M+1}$ for any distribution function $\psi(u)$ with those moments. The method is first to form the $[M/M]$ and $[M/M + 1]$ Padé approximants, then to construct $\mathcal{P}_{M+1}(z)/\mathcal{Q}_{M+1}(z)$ for the desired point z_0, and finally, to use the residues for all poles $\xi_j \leqslant z_0^{-1}$ to compute the bounds to $\psi(z_0^{-1})$.

These bounds, and arguments of this type, can be usefully applied to the construction of bounds to averages over the distributions of the sort

$$\langle \Omega \rangle = \int_{-\infty}^{+\infty} \Omega(t) \, d\psi(t) \qquad (17.67)$$

If, for example, $\Omega(t)$ is monotonic and is bounded at infinity, then integrating by parts gives

$$\langle \Omega \rangle = \Omega(\infty)g_0 - \int_{-\infty}^{+\infty} \psi(t)\, d\Omega(t) \tag{17.68}$$

Since $d\Omega$ in this example is of fixed sign, we can bound the average by using the upper bounds for ψ from (17.66).

Direct estimates, which, depending on the properties of Ω, may be bounds, can be obtained from Padé approximants via (17.56) and (17.57) as

$$\langle \Omega \rangle \cong \sum_{p=1}^{M+1} \beta_p \Omega(\gamma_p) \tag{17.69}$$

When we remember (Chapter 7) that the Padé denominators are the orthogonal polynomials with respect to a given distribution function, we recognize that the Padé approximation to the distribution function is equivalent to the usual Gaussian integration procedure for that weight function. The Padé approximant formalism forms a convenient way to locate the Gaussian points and weights.

The problem of bounding averages is an important one in applications and an extensive list is given by Wheeler and Gordon (1970), together with detailed material on the applications of these ideas to some of the problems. In particular they consider the equilibrium thermal properties of a simple harmonic solid. For this problem a large number of the moments can be computed, although the distribution itself has not been. They show how to obtain strong bounds on the quantities of interest. Ynduráin (1973) reviews a number of examples from the field of particle physics.

18

PÓLYA FREQUENCY SERIES

In the previous three chapters we exploited a particular sign pattern of the C table (Chapter 2) to derive convergence and bounding properties of the Padé approximants. An example of a function which has that sign pattern is the Euler function $\sum_n n!(-x)^n$. In this chapter we will exploit a different sign pattern. The pattern we chose is that shown by the exponential function $e^x = \sum_n x^n/n!$. We can work out directly what the sign pattern is from Gauss's continued fraction via Eq. (4.84), (4.85), (5.31), and (5.32) for half the C table, and fill in the other half by means of Hadamard's formula (4.100) since the exponential function has the property $1/e^x = e^{-x}$. We find, then, for e^x that

$$C(L/M)(-1)^{M(M-1)/2} \geqslant 0 \tag{18.1}$$

where the inequality is actually strictly greater than. Put another way, the signs of $C(L/M)$ are determined by M and are $+, +, -, -, +, +, -,$ $-, \ldots$ as $M = 0, 1, 2, \ldots$. One can work out explicitly (Arms and Edrei, 1970) that for e^x

$$C(L/M) = (-1)^{M(M-1)/2} \prod_{k=1}^{M} \frac{1}{k(k+1) \cdots (k+L-1)} \tag{18.2}$$

where $L, M \geqslant 1$. Our sample function e^x has a number of other special features which will turn out to be related to the sign pattern of the C table. As we saw in Section 5.C, the numerator and the denominator polynomials tend to entire functions in the limit as L and M tend to infinity. Also, by Eq. (5.39), $P_L(x)$ has every term positive, as does $Q_M(-x)$.

A. Characterization of Pólya Frequency Series

This class of series was introduced and studied by Schoenberg (1948). He gave the name of Pólya to these because much of his derivation of their fine structure and representation formulas depended on Pólya's investigations. An extensive exposition is given by Karlin (1968). The results on the Padé table for this class of functions are due to Arms and Edrei (1970).

We characterize the class of functions we call Pólya frequency series by Eq. (18.1). Schoenberg (1951) conjectured that it was the class (P) of functions

$$f(z) = a_0 e^{\gamma z} \left[\prod_{j=1}^{\infty} (1 + \alpha_j z) \right] \Big/ \prod_{j=1}^{\infty} (1 - \beta_j z) \qquad (18.3)$$

where

$a_0 > 0, \quad \gamma \geqslant 0,$

$$\alpha_j \geqslant 0, \quad \beta_j \geqslant 0 \quad \text{(for all} \ \ j \geqslant 1), \quad \sum_j (\alpha_j + \beta_j) < +\infty \qquad (18.4)$$

and proved that all those functions satisfied Eq. (18.1). This conjecture was proven by Edrei (1953). We will not give the proof of this conjecture, but will show, using Padé methods, why the class of functions defined by (18.3) and (18.4) satisfies Eq. (18.1). It is interesting to note that by interchanging rows Eq. (18.1) can be re-expressed by Eq. (2.1) as

$$(-1)^{M(M-1)/2} C(L/M) = \Theta(L/M)$$

$$\equiv \det \begin{vmatrix} a_L & \cdots & a_{L+M-1} \\ \vdots & \ddots & \vdots \\ a_{L-M+1} & \cdots & a_L \end{vmatrix} \geqslant 0 \qquad (18.5)$$

To verify Eq. (18.5) for the class of functions (P), we start with the obvious remark that for $f(z) = e^{\gamma z}, \lambda \geqslant$, for all $L, M \geqslant 0$, we have by the direct calculations of Section 5.c and by Eq. (18.2), the properties

$$\mathcal{P}_{L,f}(z) \quad \text{has all nonnegative terms} \qquad (18.6)$$

$$\mathcal{Q}_{M,f}(-z) \quad \text{has all nonnegative terms} \qquad (18.7)$$

$$\Theta_f(L/M) \geqslant 0 \qquad (18.8)$$

where $\mathcal{P}_L$ and $\mathcal{Q}_M$ are normalized so that $\mathcal{Q}_M(0) = 1$. Let $g(z) = (1 + \alpha z)f(z)$. Then by identity (3.54) we have

$$\mathcal{P}_{L,g}^{(J)}(z) = \mathcal{P}_{L,f}^{(J)}(z) + \alpha z \, \mathcal{P}_{L-1,f}^{(J-1)}(z)\left[\mathcal{Q}_{M,f}^{(J)}(-1/\alpha) / \mathcal{Q}_{M,f}^{(J-1)}(-1/\alpha)\right]$$

$$(18.9)$$

so that $\mathcal{P}_g$ is larger, term by term, than $\mathcal{P}_f$. If we introduce $\gg$ to mean larger term by term, then we have

$$\mathcal{P}_{L,g}^{(J)}(z) \gg \mathcal{P}_{L,f}^{(J)}(z) \tag{18.10}$$

In particular, $\mathcal{P}_g$ also satisfies (18.6). Next by identity (3.58) we have

$$\mathcal{Q}_{M,g}^{(J)}(z) = \frac{\Theta_f(L/M + 1)}{\Theta_f(L - 1/M)}$$

$$\times \sum_{j=0}^{M} \frac{(-z)^j \Theta_f(L - 1/\ M - j)\mathcal{Q}_{M-j,f}^{(J-1+j)}(-1/\alpha)\mathcal{Q}_{M-j,f}^{(J+j)}(z)}{\Theta_f(L/M + 1 - j)\mathcal{Q}_{M,f}^{(J-1)}(-1/\alpha)}$$

$$(18.11)$$

where, in case of $0/0$, identical terms are first canceled, and then for such cases as remain, $0/0 = 0$. By examining the signs we conclude from (18.7) that

$$\mathcal{Q}_{M,g}^{(J)}(-z) \gg \mathcal{Q}_{M,f}^{(J)}(-z) \tag{18.12}$$

and so (18.7) also holds for $\mathcal{Q}_g$. Finally since by identity (3.55)

$$\Theta_g(L/M) = \Theta_f(L - 1/M)\mathcal{Q}_{M,f}^{(J-1)}(-1/\alpha)\alpha^M \tag{18.13}$$

we must have

$$\Theta_g(L/M) \geqslant \Theta_f(L/M) \tag{18.14}$$

since the coefficient of $(-x)^M$ in $\mathcal{Q}_{M,f}^{(J-1)}(x)$ is $\Theta_f(L/M)/\Theta_f(L - 1/M)$, by Eqs. (3.34) and (18.5). Thus we have shown that properties (18.6)–(18.8) also hold for $g(z)$ if they do for $f(z)$.

Before proceeding further we need the following lemma.

Lemma 18.1. If $f(z) = \sum_j a_j z^j$ is a formal power series with $\Theta_f(L/M)$ ≥ 0, $a_0 > 0$, then for $g(z) = [f(-z)]^{-1}$ we have $\Theta_g(L/M) \geq 0$.

Proof: By Hadamard's formula (4.100)

$$\Theta_g(L/M) = (-1)^{\frac{1}{2}M(M-1)}C_g(L/M)$$

$$= (-1)^{\frac{1}{2}[M(M-1)+(L+M)(L+M-1)]}C_{f(-z)}(M/L)/a_0^{L+M}$$

$$= (-1)^{\frac{1}{2}[M(M-1)+(L+M)(L+M-1)]+M(L-M+1)}C_f(M/L)/a_0^{L+M}$$

$$= \Theta_f(M/L)/a_0^{L+M} \quad \blacksquare \qquad\qquad (18.15)$$

Now consider $g(z) = f(z)/(1 - \beta_1 z)$. By the duality theorem 9.2 the $[L/M]$ approximant to $g(z)$ is the $[M/L]$ to $[g(z)]^{-1} = (1 - \beta_1 z)/f(z)$. By Lemma 18.1, if $f(z)$ satisfies Eq. (18.5), then so does $[f(-z)]^{-1}$. Thus by the arguments preceding (18.9)–(18.14), $[g(-z)]^{-1}$ satisfies Eq. (18.5). Hence, by Lemma 18.1, we conclude that $g(z)$ satisfies Eq. (18.5). Thus by repeated applications of these arguments we conclude that

$$g(z) = a_0 e^{\gamma z}\left[\prod_{i=1}^{T}(1 + \alpha_i z)\right]\Big/\prod_{i=1}^{S}(1 - \beta_i z) \qquad\qquad (18.16)$$

satisfies (18.5). Since by condition (18.4) we can establish that $f(z)$ of form (18.3) has a formal power series expansion (all coefficients are finite), and by Eq. (18.14) the entries in the Θ table are monotonically increasing with T and S, we obtain at once the following result.

Theorem 18.1. If $f(z)$ belongs to the class (P) defined by Eqs. (18.3)–(18.4), then it necessarily satisfies the properties (18.6)–(18.8), and $\Theta(L/M) \geq 0$.

B. Convergence Properties

Appropriate sequences of Padé approximants to Pólya frequency series not only converge, but the numerators and denominators also converge separately as well. To establish the convergence properties, we need two more lemmas.

Lemma 18.2.   If $f(z)$ belongs to the class (P) defined by Eqs. (18.3)–(18.4), then

$$\mathcal{Q}_M^{(J)}(-z) \ll \mathcal{Q}_M^{(j)}(-z), \qquad\qquad j \leqslant J \qquad (18.17)$$

$$\mathcal{Q}_m^{(J)}(-z) \ll \mathcal{Q}_M^{(J+m-M)}(-z), \qquad m \leqslant M \qquad (18.18)$$

$$\mathcal{P}_L^{(J)}(z) \ll \mathcal{P}_L^{(j)}(z), \qquad\qquad j \leqslant J \qquad (18.19)$$

$$\mathcal{P}_l^{(J)}(z) \ll \mathcal{P}_L^{(J+l-L)}(z), \qquad\qquad l \leqslant L \qquad (18.20)$$

Proof: If we rewrite the three-term identities, (3.24) and (3.25), using Eq. (3.34) and definition (18.5) in the notation of this chapter, they become

$$\mathcal{S}(L + 1/M) - \mathcal{S}(L/M)$$

$$= z\mathcal{S}(L/M - 1)\, \frac{\Theta(L + 1/M + 1)\Theta(L/M - 1)}{\Theta(L + 1/M)\Theta(L/M)} \qquad (18.21)$$

$$\mathcal{S}(L/M + 1) - \mathcal{S}(L/M)$$

$$= -z\mathcal{S}(L - 1/M)\, \frac{\Theta(L + 1/M + 1)\Theta(L - 1/M)}{\Theta(L/M)\Theta(L/M + 1)} \qquad (18.22)$$

where

$$\mathcal{S}(L/M) = G(z)\mathcal{P}_L^{(J)}(z) + H(z)\mathcal{Q}_M^{(J)}(z) \qquad (18.23)$$

with G and H arbitrary. The conclusions in Eqs. (18.17)–(18.20) for a difference of one in L or M follow directly by the specializations $G = 1$, $H = 0$ and $G = 0$, $H = 1$ from a term-by-term comparison of the signs implied by Theorem 18.1. Since we can repeat the comparison one step at a time until we have reached each of the relations (18.17)–(18.20), the lemma follows by induction. ∎

Lemma 18.3.   Let $f(z)$ be any function for which the numerators and denominators of the sequences, $l = 1, \ldots, T$ and $m = 1, \ldots, S$ of Padé approximants $[L_k - l/M_k - m]$ separately converge to an entire function as $k \to \infty$. Then the sequence $[L_k/M_k]$ for

$$g(z) = f(z)\left[\prod_{i=1}^{T} (1 + \alpha_i z) \right] \Big/ \prod_{i=1}^{S} (1 - \beta_i z) \qquad (18.24)$$

also converges, provided only that no α_i is a pole nor β_i a zero of $f(z)$, and $T, S < \infty$.

Proof: First let us consider $h(z) = (1 + \alpha_i z)f(z)$. If we take the limit of Eq. (18.9), then under the assumptions of the theorem we have directly that

$$\lim_{k \to \infty} \mathscr{P}_{L,h}^{(J)}(z) = (1 + \alpha_i z) \lim_{k \to \infty} \mathscr{P}_{L,f}^{(J)}(z) \qquad (18.25)$$

If we use identity (3.60), we conclude in like manner

$$\lim_{k \to \infty} \mathscr{Q}_{M,h}^{(J)}(z) = \lim_{k \to \infty} \mathscr{Q}_{M,f}^{(J)}(z) \qquad (18.26)$$

Thus by repeating this argument we can find (18.24) for any $T < \infty$ and $S = 0$. If we now use the duality theorem 9.2, we can by the same argument extend our conclusions to any $S < \infty$, since

$$[g(-z)]^{-1} = \left[\prod_{i=1}^{S} (1 + \beta_i z) \right] \left\{ 1 / \left[f(-z) \prod_{i=1}^{T} (1 - \alpha_i z) \right] \right\} \qquad (18.27)$$

and the part in the curly braces plays the role of $f(z)$ in the foregoing argument. ∎

We remark that this lemma is valuable in its own right. It can be used to extend the results previously obtained when $f(z)$ is a member of one of the other classes of functions we have studied. For example, it can be applied to certain of the confluent hypergeometric functions of Section 5.C, and to meromorphic series of Stieltjes. Likewise, we can use it to extend the results we will obtain for Pólya frequency series.

We can now prove the convergence theorem (Arms and Edrei, 1970) for Pólya frequency series.

Theorem 18.2   (Arms and Edrei).   If $f(z)$ belongs to the class (P) defined by Eqs. (18.3)–(18.4), then if both L and M tend to infinity, while $\lim (M/L) = \omega$,

$$\mathscr{P}_L^{(J)}(z) \to a_0 \exp[\gamma z / (1 + \omega)] \prod_{j=1}^{\infty} (1 + \alpha_j z) \qquad (18.28)$$

$$\mathscr{Q}_M^{(J)}(z) \to \exp[-\gamma \omega z / (1 + \omega)] \prod_{j=1}^{\infty} (1 - \beta_j z) \qquad (18.29)$$

uniformly in any closed and bounded region of the complex plane. The limits of $\mathcal{P}$ and $\mathcal{Q}$ are entire.

Proof: That the limits (18.28) and (18.29) are entire follows directly from the restriction (18.8), which prevents the accumulation of an infinite number of zeros in any finite region of the complex z plane.

Since the functions (18.3) are meromorphic, we can conclude by Montessus's theorem 11.1 and its corollary 11.3 that

$$\lim_{L \to \infty} [L/M] = \frac{a_0 e^{\gamma z} \left[\prod_{j=M+1}^{\infty} (1 - \beta_j z)^{-1} \right] \prod_{j=1}^{\infty} (1 + \alpha_j z)}{\prod_{j=1}^{M} (1 - \beta_j z)} \tag{18.30}$$

inside $|z| < \beta_{M+1}$, and

$$\lim_{M \to \infty} [L/M] = \frac{a_0 \prod_{j=1}^{L} (1 + \alpha_j z)}{e^{-\gamma z} \left[\prod_{j=L+1}^{\infty} (1 + \alpha_j z)^{-1} \right] \prod_{j=1}^{\infty} (1 - \beta_j z)} \tag{18.31}$$

inside $|z| < \alpha_{L+1}$. For convenience we have ordered

$$\alpha_1 \geqslant \alpha_2 \geqslant \alpha_3 \cdots, \qquad \beta_1 \geqslant \beta_2 \geqslant \beta_3 \cdots \tag{18.32}$$

In (18.30) and (18.31) the limits of the numerators and denominators are indicated above and below the lines, respectively. By combining the results of Lemma 18.2 and Montessus's theorem [(18.30)–(18.31)], we obtain the bounds

$$(1 + \beta_1 z)(1 + \beta_2 z) \cdots (1 + \beta_M z)$$

$$\ll \mathcal{Q}_M^{(J)}(-z) \ll e^{\gamma z} \frac{\prod_{j=1}^{\infty} (1 + \beta_j z)}{\prod_{j=M+1}^{\infty} (1 - \alpha_j z)} \tag{18.33}$$

for $0 \leqslant z < 1/|\alpha_{M+1}|$, and

$$a_0 (1 + \alpha_1 z)(1 + \alpha_2 z) \cdots (1 + \alpha_L z)$$

$$\ll \mathcal{P}_L^{(J)}(z) \ll a_0 e^{\gamma z} \frac{\prod_{j=1}^{\infty} (1 + \alpha_j z)}{\prod_{j=L+1}^{\infty} (1 - \beta_j z)} \tag{18.34}$$

for $0 \leqslant z < 1/|\beta_{L+1}|$.

Since the coefficients in the upper bounding series in (18.33) and (18.34) are all positive, we can use them to bound the absolute values of $\mathcal{Q}_L^{(J)}$ and

$\mathcal{P}_L^{(J)}$ for any $|z| < \min(|\alpha_{M+1}|^{-1}, |\beta_{L+1}|^{-1})$. But since α_j and β_j go to zero as $j \to \infty$, this property means that for L, M large enough we can bound $|\mathcal{P}_L^{(J)}(z)|$ and $|\mathcal{Q}_M^{(J)}(z)|$ uniformly for all $|z| \leqslant R < \infty$ and $M > M_0(R)$, $L > L_0(R)$. Thus by the same arguments as we used to establish the more general theorem 12.2 we conclude that there must exist at least a subsequence of the given sequence of Padé approximants which converges to a limit function which is entire. To complete the proof of this theorem, we need to show that the condition on the sequence that $\lim(M/L) = \omega$ suffices to ensure that the entire sequence converges and to demonstrate that the limits are given correctly by (18.28) and (18.29).

If we combine Eq. (18.9) with inequalities (18.17) and (18.20) of Lemma 18.2, then we have as an extension of Eq. (18.10) for $u(z) = (1 + \alpha z)v(z)$

$$\mathcal{P}_{L,v}^{(J)}(z) \ll \mathcal{P}_{L,u}^{(J)}(z) \ll \mathcal{P}_{L,v}^{(J)}(z)(1 + \alpha z) \qquad (18.35)$$

for $0 \leqslant z$, $\alpha < \infty$. If we apply Lemma 18.3 to $w(z) = a_0[\prod_{i=1}^{N}(1 + \alpha_i z)]e^{\gamma z}$, we have by Padé's results (5.40) for the exponential function

$$\lim \mathcal{P}_{L,w}^{(J)}(z) \to a_0 \left[\prod_{i=1}^{N} (1 + \alpha_i z) \right] \exp[\gamma z/(1 + \omega)] \qquad (18.36)$$

since $\lim(M/L) = \omega$ by assumption. By using Eq. (18.35) for the successive application of the factor $(1 + \alpha_i z)$, $i > N$, we can establish, by taking the limit as L, $M \to \infty$,

$$a_0 \left[\prod_{i=1}^{N} (1 + \alpha_i z) \right] \exp[\gamma z/(1 + \omega)] \ll \lim \mathcal{P}_{L,f}^{(J)}(z)$$

$$\ll a_0 \left[\prod_{i=1}^{\infty} (1 + \alpha_i z) \right] \exp[\gamma z/(1 + \omega)]$$

$$(18.37)$$

However, since N can be chosen as large as we please, Eq. (18.37) implies (18.28). Thus the entire sequence of numerators converges for $0 \leqslant z < \infty$. By our previous argument this convergence extends throughout the whole complex plane. However, by Lemma 18.1 we can repeat the argument for $1/f(-z)$ if we interchange the role of the α's and β's. Thus the whole sequence of denominator polynomials also converges to the entire function, which is given by Eq. (18.29). ∎

Corollary 18.1.   Let the assumptions of Theorem 18.2 hold, except that a finite number of the α_i and β_i are unrestricted provided that in no case is $\alpha_i = -\beta_j$. The conclusions of that theorem remain valid.

Proof: Apply Theorem 18.2 to the function formed by the nonexceptional α's and β's. This procedure gives a function $f(z)$ to which to apply Lemma 18.3. Select the exceptional α_i and β_i as those to be treated by Eq. (18.24). This corollary then follows by Lemma 18.3.   ■

PART IV

GENERALIZATIONS AND APPLICATIONS

19

GENERALIZED PADÉ APPROXIMANTS

The presentation in this part will differ from that in previous parts in that the emphasis will be on some of the large number of fruitful ideas which have grown out of the study of Padé approximants. The details of the proofs will not be given, but only the general ideas. It is hoped that this part will give the reader an appreciation for the breadth of the current lines of work in which the underlying ideas are now embodied. The results presented here by no means exhaust those generalizations and applications currently known. In fact, an entire book (Baker and Gammel, 1970) was devoted to this area and was admittedly incomplete then. One may anticipate with confidence a further expansion of these methods in the future.

In this chapter we will discuss one way to extend the scope of the bounding approximation procedures for series of Stieltjes which were discussed in Part III. In connection with the study of the Padé approximant the idea of this sort of approximant was first proposed by Gammel *et al.* (1967) and the criteria for and proofs of the bounding properties were given by Baker (1967). [See Baker (1970) for a more complete discussion.] Our starting point is the observation of Section 17.D that for a series of Stieltjes, the Padé approximant [see Eq. (17.56)] approximates the weight function of $d\psi(u)$ by a discrete sum of Dirac delta functions whose strengths and locations are determined by the Padé equations. The delta functions are all located on the negative real axis and have positive weights for series of Stieltjes. We seek to relax the forms (16.40) of the functions and still retain the desirable properties of the approximants. To this end, let us consider

$$g(z) = \int_0^\infty b(z, s) \, d\varphi(s) \qquad (19.1)$$

where $d\varphi \geqslant 0$. We introduce approximants of the form

$$B_{n,j}(x) = \sum_{m=1}^{n} \alpha_m b(x, \sigma_m) + \sum_{k=0}^{j} \frac{\beta_k}{k!} \left[\left(\frac{\partial}{\partial s} \right)^k b(x, s) \Big|_{s=0} \right] \quad (19.2)$$

where $j \geqslant -1$. If $j = -1$, the second sum is entirely omitted. We obtain the $[M + J/M]$ Padé approximant as a special case of Eq. (19.2) when we pick

$$b(x, s) = 1/(1 + xs) \quad (19.3)$$

as can be easily checked.

The defining equations for the α, β, and σ of (19.2) are obtained as follows. Let us assume $b(x, s)$ is such that we can expand it as

$$b(x, s) = \sum_{m=0}^{\infty} b_m(x)(-s)^m \quad (19.4)$$

and denote

$$c_n = \int_0^{\infty} s^n \, d\varphi(s) \quad (19.5)$$

Then by equating (19.1) and (19.2) we obtain

$$\sum_{m=0}^{\infty} b_m(x) \sum_{l=1}^{n} \alpha_l(-\sigma_l)^m + \sum_{m=0}^{j} b_m(x)\beta_m(-1)^m = \sum_{m=0}^{\infty} b_m(x)(-1)^m c_m \quad (19.6)$$

Let us assume that $b(x, s)$ satisfies a "solvability" condition so that (19.6) can be made equivalent to the set of equations obtained by equating the coefficients of the $b_m(x)$. For example, if $b_m(x) \propto x^m$ as x goes to zero, then we could obtain this equivalence. We therefore take the defining equations to be

$$\sum_{l=1}^{n} \alpha_l(\sigma_l)^k + \beta_k = c_k, \qquad 0 \leqslant k \leqslant j$$

$$\sum_{l=1}^{n} \alpha_l(\sigma_l)^k = c_k, \qquad j < k \leqslant 2n + j \quad (19.7)$$

where there are $2n + j + 1$ equations in the same number of parameters and the c_k are determined by the solution of

$$g(z) = \sum_{m=0}^{\infty} (-1)^m b_m(z) c_m \quad (19.8)$$

If we identify the $(-\sigma_l)^{-1}$ as the location of the poles and α_l/σ_l as the respective residues, then Eq. (19.7) are exactly the equations for the $[n + j/n]$ Padé approximants to the function

$$C(z) = \sum_{k=0}^{\infty} c_k(-z)^k \qquad (19.9)$$

Thus for suitably restricted $b(x, s)$ we can reduce the study of this more general approximation procedure to the study of the Padé approximants to a transformed function. The series $C(z)$ will be a series of Stieltjes for $g(z)$ of the form (19.1).

We are now in a position to give several theorems analogous to those for series of Stieltjes.

Theorem 19.1. If $b(x, s)$ is regular in a uniform neighborhood of the positive real s axis and $(\ln s)^{1+\eta} b(x, s)$ is bounded as $s \to +\infty$ for some $\eta > 0$, then the approximants $B_{n,j}(x)$ converge as $n \to \infty$ for $g(z)$ of the form (19.1).

The proof follows from a consideration of the contour integral representation of the approximants

$$(1/2\pi i) \int_C b(x, s)[n + j/n](-1/s) \, ds/s \qquad (19.10)$$

and the convergence theorems for series of Stieltjes.

As the next theorem will show, a remarkably simple additional property of the function $b(x, s)$ is both necessary and sufficient to ensure that the $B_{n,j}(x)$ to functions of form (19.1) have the same bounding properties as do the $[n + j/n]$ Padé approximants to series of Stieltjes.

Theorem 19.2. The $B_{n,j}(x)$ for a function of the form (19.1) obey the following inequalities, where x is real and nonnegative,

$$(-1)^{1+j}\{B_{n+1,j}(x) - B_{n,j}(x)\} \geqslant 0 \qquad (19.11a)$$

$$(-1)^{1+j}\{B_{n,j}(x) - B_{n-1,j+2}(x)\} \geqslant 0 \qquad (19.11b)$$

$$B_{n,0}(x) \geqslant g(x) \geqslant B_{n-1}(x) \qquad (11.11c)$$

where $j \geqslant -1$, if and only if

$$(-\partial/\partial s)^k b(x, s) \geqslant 0 \qquad (19.12)$$

for all real, nonnegative x and s and $k = 0, 1, 2, \ldots$. These inequalities imply that (19.11c) gives the closest bounds for a given fixed number of coefficients.

That the derivative condition is sufficient can be established by showing that the inequalities are related to a high difference operator acting on $b(x, s)$, which in turn is related to (19.12) by a mean-value theorem. The necessity can be established by induction and by varying $g(x)$.

One can characterize the class of functions which satisfy (19.12) by using the following theorem due to Bernstein (1928).

Theorem 19.3. In the class of functions which are regular in the right-half s plane and go to zero faster than s^{-k}, $k > 1$, for some k, as s goes to infinity there the following statements are equivalent:

$$(-\partial/\partial s)^j f(s) \geqslant 0, \qquad 0 \leqslant s < \infty \tag{19.13}$$

for all $j \geqslant 0$, and

$$f(s) = \int_0^\infty e^{-st}\, d\varphi(t) \tag{19.14}$$

with $d\varphi \geqslant 0$.

That (19.14) implies (19.13) follows by differentiation under the integral sign. The relation in the other direction can be established by the existence of the inverse Laplace transform and the observation that the right limit of j and s going to infinity can select any t in (19.14) very accurately.

We give two sample kernels which have the required properties. First

$$b(x, s) = L_\xi(xs) = \sum_{n=0}^\infty \frac{\Gamma(1 + \xi n)}{\Gamma(1 + n)} (-x)^n, \qquad 0 \leqslant \xi < 1 \tag{19.15}$$

which is entire for $0 \leqslant \xi < 1$ and has the special cases $L_0(x) = e^{-x}$ and $L_1(x) = (1 + x)^{-1}$. The other example is

$$b(x, s) = [1 + xs/(n + 1)]^{-n}, \qquad 0 < n < \infty \tag{19.16}$$

The range $n = 1$ to ∞ again interpolates between the Padé and exponential kernels. Some examples of the applications of these approximants will be given in later chapters.

SERIES WITH INFINITE COEFFICIENTS

Let us consider the example

$$F(z) = \int_0^\infty \frac{\varphi(u)\,du}{1 + zu} \tag{20.1}$$

where φ is a bounded, continuous function with the properties $\varphi(u) > 0$ and $\varphi(u) \propto K/u^{1+\alpha}$ ($0 < \alpha \leqslant 1$; $k > 0$). This function is not a Stieltjes function (Part III), because

$$\int_0^\infty u^n \varphi(u)\,du = \infty, \qquad n = 1, 2, \ldots \tag{20.2}$$

so it has no formal power series expansion about the origin. However, if we introduce a cutoff ρ, then we can define

$$F_\rho(z) = \int_0^\rho \frac{\varphi(u)\,du}{1 + zu} = \sum_{n=0}^\infty f_n(\rho)(-z)^n \tag{20.3}$$

where

$$\lim_{\rho \to \infty} F_\rho(z) = F(z) \tag{20.4}$$

in the cut ($-\infty < z < 0$) complex z plane. The functions $F_\rho(z)$ are Stieltjes functions with radius of convergence ρ^{-1}. From the theory of Stieltjes functions we then have, for any z in the cut, complex plane

$$F(z) = \lim_{\rho \to \infty} \lim_{M \to \infty} [M + J/M](z, \rho) \tag{20.5}$$

This method of approach to the problem of summing series with formally divergent series has been studied by Garibotti *et al.* (1970), Fogli *et al.*

267

(1971), and Villani (1972). The fundamental idea rests on Villani's limit theorem, which reduces the difficult double limiting process (20.5) to a single limit.

Theorem 20.1 (Villani). Let there be a sequence of functions $G(m, \rho)$ with the properties:

(a)   Convexity, $2G(m, \rho) - G(m, 0) - \lim_{\rho \to \infty} G(m, \rho) > 0$, $0 < \rho < \infty$.

(b) $\lim_{m \to \infty} G(m, \rho) = G(\rho), 0 \leqslant \rho < \infty, \lim_{\rho \to \infty} G(\rho) = \Lambda$.

(c) $G(\rho_1) \geqslant G(\rho_2), \rho_1 \geqslant \rho_2$.

(d) $G(\rho) \geqslant G(m, \rho), 0 \leqslant \rho < \infty$.

Then there exists an infinite sequence of maxima in ρ of $G(m, \rho)$, ρ_m, such that

$$\lim_{m \to \infty} \rho_m = \infty, \qquad \lim_{m \to \infty} G(m, \rho_m) = \Lambda \qquad (20.6)$$

where by properties (c) and (d), $G(m, \rho_m) \leqslant \Lambda$ for all m.

Proof: By properties (a)–(c), for m sufficiently large there must exist at least one maximum of $G(m, \rho)$, since $G(m, \rho)$ can be made to agree arbitrarily closely to a monotonically increasing function over any preassigned range $0 \leqslant \rho \leqslant R < \infty$. By choosing m large enough we can have the $\rho_m > R$ for any R we please, which implies that the ρ_m tend to infinity. For any $\epsilon > 0$ we can find an $R(m, \epsilon)$ such that $|G(\rho) - G(m, \rho)| < \epsilon$ for $0 \leqslant \rho \leqslant R(m, \epsilon)$. It must be by properties (c) and (d) that

$$G(R) - \epsilon \leqslant G(m, R) \leqslant G(m, \rho_m) \leqslant \Lambda \qquad (20.7)$$

but since $R(m, \epsilon) \to \infty$ as $m \to \infty$, Eq. (20.7) implies an error of at most ϵ in (20.6). Since $\epsilon > 0$ is arbitrary, however, we conclude that the conditions (20.6) hold and so does the theorem. ∎

Now let us consider the $[M - 1/M](z, \rho)$ approximant to the series (20.3). For $\rho = 0$, $F_\rho(z)$ is identically zero and the Padé approximant is, too.

One can conclude for ρ large that since

$$f_n(\rho) \approx k\rho^{n-\alpha}/(n - \alpha) \qquad (20.8)$$

then, by Eq. (1.27), $[M - 1/M](z, \infty) = 0$. Thus the $[M - 1/M](z, \rho)$ satisfy property (a) of Theorem 20.1. By the theory of series of Stieltjes (Part III) and Eq. (20.4) we can satisfy the rest of the theorem.

Thus we have the following recipe for dealing with the calculation of $F(z)$ of Eq. (20.1) given the series coefficients $f_n(\rho)$ [Eq. (20.3)]. Form the $[M - 1/M](z, \rho)$ Padé approximant and find the (largest) maximum

$\rho_M(z)$ with respect to ρ for M and z fixed. Since this is a lower bound to $F(z, \rho)$, which is in turn a lower bound to $F(z)$, we have in $[M - 1/M]$ $(z, \rho_M(z))$ a sequence of converging lower bounds to $F(z)$ for real and positive. If $f_0(\infty)$ is finite, we can give a corresponding sequence of upper bounds formed by selecting $\tilde{\rho}_M(z)$ to minimize $[M/M](z, \rho)$, which is an upper bound to $F(z, \rho)$. In this minimization procedure, however, $f_0(\rho)$ is taken equal to the fixed value $f_0(\infty)$.

These procedures can be extended (Fogli *et al.*, 1971) to give results for complex z and to give results even if $f_0(\infty)$ is not finite. These procedures have also been applied to a number of problems of interest in physics. Fogli *et al.* (1972) have used them to greatly simplify the problem of the scattering of electrons by an electromagnetic field in quantum electrodynamics. Baker (1972) has used the procedure to provide converging bounds in certain statistical mechanical problems.

By way of illustration, we give Villani's (1972) example. Let

$$F(z) = \int_0^\infty \frac{dt}{(1 + t^2)(1 + zt)} = \frac{1}{1 + z^2} \left(\frac{\pi}{2} + z \ln z \right) \quad (20.9)$$

where

$$f_n(\rho) = [\rho^{n-1}/(n - 1)] - f_{n-2}(\rho) \quad (20.10)$$

by direct integration with

$$f_0(\rho) = \tan^{-1}(\rho), \qquad f_1(\rho) = \tfrac{1}{2} \ln (1 + \rho^2) \quad (20.11)$$

The results for some low-order approximants are given in Table 20.1.

TABLE 20.1 $A_M = [M - 1/M](z, \rho_n(z))$ for Example (20.9)

M	$z = 0.01$		$z = 0.1$	
	$\rho_M(z)$	A_M	$\rho_M(z)$	A_M
1	1×10^2	1.5161	1.4×10^1	1.2749
2	2×10^2	1.5229	2.5×10^1	1.3153
3	3×10^2	1.5240	5×10^1	1.3226
4	4×10^2	1.5243	6×10^1	1.3250
5	5×10^2	1.5244	1×10^2	1.3259
∞	∞	1.5246	∞	1.3273

We see from Table 20.1 that the smaller z is, the better is the convergence and the larger are the $\rho_M(z)$. This method of handling series with formally infinite coefficients appears to give a better result for smaller z, just as we have been accustomed to for finite coefficients.

21

MATRIX PADÉ APPROXIMANTS

Instead of forming Padé approximants to a single function of a complex variable z, we wish in this chapter to consider the case of a given matrix function of z. In particular, we assume that we have the formal expansion

$$\mathbf{f}(z) = \sum_{j=0}^{\infty} \mathbf{f}_j z^j \tag{21.1}$$

where the $\mathbf{f}_j$ are matrices, or more generally elements of some, possibly noncommutative, algebra. This study seems to have been started by Gammel and McDonald (1966) and substantially developed by Basdevant *et al.* (1969) and Zinn-Justin (1971a). A good summary is given by Bessis (1973), which we follow.

One may define two types of Padé approximants, since the $\mathbf{f}_j$ need no longer commute, by the equations

$$\mathbf{f}(z) - \mathbf{P}_L(z)\mathbf{Q}_M^{-1}(z) = O(z^{L+M+1}) \tag{21.2}$$

or

$$\mathbf{f}(z) - \mathfrak{Q}_M^{-1}(z)\mathfrak{P}_L(z) = O(z^{L+M+1}) \tag{21.3}$$

However, one can show that both these Padé approximants are identical, although $\mathbf{P}_L \neq \mathfrak{P}_L$ and $\mathbf{Q}_M \neq \mathfrak{Q}_M$. If we subtract Eq. (21.2) from (21.3) and multiply on the left by $\mathfrak{Q}_M(z)$ and on the right by $\mathbf{Q}_M(z)$, we obtain

$$\mathfrak{Q}_M(z)\mathbf{P}_L(z) - \mathfrak{P}_L(z)\mathbf{Q}_M(z) = O(z^{L+M+1}) \tag{21.4}$$

But, since the left-hand side is a polynomial in z of maximum degree $L + M$, it must be identically zero. Therefore, if they both exist, we must

270

have the unique definition

$$[L/M]_f \equiv \mathbf{P}_L(z)\mathbf{Q}_M^{-1}(z) \equiv \mathfrak{Q}_M^{-1}(z)\mathfrak{P}_L(z) \qquad (21.5)$$

for the matrix Padé approximants. The defining equations can be taken as

$$\mathbf{f}(z)\mathbf{Q}_M(z) - \mathbf{P}_L(z) = O(z^{L+M+1}), \qquad \mathbf{Q}_M(0) = \mathbf{I} \qquad (21.6)$$

where $\mathbf{I}$ is the multiplicative identity. Depending on the nature of the $\mathbf{f}_j$, the solution of the $L + M$ linear matrix equations (21.6) may be a complex problem. We will not treat conditions for this solution to exist but simply assume that a unique solution does exist, and go on to study some of its properties.

The matrix Padé approximants have the invariance properties of the ordinary Padé approximants described in Chapter 9, both argument transformations and value transformations. In addition, they possess a few additional transformation properties which specifically relate to their matrix, or more general character.

The duality property (Theorem 9.2) directly extends, so that

$$[L/M]_{\mathbf{T}^{-1}} = [M/L]_{\mathbf{T}}^{-1} \qquad (21.7)$$

where $\mathbf{T}^{-1}$ and everything in (21.7) exist. The proof is as in Chapter 9.

If $\mathbf{A}$, $\mathbf{B}$, $\mathbf{C}$, and $\mathbf{D}$ are constant matrices independent of z, then for the diagonal $[M/M]$ we have the result

$$[M/M]_{(\mathbf{A}+\mathbf{B}f)(\mathbf{C}+\mathbf{D}f)^{-1}} = (\mathbf{A} + \mathbf{B}[M/M]_f)(\mathbf{C} + \mathbf{D}[M/M]_f)^{-1} \qquad (21.8)$$

which can be proved by the same method as used in Chapter 9, by using the identity

$$(\mathbf{A} + \mathbf{B}\mathbf{T})(\mathbf{C} + \mathbf{D}\mathbf{T})^{-1} = \mathbf{B}\mathbf{D}^{-1} + (\mathbf{A} - \mathbf{B}\mathbf{D}^{-1}\mathbf{C})(\mathbf{T} + \mathbf{D}^{-1}\mathbf{C})^{-1}\mathbf{D}^{-1}$$

$$(21.9)$$

and the property [analogous to Eqs. (3.45)]

$$[L/M]_{f+R} = \mathbf{R}(z) + [L/M]_f \qquad (21.10)$$

where $\mathbf{R}(z)$ is a polynomial in z of degree less than or equal to $L - M$. Equation (21.10) follows easily from uniqueness since

$$\mathbf{R}(z) + \mathbf{P}_L(z)\mathbf{Q}_M^{-1}(z) = \{\mathbf{R}(z)\mathbf{Q}_M(z) + \mathbf{P}_L(z)\}\mathbf{Q}_M^{-1}(z) \qquad (21.11)$$

is of the required form.

The argument invariance theorem 9.1 again holds, as in Chapter 9, as can be seen by manipulations on

$$[M/M]_f\left(\frac{Ay}{1 + By}\right)$$

$$= \mathbf{P}_M\left(\frac{Ay}{1 + By}\right)\mathbf{Q}_M^{-1}\left(\frac{Ay}{1 + By}\right)$$

$$= \left\{(1 + By)^M \mathbf{P}_M\left(\frac{Ay}{1 + By}\right)\right\}\left\{(1 + By)^M \mathbf{Q}_M\left(\frac{Ay}{1 + By}\right)\right\}^{-1}$$

(21.12)

which is the quotient of one matrix polynomial of degree M by another. By uniqueness, then, (21.12) gives the Padé approximant to $\mathbf{g}(y) = \mathbf{f}(Ay/(1 + By))$.

In addition, we have the following properties. Let $\mathbf{A}$ be a matrix independent of z; then if $\mathbf{A}^{-1}$ exists,

$$[L/M]_{\mathbf{AfA}^{-1}} = \mathbf{A}[L/M]_f \mathbf{A}^{-1} \tag{21.13}$$

as we can rewrite Eq. (21.2), by multiplying on the right and left by $\mathbf{A}$ and $\mathbf{A}^{-1}$, as

$$\mathbf{Af}(z)\mathbf{A}^{-1} - [\mathbf{AP}_L(z)\mathbf{A}^{-1}][\mathbf{AQ}_M(z)\mathbf{A}^{-1}]^{-1} = O(z^{L+M+1}) \quad (21.14)$$

which is again of the correct form. Consequently, if all the $\mathbf{f}_j$ commute so that there exists an $\mathbf{A}$ which simultaneously diagonalizes them, then by (21.14) we see that the matrix Padé approximant equations have a solution (assumed unique) for that $\mathbf{A}$ so that $\mathbf{APA}^{-1}$ and $\mathbf{AQA}^{-1}$ are diagonal matrices whose elements are the ordinary Padé approximants to the ordinary formal power series which form the diagonal elements of $\mathbf{AfA}^{-1}$. Generally speaking, if $\mathbf{f}(z)$ can be decomposed into a direct sum of commuting matrices (or operators), then the corresponding matrix Padé approximants are the direct sum of the corresponding Padé approximants computed separately.

If we denote $\mathbf{A}^\dagger$ as the Hermitian conjugate of $\mathbf{A}$ (i.e., take the complex conjugate of the transposed matrix), then by following the same argument that we used at Eq. (21.4) we can conclude that

$$[L/M]_{f^\dagger} = [L/M]^\dagger \tag{21.15}$$

We are now in a position to derive the results of Gammel and McDonald (1966), which make the matrix Padé approximant particularly useful in problems of scattering physics. That is, in addition to preserving the analytic properties of the power series, this diagonal matrix Padé approximant to a unitary operator is unitary.

Suppose $S(z)$ is unitary; then by definition

$$S(z)S^{\dagger}(z^*) = S^{\dagger}(z^*)S(z) = I \tag{21.16}$$

so that $S^{-1}(z)$ exists and

$$S^{-1}(z) = S^{\dagger}(z^*) \tag{21.17}$$

Thus we can write by duality, Eq. (21.17), and Eq. (21.15)

$$[L/M]_{S(z)}^{-1} = [M/L]_{S^{-1}(z)}(z)$$

$$= [M/L]_{S^{\dagger}(z^*)}(z)$$

$$= [M/L]_{S(z)}^{\dagger}(z^*) \tag{21.18}$$

Hence

$$[M/L]_{S}^{\dagger}(z^*)[L/M]_{S}(z) = I \tag{21.19}$$

and so for $L = M$ the diagonal matrix Padé approximants are unitary.

22

CRITICAL PHENOMENA

In this chapter we discuss practical approaches to certain problems of statistical physics as an application of Padé approximants. For a fuller discussion we mention that there are two recent good reviews of this area by Gaunt and Guttmann (1973) and Hunter and Baker (1973). In his history of a part of the theory of critical phenomena Brush (1967) has pointed out that with the Padé approximant method it was possible to obtain significant results where previous methods had failed.

We gave in Section 1.E an example of a typical problem encountered in this area. There is a real function for which the first few, typically 6 to 40, Taylor series coefficients can be computed. From the physics of the problem we expect this function to have an analytic singularity at some real point, either a divergence, a zero, or a cusp. In the framework of Padé approximants the following methods of manipulating original series have been used extensively. First, in combination with the other methods which follow, an Euler transformation

$$f(z) \to f(aw/(1 + bw)) \tag{22.1}$$

has been employed to diminish the effects of competing singularities for nondiagonal Padé approximants. The diagonal Padé approximants $[M/M]$ are invariant under such procedures, but (a) the nondiagonal Padé approximants $[L/M]$, $L \neq M$, are not, and (b) even the diagonal approximants are not for many of the following series manipulations.

We particularly consider the case where the asymptotic form

$$f(x) \sim A(1 - yx)^{-\gamma} + B \tag{22.2}$$

holds as $x \to y^{-1}$.

(i) Form Padé approximants to

$$F_1(x) = \frac{d}{dx} \ln f(x) \sim \frac{-\gamma}{x - y^{-1}} \tag{22.3}$$

and obtain unbiased estimates of y and γ by choosing the appropriate zero of the denominator and calculating the residue at that point. An example of this procedure was described in Section 1.E and results were given in Table 1.2. The γ estimate is unbiased in the sense that no assumed value of y is used.

(ii) Form Padé approximants to

$$F_2(x) = (y^{-1} - x)(d/dx) \ln f(x) \sim \gamma \tag{22.4}$$

for an assumed value of y, and obtain a biased estimate of γ by evaluating the Padé approximants at that assumed value $x = y^{-1}$.

(iii) Form Padé approximants to

$$F_3(x) = [f(x)]^{1/\gamma} \sim A^{1/\gamma}/(1 - yx) \tag{22.5}$$

by assuming a value of γ, and obtain biased estimates of A and y from the roots and residues of the Padé approximants.

(iv) Form Padé approximants to

$$F_4(x) = (1 - yx)^\gamma f(x) \sim A \tag{22.6}$$

by assuming values of y and γ and obtain biased estimates of A by evaluating the approximants at $x = y^{-1}$.

(v) Form Padé approximants to

$$F_5(x) = \left[\frac{d}{dx} \ln \frac{d}{dx} f(x) \right] \left[\frac{d}{dx} \ln f(x) \right]^{-1} \sim 1 + \frac{1}{\gamma} \tag{22.7}$$

and evaluate at an assumed value $x = y^{-1}$. While this is technically a biased procedure, it is in practice quite insensitive to the choice of y. One drawback is that for large γ the quantity calculated is relatively insensitive to γ as well.

(vi) If $f(x) = \sum_j f_j x^y$ has the asymptotic form (22.2) and $g(x) = \sum_j g_j x^j$ has the asymptotic form

$$g(x) \sim C(1 - yx)^{-\varphi} + D \tag{22.8}$$

for the same value of y, then

$$h(x) = \sum_j \frac{f_j}{g_j} x^j \sim \frac{A}{C} (1 - x)^{-(\gamma - \varphi + 1)} \tag{22.9}$$

for x near one. Thus, forming Padé approximants to

$$F_6(x) = (1 - x)(d/dx) \ln h(x) \sim \gamma - \varphi + 1 \tag{22.10}$$

we obtain an unbiased estimate of $\gamma - \varphi$ by evaluating the Padé approximants at $x = 1$. This technique is called "critical-point renormalization" (Fisher and Hiley, 1961; Fisher and Burford, 1967; Moore et al., 1969; and Ferer et al., 1971) because the variable x is normally a temperature variable and its scale is changed so that the singularity or "critical point" or "Curie point" occurs at $x = 1$. This method is useful in the study of critical phenomena because usually the series for several different physical properties are known, all of which are expected to be singular at the same (unknown) point.

As an example of the application of some of these procedures we give the results of Baker et al. (1967) on the analysis of the high-temperature magnetic susceptibility χ (discussed in Section 1.E) for the spin-$\frac{1}{2}$ Heisenberg model on the three space-dimensional, face-centered cubic lattice. The series is

$$\chi(x) = 1 + 12x + 240x^2 + 6,624x^3 + 234,720x^4$$

$$+ 10,208,832x^5 + 526,810,176x^6 + 31,434,585,600x^7$$

$$+ 2,127,785,025,024x^8 + 161,064,469,168,128x^9 + \cdots \tag{22.11}$$

The results are displayed in Fig. 22.1. We observe that methods (ii) and (iii) are very closely correlated and quite consistent with each other over a range of values, and begin to diverge from one another at the same time that the spread, indicated by the error bars, of the highest-order near-diagonal Padé approximants begins to grow markedly. The approximation to $F_5(x)$ of method (v) is also shown; the fluctuations between the highest-order near-diagonal Padé approximants is shown by the shading. The results of method (i) lie along the curve traced out by methods (ii) and (iii), with the results from the highest-order near-diagonal Padé approximants clustering at the intersection of that curve and the curve of method (v). Since there are no rigorous means to access the errors, the self-consistency has been used to estimate the errors. In this case an estimate of $y^{-1} \equiv x_c = 0.2492 \pm 0.001$ and $\gamma = 1.43 \pm 0.01$ was made.

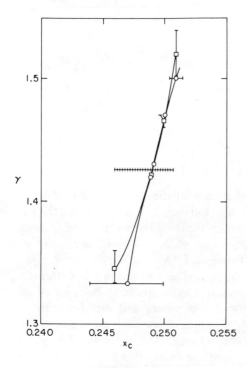

Fig. 22.1 **Fig. 22.1** Locus of points in the x_c, γ plane consistent with methods (ii) and (iii). ($\square$) γ from x_c, (O) x_c from γ; barred horizontal indicates γ from method (v).

In order to understand the structure of the errors in this example, which is typical of many such applications, we give the following general analysis (Hunter and Baker, 1973). Let us suppose we are trying to approximate the sum of a series which represents a function with a singularity

$$A(1 - Yx)^{-G} \tag{22.12}$$

in which we are particularly interested. The approximate solution

$$a(1 - yx)^{-g} \tag{22.13}$$

will represent, together with the rest of our approximation scheme, the coefficients of the function being approximated, so that we have the equations

$$a\binom{-g}{j}y^j = A\binom{-g}{j}Y^j(1 + \eta_j), \quad j = J, J + 1, J + 2 \tag{22.14}$$

The η_j are considered to be small percentage errors. If we now expand

$$a = A + \Delta A, \quad g = G + \Delta G, \quad y = Y + \Delta Y \tag{22.15}$$

we can easily solve Eq. (22.14) for

$$\frac{\Delta Y}{Y} = (2G + 2J + 1)\eta_{J+1} - (G + J + 1)\eta_{J+2} - (G + J)\eta_J$$

$$\Delta G = (G + J)\left(\eta_{J+1} - \eta_J - \frac{\Delta Y}{Y}\right)$$

$$\frac{\Delta A}{A} = \eta_J - J\frac{\Delta Y}{Y} - \left(\sum_{k=0}^{J-1}\frac{1}{G + k}\right)\Delta G \qquad (22.16)$$

From Eq. (22.16) we see that, barring unusual cancellations, $\Delta Y/Y$ is of the order of J times the size of the η's. Further, ΔG is of the order of J times $\Delta Y/Y$. It is this connection which explains the steep slope (J is of order ten) in Fig. 22.1. Finally, the value of $\Delta A/A$ is of the order of $\ln(G + J)$ times ΔG or $J\ln(G + J)$ times $\Delta Y/Y$.

Obtaining accurate estimates of the magnitude of the η's is extremely difficult except in special cases. Empirically, estimates have been based on a combination of the ideas used for Taylor series and the theorems of Chapter 11. If one examines a table of values of successive Padé approximants, one notices that their difference behaves, at least for small x, like a high power of x. At some distance from the origin this law breaks down and the magnitude of the fluctuations remains relatively constant beyond that point. A conservative guide would be to extrapolate the small-x error law to x_0, the point of interest. Since a much higher degree of self-consistency is frequently seen, as indicated before, the temptation has been irresistible to quote much smaller, purely ad hoc errors.

According to identities (3.5) and (3.7)–(3.9), the empirically observed behavior is what is to be expected if the roots of the denominators are all of the same general order of magnitude. Inside the region of a high-power-law error one presumes that there is a denominator of high enough degree so that one can apply Wilson's theorem 11.2 or Montessus's theorem 11.1 or one of their generalizations or corollaries to argue that the Padé approximant is in the relatively converged part of a convergent vertical sequence in the Padé table. Using the magnitude of the difference between the $[M + j/M]$ and the $[M + j - 1/M]$ Padé approximants to estimate the magnitude of the errors η_j by this procedure, we obtain, from identity (3.7),

$$A\begin{pmatrix} -G \\ 2M + j \end{pmatrix}Y^{2M+j}\eta \approx C(M + j/M + 1)/C(M + j - 1/M) \quad (22.17)$$

in the notation of Chapter 2. The estimates of the errors for $\Delta Y/Y$, ΔG, and $\Delta A / A$ would be J, J^2, and $J^2 \ln J$ times as great, respectively. This method of assessment of the error has been tested on an 18-term expansion of $(1 - x)^{-1.5}(1 - \frac{1}{2}x)^{1.5} + e^{-x}$ with good results near $x = 1$.

The method just described for error assessment assumes that one has reached such an order of approximation that the convergence theorems of Chapter 11 hold. Prima facie evidence that that assumption is false would be the appearance of a pole in a previously converging region of the complex plane where there is no singularity of the approximated function. As a practical matter one frequently observes such poles, and we have discussed them in Chapters 13 and 14. By the theorems of Chapter 14, they must have very small residues. I call them *defects*. A criterion which is fully in accord with the general theorems of Chapter 12 is to simply discard from consideration those Padé approximants that have a pole closer to the origin than the physically interesting point with a residue less in absolute value than some fixed value (0.003 has served well for applications in this area.) This criterion can be thought of as an approximate, but servicable version of the necessary and sufficient criterion of Marty (1931). It is further worthwhile to point out that, as a practical matter, the occurrence of a defect seems to sometimes cause the difference between the values of successive Padé approximants to be anomalously small, and one can be, and some have been, misled as to the rate of convergence if this effect is not taken into account. The occurence of defects can be thought of as a near miss at forming a block (see Chapter 2) in the Padé table, where the determinant did not actually vanish, but was unusually small.

For a wide variety of models of physical systems the generalized Padé approximants of Chapter 19 can be used, together with the knowledge that all the zeros of the partition function (fundamental mathematical function which describes the system) lie on the unit circle in a complex plane related to a magnetic field, to provide converging upper and lower bounds on several of the thermodynamic properties. These properties establish, with the right generalized kernel, a nonnegative weight function, and so lead to series of Stieltjes-type results [see Baker (1967, 1968, 1971), Griffiths (1972)].

In addition, by using the structure of the partition function

$$Z_N = \int e^{-\beta E} \, d\rho_N(E) \tag{22.18}$$

where β is the reciprocal of Boltzmann's constant times the absolute temperature, E is the energy of a given state, and $\rho_N(E)$ is the number of states of a system of size N with energy less than or equal to E, we can, by using the techniques of Chapter 20 and the high-temperature series expansion for a system of size N compute a different set of bounds on the thermodynamic limit ($N \to \infty$) [see Baker (1972)].

23

SCATTERING PHYSICS

In this chapter we shall be concerned with various kinds of scattering. First we consider nonrelativistic, quantum mechanical scattering by a fixed potential source. This scattering is governed by the Schrödinger equation, which is

$$-\nabla^2\psi(\mathbf{r}) + gV(\mathbf{r})\psi(\mathbf{r}) = k^2\psi(\mathbf{r}) \tag{23.1}$$

where ∇^2 is the Laplacian operator, ψ is the wave function, $\mathbf{r}$ is a three-dimensional vector, g is called the coupling constant, $V(\mathbf{r})$ is called the potential energy, and k^2 is the energy. To complete the description of the scattering problem, we must specify the boundary conditions for (23.1). The standard ones are that at large distances ψ should look like an incoming plane wave plus an outgoing scattered wave. That is,

$$\psi(\mathbf{r}) \approx \exp(i\mathbf{k} \cdot \mathbf{r}) - (1/4\pi r)[\exp(ik'r)]T(\mathbf{k}', \mathbf{k}) \tag{23.2}$$

where $\mathbf{k}$ is the wave vector of the incoming wave and $\mathbf{k}' = |\mathbf{k}|\mathbf{r}/|\mathbf{r}|$ is thought of as the wave vector describing the outgoing wave. One can derive from the boundary conditions and Eq. (23.1) the Lippmann–Schwinger equation for the scattering equation for the scattering amplitude $T(\mathbf{k}', \mathbf{k})$:

$$T(\mathbf{k}', \mathbf{k}) = gV(\mathbf{k}', \mathbf{k}) + \frac{g}{(2\pi)^3} \int V(\mathbf{k}', \mathbf{q}) \frac{d\mathbf{q}}{k^2 - q^2 + i\epsilon} T(\mathbf{q}, \mathbf{k}) \tag{23.3}$$

where ϵ is a positive infinitesmal and the integral is over the whole three-dimensional momentum space. We have defined

$$V(\mathbf{k}', \mathbf{k}) = \int \exp(-i\mathbf{k}' \cdot \boldsymbol{\rho})V(\boldsymbol{\rho}) \exp(i\mathbf{k} \cdot \boldsymbol{\rho}) \, d\boldsymbol{\rho} \tag{23.4}$$

and T is related to ψ by

$$T(\mathbf{k}', \mathbf{k}) = g \int \exp(-i\mathbf{k}' \cdot \boldsymbol{\rho}) V(\boldsymbol{\rho}) \psi(\boldsymbol{\rho}) \, d\boldsymbol{\rho} \qquad (23.5)$$

We will next see that the Padé approximants to the scattering amplitude have a number of interesting properties. To simplify the presentation, we will restrict the potential function $V(\mathbf{r})$ to depend on $r = |\mathbf{r}|$ only. Further we will assume that it is short ranged and well behaved at $r = 0$, so that

$$\int_0^\infty e^{2\alpha r} |V(r)| \, dr < \infty \qquad (23.6)$$

for some $\alpha > 0$. These restrictions can be relaxed in various ways and the application of Padé approximants to this type of problem has been discussed by a number of authors [e.g., Chisholm (1963), Baker (1965), Tani (1965, 1966a, b), Nuttall (1967), Garibotti and Villani (1969a, b), Garibotti (1972), and Graves-Morris (1973)]. First, if we think of $\mathbf{T}$, $\mathbf{V}$, and $\mathbf{G}(\mathbf{q}, \mathbf{q}') = -\delta(\mathbf{q} - \mathbf{q}')/(k^2 - q^2 + i\epsilon)$ as operators, we can write as a formal power series in the coupling constant

$$\mathbf{T} = g\mathbf{V} - g^2\mathbf{VGV} + g^3\mathbf{VGVGV} - \cdots$$

$$= g\mathbf{V} - g^2 V \mathbf{G}^{1/2} \Big[\mathbf{I} - g(\mathbf{G}^{1/2}\mathbf{VG}^{1/2}) + g^2(\mathbf{G}^{1/2}\mathbf{VG}^{1/2})^2 + \cdots \Big] \mathbf{G}^{1/2}\mathbf{V}$$

$$(23.7)$$

If we choose k^2 a negative real, then $G^{1/2}$ is a well-defined real Hermitian operator, and therefore $\mathbf{G}^{1/2}\mathbf{VG}^{1/2}$ is Hermitian, as $\mathbf{V}$ is. Therefore $\omega = \mathbf{G}^{1/2}\mathbf{VG}^{1/2}$ has real eigenvalues, by standard theorems. If we take the diagonal matrix elements of $\mathbf{T}$ then, by expanding

$$\mathbf{G}^{1/2}\mathbf{V}|\mathbf{k}\rangle = \Sigma \, a_k u_k \qquad (23.8)$$

in eigenstates of ω we can rewrite (23.7) as

$$\langle \mathbf{k}|\mathbf{T}|\mathbf{k}\rangle = gV(\mathbf{k}, \mathbf{k}) - g^2\Sigma \, |a_k|^2(1 - \omega_k g + \omega_k^2 g^2 + \cdots) \qquad (23.9)$$

By our assumptions on the potential and the normalization of $|\mathbf{k}\rangle$ (Tani, 1966b) there must exist

$$\varphi(u) = \sum_k |a_k|^2, \qquad \omega_k \leqslant u \qquad (23.10)$$

where $d\varphi \geq 0$ necessarily. Thus, formally summing (23.9), we obtain

$$T(\mathbf{k}, \mathbf{k}) = gV(\mathbf{k}, \mathbf{k}) - g^2 \int_{-\infty}^{+\infty} \frac{d\varphi(u)}{1 + ug} \tag{23.11}$$

Our assumptions on the potential suffice to prove that the spectrum of ω is bounded for $k^2 < 0$. Thus by means of transformations (17.36) we can construct a series of Stieltjes from (23.11), and by the results of Part III construct converging inclusion regions for $T(\mathbf{k}, \mathbf{k})$ in its regions of analyticity in g (for $k^2 < 0$). Without making this transformation, by Corollary 9.1 and Theorem 16.2, the $[M + 1/M]$ Padé approximants to T will converge smoothly in an appropriately cut complex g plane. This conclusion is in accord with Theorem 16.6. By Lemma 18.3 we can extend this result to any diagonal sequence $[M + J/M]$, $J \geq 0$.

Our assumptions on the potential V are also sufficient to ensure that the spectrum of eigenvalues is discrete. Thus the integral in Eq. (23.11) is actually an (infinite) sum of pole terms. Hence since $gV(\mathbf{k}, \mathbf{k})$ is entire, $T(\mathbf{k}, \mathbf{k})$ is a meromorphic function of g. If the potential is of definite sign, then by considering the Hermitian operator $\mathbf{v} = \mathbf{V}^{1/2}\mathbf{G}\mathbf{V}^{1/2}$ and by a series of arguments similar to (23.7)–(23.11), we conclude that the forward scattering amplitude

$$T(\mathbf{k}, \mathbf{k}) = g \int_0^\infty \frac{d\omega(u)}{1 + ug} \tag{23.12}$$

is a series of Stieltjes directly, and hence conclude $(k^2 < 0)$ from the theorems of Chapter 16 the convergence of every diagonal-type sequence of Padé approximants.

In order to make further progress in the discussion of the properties of the Padé approximant to the scattering amplitude, it will be convenient to expand it in partial waves. We will then treat the partial waves separately. This procedure is equivalent to forming the matrix Padé approximants (Chapter 21). An important property of the scattering amplitude is that the associated matrix

$$\mathbf{S}(g) = \mathbf{I} + 2i\mathbf{T}(g) \tag{23.13}$$

is unitary, i.e., $\mathbf{S}\mathbf{S}^\dagger = \mathbf{S}^\dagger\mathbf{S} = \mathbf{I}$. This property expresses conservation during the scattering process, i.e., everything that goes in must come out. As we saw in Chapter 21, the matrix Padé approximants preserve this property, by Eq. (21.9), (21.10), and (21.18). Thus let us define the projections

$$T_l(k', k) = \frac{1}{8\pi} \int_{-1}^{+1} T(\mathbf{k}', \mathbf{k}) P_l(\cos\theta) \, d\cos\theta$$

$$= g \int_0^\infty j_l(k'\rho) V(\rho) \psi_{k, l}(\rho) \rho^2 \, d\rho$$

$$V_l(k', k) = \frac{1}{8\pi} \int_{-1}^{+1} V(\mathbf{k}', \mathbf{k}) P_l(\cos\theta) \, d\cos\theta$$

$$= \int_0^\infty j_l(k'\rho) V(\rho) j_l(k\rho) \rho^2 \, d\rho \qquad (23.14)$$

where θ is the angle between k and k', and we have expanded $\psi(\rho)$ $= \Sigma(2l + 1) i^l P_l(\cos \theta) \psi_{k, l}(\rho)$. In terms of these variables we can transform Eq. (23.3) into the decoupled set of equations for $l = 0, 1, \ldots$:

$$T_l(k', k) = g V_l(k', k) + \frac{2g}{\pi} \int_0^\infty V_l(k', q) \frac{q^2 \, dg}{k^2 - q2 + i\epsilon} T_l(q, k) \quad (23.15)$$

Again this equation has the structure of (23.3) and by the same arguments we can prove for $k^2 < 0$ (there we can set $\epsilon = 0$ directly) that again we have a Hamburger moment problem with bounded spectrum, so that the $[L + J/M]$ Padé approximants in the coupling constant must converge in the appropriately cut g plane.

There is one important advantage to treating the partial-wave amplitudes rather than the forward scattering amplitudes. That is, that the operator $\mathbf{v}_l$ corresponding to $\mathbf{v}$ is of trace class C_1. That is, $\text{Tr}[(\mathbf{v}_l^\dagger \mathbf{v}_l)^{1/2}]$ $< \infty$ and $\text{Tr}(\mathbf{v}_l^\dagger \mathbf{v}_l) < \infty$ for $0 < ik < \alpha$. In this case we can approximate $\mathbf{v}_l$ by a sequence of finite-rank operators and expect convergence in the limit as the rank becomes infinite. If we choose as a subspace of the Hilbert space of wave functions the space spanned by the wave functions

$$\{ \mathbf{V}^{1/2} P_l(\cos\theta) j_l(kr), \mathbf{V}^{1/2} \mathbf{GV} P_l(\cos\theta) j_l(kr),$$

$$\ldots, \mathbf{V}^{1/2} (\mathbf{GV})^{M-1} P_l(\cos\theta) j_l(kr) \} \qquad (23.16)$$

then the solution to the $M \times M$ matrix equation for the scattering amplitude in this subspace is exactly of the form of a polynomial in g of degree less than or equal to $M + 1$, divided by another polynomial of degree less than or equal to M, as can be seen by a direct determinantal solution. However, since it can be shown for a single-signed potential (Garibotti and Villani, 1969b) that the exact solution has all the properties of the $[M + 1/M]$ Padé approximant, it is, by the uniqueness theorem, equal to it. Garibotti (1972) extends this result to the $[M + J/M]$.

By virtue of being the exact solution to this truncated problem, it follows that the denominator of the Padé approximant is the Jost function (Newton, 1960; Goldberger and Watson, 1964), $f_l(-k, g)$, about which considerable is known. In particular, for $|\text{Im}\,(k)| < \alpha$, it is an entire function of g. One can prove (Garibotti and Villani, 1969b) the existence of uniform bounds in M for each coefficient in the series expansion in powers of g, from our assumptions on the potential. In the range $-\alpha < \text{Im}\,(k) \leqslant 0$ the scattering amplitude is meromorphic. Its only singularities are simple poles on the negative imaginary k axis, which correspond to bound states. Thus for fixed g there exists a range of k in $0 < ik < \alpha$ for which, by the properties of Padé approximants to this Hamburger moment series, we have convergence of the $[M + J/M]$ Padé approximants. The entire elastic scattering amplitude can be written in terms of the Jost function as

$$T_l(k, k) = \tfrac{1}{2} i [1 - f_l(k)/f_l(-k)] \tag{23.17}$$

By means of the properties of the Jost function, if $f_l(-k_0) = 0$, we can prove a lower bound on $f_l(k_0)$ as well as an upper bound to $f_l'(-k_0)$ by using restriction (23.6) on the potential, and hence a lower bound on the residue at the poles, $|\text{Im}\,(k_0)| < \alpha$, of $T_l(k, k)$. From this result one can easily get equicontinuity on the sphere (Chapter 12) of the $T_l(k, k)$ in any closed, bounded subregion of $|\text{Im}\,(k)| < \alpha$. Thus by applying the convergence continuation theorem 12.4 in k we obtain the following result.

Theorem 23.1. The $[M + J/M]$ sequence of Padé approximants $(M \to \infty, J \geqslant 0)$ in the coupling constant g to the partial-wave scattering amplitude $T_l(k, k)$ [Eq. (23.14)] converges on the sphere in any closed region $\mathcal{R}$ interior to $|\text{Im}\,(k)| < \alpha$ to a function which is continuous on the sphere and meromorphic in k in the interior of $\mathcal{R}$, provided the potential energy is of one sign and satisfies restriction (23.6).

Masson (1967) has shown that for V of definte sign and k real, $T_l(k, k)$ is a Hamburger series in g.

We remark that the $[M/M]$ sequence is automatically unitary (see Chapter 21), which is a physically important property. In addition, the bound states are correctly characterized as simple poles in the scattering amplitude, at least for $ik < \alpha$, and probably more extensively. An interesting example has been worked out by Basdevant and Lee (1969) for an exponential potential. They picked $V(r) = -v_0 e^{-r/d}$; then in terms of $g = d^2 v_0$ and $\nu = 2idk$ they computed for the s-wave ($l = 0$) scattering amplitude

$$S_0(g, \nu) = 1 + 2iT_0(g, \nu) = g^{-\nu} \frac{J_\nu(2g^{1/2})\Gamma(\nu + 1)}{J_{-\nu}(2g^{1/2})\Gamma(-\nu + 1)} \qquad (23.18)$$

where Γ and J_ν are the usual gamma and Bessel functions, and by series expansion one has

$$S_\nu(g, \nu) = 1 + g t_1 + g^2 t_2 + \cdots, \qquad t_1 = \frac{2\nu}{\nu^2 - 1},$$

$$t_2 = \frac{1}{2}\left(\frac{1}{\nu + 2} \frac{1}{\nu + 1} - \frac{1}{\nu - 2} \frac{1}{\nu - 1} \right) + \left(\frac{1}{\nu - 1} \right)^2 + \frac{1}{\nu^2 - 1}$$

$$(23.19)$$

The conclusions of this section are illustrated in their results; Casar *et al.* (1969) have investigated the same problem using the Yukawa potential $(e^{-\mu r}/r)$.

In the special case of a potential of definite sign there are special methods based on the properties of Padé approximants to series of Stieltjes which give bounds for the non-forward scattering amplitude. They are due to Nuttall (Baker, 1970) and Masson (1970a). In connection with his study of the convergence of Padé approximants to the solution of the Bethe–Salpeter equation (this equation bears a strong formal resemblance to the Lippmann–Schwinger equation for a potential of definite sign), Nuttall (1967) found that the $[M/M]$ Padé approximants were solutions of the Schwinger variational principle, given an appropriate trial function. A recent summary of the relation of Padé approximants to variational principles is given by Bessis (1973).

There is another aspect in the application of Padé approximants to scattering physics. This aspect is the structure of the scattering amplitude as a function of energy (k^2). Baker (1965) pointed out that if the scattering amplitude A, for example, for pion–nuclear scattering, has the Mandelstam (1958) representation

$$A = \frac{g^2}{m^2 - s} + \frac{g^2}{\mu^2 - s_c} + \frac{1}{\pi^2} \int_{(M+\mu)^2}^{\infty} ds' \int_{4\mu^2}^{\infty} dt' \frac{A_{13}(s', t')}{(s' - s)(t' - t)}$$

$$+ \frac{1}{\pi^2} \int_{(M+\mu)^2}^{\infty} ds'_c \int_{4\mu^2}^{\infty} dt' \frac{A_{23}(s'_c, t')}{(s'_c - s_c)(t' - t)}$$

$$+ \frac{1}{\pi^2} \int_{(M+\mu)^2}^{\infty} ds' \int_{(M+\mu)^2}^{\infty} ds'_c \frac{A_{12}(s', s'_c)}{(s' - s)(s'_c - s_c)} \tag{23.20}$$

where the energy and momentum transfer variables satisfy $s + s_c + t = 2(M^2 + \mu^2)$, then in terms of any one of these three variables (the other independent one is fixed), by some obvious manipulations and by identifying the "absorptive" or positive amplitudes, we find that A is of the Hamburger form with finite radius of convergence. Thus Padé approximants in the energy variables (as well as the coupling constant) can be expected to converge. This feature has been exploited to practical advantage using N-point Padé approximants in phase shift calculations to the Bethe–Salpeter equation (Haymaker and Schlessinger, 1970). Additional results in this area have been obtained on the solution of the N/D equations by Padé approximants [see Mason (1970b), Common (1970)].

An important reason why the application of Padé approximants to potential scattering has been studied as thoroughly as it has is that it offers a proving ground for the possible application of the same techniques to quantum field theory. The direct solution of the Schrödinger equation for potential scattering is, of course, feasible with modern computers for the two-body problem, though very difficult for the three-body problem, where the Padé technique can indeed be a practical tool (Tjon, 1973). The quantum field theory problem differs from the potential scattering theory problem in a number of fundamental ways, even though experimentally excellent results are obtained from potential theory at low energies. The quantum field theory is described by an infinite number of coupled, nonlinear integrodifferential equations. It is not definitely known if a solution exists. The theory does produce a formal power series expansion in the coupling constant; unfortunately, every term is given by a divergent integral. An elaborate theory, renormalization, has been developed to redefine these series. Another difficulty is that once the series is redefined it most likely is at best asymptotic. A possible alternative is the approach of Chapter 20 of "intermediate regularization" to make the integrals finite until the limit is taken. Consequently, the investigation of powerful methods of analytic continuation based on quantitative information at a point has seemed to many researchers in this area to be a desirable approach. The thrust of these investigations has been twofold: (i) to study the analytic structure and the suitability of Padé methods to handle it, and (ii) to study empirically the application of Padé methods to physical data.

In the first area Baker and Chisholm (1966) investigated the simple Peres model field theory described by the Hamiltonian

$$H = \tfrac{1}{2}\left(p_x^2 + p_y^2 + x^2 + y^2 \right) + g\,\delta(t)x^2y \qquad (23.21)$$

which couples two harmonic oscillator "fields" by an instanteous interaction. It shares with field theory the property that the series expansion in g^2 of the elements of the S matrix is divergent. In this case every element is the difference of two series of Stieltjes, and as a practical matter the Padé approximants converge rapidly. The separation and approximation of such series have been discussed by Baker and Gammel (1971). Another interesting and thorough investigation has been that of the eigenvalues of the anharmonic oscillator. The Hamiltonian is

$$H = p^2 + x^2 + gP_{2m}(x), \qquad m = 2, 3, \ldots \qquad (23.22)$$

where $P_{2m}(x)$ is a polynomial of degree $2m$ in x. This problem is discussed in detail by Simon (1970) and Loeffel et al. (1969). For $m = 2, 3$ the asymptotic series generated are rigorously summed to the correct value by the Padé approximants. For $m > 3$ a Padé–Borel summation procedure of the type treated in Chapter 19 is required (Graffi et al., 1970), with

$$b(z, s) = \int_{\infty}^{\infty} \frac{e^{-t}\,dt}{1 + szt^m} \qquad (23.23)$$

used for the kernel in Eq. (19.1).

In the second area Bessis and Pusterla (1967, 1968) and Copley and Masson (1967) pioneered the application of Padé approximants to realistic field theories. A large number of such calculations have been made, and many of them are reviewed by Basdevant (1973).

24

ELECTRICAL CIRCUITS AND
SEVERAL OTHER APPLICATIONS

In this chapter we will discuss briefly some of the relations between the analysis of electrical networks and Padé approximants. For a more detailed treatment one can consult the standard works of Gillemin (1957), Weinberg (1962), and Newcomb (1966). At the end of this chapter we will list several additional areas where the Padé approach is being usefully applied.

Specifically, we will consider passive, linear, lumped, reciprocal networks. Such a network consists of resistors, capacitors, and inductors connected in any desired way. The voltage v across any single element is governed by the following relations to the current j flowing through it:

$$\begin{aligned} \text{resistor:} \quad & v(t) = Rj(t) \\ \text{inductor:} \quad & v(t) = L\,dj(t)/dt \\ \text{capacitor:} \quad & v(t) = v(0) + (1/C)\int_0^t j(t')\,dt' \end{aligned} \tag{24.1}$$

These equations relating the instantaneous current and the instantaneous voltage hold, provided the current variations are sufficiently slow so that the corresponding electric and magnetic fields are essentially the same as would be produced in a steady state. These assumed conditions are called quasi-stationary.

A convenient way to analyze the response of a circuit is to consider a sinusodial, oscillating current of frequency ω. Since the system is linear, we can study any current as a superposition of sine waves. Mathematically, it is convenient to use the complex current $j(t) = J(i\omega)e^{i\omega t}$. Then Eq. (24.1) become

$$\text{resistor:} \qquad v(t) = Rj(t)$$
$$\text{inductor:} \qquad v(t) = i\omega Lj(t) \qquad\qquad\qquad (24.2)$$
$$\text{capacitor:} \qquad v(t) = j(t)/(i\omega C)$$

where we have chosen $v(0) = -i/\omega C$. In each case the induced voltage is of the same form, though perhaps shifted in phase. Kirchhoff's rules for the behavior of a network under these conditions tell us that the sum of the (complex) currents entering any node is zero and the sum of the voltage drops around any closed loop is zero. Now we can decompose any circuit into a set of independent loops. The current flowing in any branch is the sum of the independent loop currents. The equations then are

$$\sum_{l=1}^{\mathcal{L}} Z_{kl}(i\omega) J_l(i\omega) = V_k(i\omega), \qquad k = 1, \ldots, \mathcal{L} \qquad (24.3)$$

where we have canceled the common factors of $e^{i\omega t}$ and written an eqation for the amplitudes for each of the $\mathcal{L}$ independent loops. The V_l here are the amplitude of such impressed external voltages as there may be in each loop. The Z_{kl} is the impedance encountered by the lth current in the kth loop and is the sum $R_{kl} + i\omega L_{kl} + (1/i\omega C_{kl})$ of the impedances of the respective resistors, inductors, and capacitors. For simplicity suppose $V_k = 0$, $k > 1$; the solution, by Cramer's rule, must necessarily be of the form of a rational fraction in ω, the degree of neither the numerator nor denominator exceeding $2\mathcal{L}$. If we write this solution

$$V_1(i\omega) = Z(i\omega) J_1(i\omega) \qquad\qquad\qquad (24.4)$$

then it can be shown from the structure of the equations, R, L, and C necessarily nonnegative, that $Z(p)$ is regular in the open right-half p plane and there has the properties

$$Z(p) \quad \text{real}, \quad p \quad \text{real}, \quad \text{Re}\,[Z(p)] > 0 \quad \text{for} \quad \text{Re}\,(p) > 0 \quad (24.5)$$

These conditions on the impedance function $Z(p)$ are both necessary and sufficient for $Z(p)$ (rational) to be realizable as a network.

In the special case of a lossless (no resistors present) network the impedance can be expanded as

$$Z(p) = -L_0 p + \frac{C_0}{p} + \sum \frac{L_i p}{p^2 + L_i C_i} \qquad (24.6)$$

a partial fraction decomposition. The form (24.6) is clearly related to series of Stieltjes as discussed in Part III, and given such a desired function $Z(p)$, one can approximate it by successive Padé approximants. A circuit which corresponds to Eq. (24.6) is shown in Fig. 24.1.

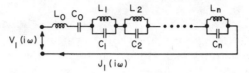

Fig. 24.1 A circuit with the impedance of Eq. (24.6).

If we use the formula

$$Z = Z_1 + \frac{Z_2}{1 + Z_2/Z_3} \tag{24.7}$$

for the circuit in Fig. 24.2 repeatedly, we can derive a continued fraction expansion

$$Z = Z_1 + \cfrac{Z_2}{1 + \cfrac{Z_2}{Z_3 + \cfrac{Z_4}{1 + Z_4 \cfrac{\ddots}{\quad} + \cfrac{Z_{2n-2}}{Z_{2n-1} + Z_{2n}}}}} \tag{24.8}$$

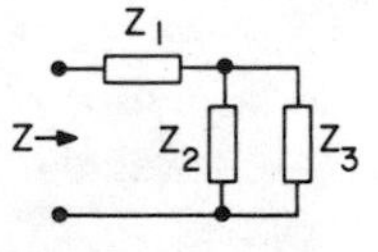

Fig. 24.2 A circuit with the impedance of Eq. (24.7).

This means that a network (see Fig. 24.3) can be found for any function $Z(i\omega)$ with a continued fraction expansion (see Chapter 4) of the form (24.8), provided $\operatorname{Re} Z_i \geqslant 0$. This condition is always possible to meet, provided the original circuit was composed of only two types of circuit elements ($L - C$, $R - L$, $R - C$). There is a substantial body of theory dealing with a great variety of problems and situations in this area. We have only indicated here a tiny sample.

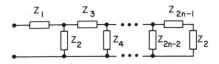

Fig. 24.3 A circuit with the impedance of the continued fraction (24.8).

We conclude by mentioning very briefly a number of additional applications and generalizations of the Padé approach. The first idea was studied by Common (1969) and Fleischer (1972, 1973), though not stated in just this way. Instead of assuming an original expansion about a point (the Padé case), or even about a finite set of points (Chapter 8), let us assume an expansion appropriate to an interval or region. For example, an expansion in Legendre polynomials would be appropriate to the interval -1 to $+1$ with a uniform weight function. Many important results have been obtained on the extension of the region of convergence of such expansions outside their natural ellipse with foci at ± 1. We remark that the formulas

$$f(z) - [\Sigma\, a_l P_l(z)/\Sigma\, b_l P_l(z)] = O[P_{L+M+1}(z)] \qquad (24.9)$$

and

$$f(z)\, \Sigma\, b_l P_l(z) - \Sigma\, a_l P_l(z) = O[P_{L+M+1}(z)] \qquad (24.10)$$

are not in general equivalent since

$$P_l(z)P_n(z) \neq P_{l+n}(z) \qquad (24.11)$$

The expansion

$$\{f(z) - [\Sigma\, a_l P_l(z)/\Sigma\, b_l P_l(z)]\} / [\Sigma\, b_l P_l(z)]^2 = O[P_{L+M+1}(z)] \quad (24.12)$$

corresponds to the minimum problem over the interval, and has not yet been studied.

Another application has been reviewed by Langhoff and Karplus (1970). In this application the dynamic dipole polarizability for an atomic or molecular system was studied. In this case the function being approximated is a series of Stieltjes. The moments are of the oscillator-strength distribution and are the Cauchy coefficients, which are obtainable from measured dispersion and absorption data or theoretical calculations. Rapid bounding convergence is obtained. This procedure furnishes a new method for performing bounded extrapolation of measured optical refractivity from the visible into the vacuum ultraviolet.

Another large area which we have hardly touched on is the field of numerical analysis. Here various Padé techniques have been used to accelerate the evaluation of numerical integrals [see, for instance, Genz (1973)]. Padé approximants to e^t have entered in an important way in the solution of ordinary and partial differential equations (Varga, 1961; Legras, 1966). Padé approximant methods have been applied [see Longman (1973) and others] to the inversion of Laplace transforms. The Padé approximant to the function can be expanded as a partial fraction and directly inverted by the standard result

$$\mathcal{L}^{-1}[a/(s - b)] = ae^{bt} \tag{24.13}$$

Good results have been obtained by this approach in a number of cases.

The development of efficient calculational procedures for special functions has been studied in detail. In a two-volume work on this subject Luke (1969) gives extensive results of Padé approximants to the functions of Chapter 5, and their generalizations, together with explicit error analysis.

The N-point Padé approximant (Basdevant et al., 1969) has been used to accelerate the solution of the equation $f(z) = 0$. The technique is to determine a function of the form $(A + Bx)/(1 + Cx)$ from the value at three points and compute a new approximation for the zero as $-A/B$. In the neighborhood of a simple zero (taken at $x = 0$ for convenience) the error of the next approximation is proportional to $x_1 x_2 x_3$, and near a multiple zero its error decays geometrically.

The final topic is the application of the Padé approximant to power series in more than one variable. A very natural way to do this application has been suggested by Chisholm (1973) for the two-variable case. These approximants were generalized by Chisholm and McEwan (1974), Common and Graves-Morris (1974), Graves-Morris et al. (1974), and Jones (1974). We use the definitions

$$[L_1, L_2, \ldots, L_n / M_1, M_2, \ldots, M_n]$$

$$\equiv [\mathbf{L/M}] \equiv \left[\sum_{l_1=0}^{L_1} \sum_{l_2=0}^{L_2} \cdots \sum_{l_n=0}^{L_n} p_{l_1, l_2, \ldots, l_n} x_1^{l_1} x_2^{l_2} \cdots x_n^{l_n} \right]$$

$$\times \left[\sum_{m_1=0}^{M_1} \sum_{m_2=0}^{M_2} \cdots \sum_{m_n=0}^{M_n} q_{m_1, m_2, \ldots, m_n} x_1^{m_1} x_2^{m_2} \cdots x_n^{m_n} \right]^{-1} \tag{24.14}$$

$$f(x_1, x_2, \ldots, x_n) \approx [\mathbf{L/M}] \tag{24.15}$$

The essential feature of Eq. (24.14) is that the same coefficients are involved in the multivariate Padé approximant as in the product

$$\prod_{i=1}^{n} [L_i/M_i] \tag{24.16}$$

of Padé approximants. The sign $\approx$ in Eq. (24.15) means that

$$\mathfrak{N} = \prod_{i=1}^{n} (L_i + 1) + \prod_{i=1}^{n} (M_i + 1) - 1 \tag{24.17}$$

power series coefficients are equal in a region of coefficient index space such that if $(k_1, k_2, \ldots, k_n)$ is included the region, then so, too, is $(K_1, K_2, \ldots, K_n)$, provided $K_i \leqslant k_i$, for $i \leqslant n$. As an example for $n = 2$ we have

$$f(x, y) \approx \frac{p_{00} + p_{10}x + p_{01}y + p_{11}xy}{1 + q_{10}x + q_{01}y + q_{11}xy} = [1, 1/1, 1] \tag{24.18}$$

The problem has been to pick the defining equations in a symmetrical way. There are seven coefficients to be chosen, but the "natural" triangle $x^\gamma y^\delta$, $\gamma + \delta \leqslant 2$, contains only six coefficients. The proposed solution is to permit the equating of sums of the equations corresponding to the boundary of the region in coefficient index space. If we specify that the points

$$(L_1 + M_1, 0, \ldots, 0), \quad (0, L_2 + M_2, \ldots, 0), \ldots, (0, 0, \ldots, L_n + M_n)$$

are included in the region as corners, and further that the projection on any subspace has the correct number of points, then the resultant approximants have the following properties.

(i) *Projection.* If k $(< n)$ of the variables are set to zero, then the approximants reduce to the corresponding approximants in the remaining $n - k$ variables formed from the power series with the same k variables set to zero.

(ii) *Homographic invariance.* The definition is invariant under the transformation $x_j = Ay_j/(1 + B_j Y_j)$, provided $L_j = M_j$ whenever $B_j \neq 0$.

(iii) *Duality.* $[L/M]_f^{-1} = [M/L]_{f^{-1}}$.

(iv) *Factorization.* If the given series is the product of two power series, one in k variables and the other in $n - k$, then an approximant is the product of the corresponding approximants to the two-power series.

(v) *Additivity.* If the given series is the sum of two power series, one in k variables and one in the other $n - k$, then an approximant is the sum of the corresponding approximants to the two-power series.

REFERENCES

Arms, R. J., and Edrei, A. (1970). The Padé tables and continued fractions generated by totally positive sequences. *In* "Mathematical Essays dedicated to A. J. Macintyre," pp. 1–21. Ohio Univ. Press, Athens, Ohio.

Baker, G. A., Jr. (1961). Application of the Padé method to the investigation of some magnetic properties of the Ising model. *Phys. Rev.* **124**, 768–774.

Baker, G. A., Jr. (1965). The theory and application of the Padé approximant method. *Advan. Theor. Phys.* **1**, 1–58.

Baker, G. A., Jr. (1967). Convergent, bounding approximation procedures with applications to the ferromagnetic Ising model. *Phys. Rev.* **161**, 434–445.

Baker, G. A., Jr. (1968). Some rigorous inequalities satisfied by the ferromagnetic Ising model in a magnetic field. *Phys. Rev. Lett.* **20**, 990–992.

Baker, G. A., Jr. (1969). Best error bounds for Padé approximants to convergent series of Stieltjes. *J. Math. Phys.* **10**, 814–820.

Baker, G. A., Jr. (1970). The Padé approximant method and some related generalizations. *In* "The Padé Approximant in Theoretical Physics" (G. A. Baker, Jr., and J. L. Gammel, eds.), pp. 1–39. Academic Press, New York.

Baker, G. A., Jr. (1971). Inequalities among the Ising-Heisenberg model critical indices. *In* "Critical Phenomena in Alloys, Magnets, and Superconductors" (R. E. Mills, E. Ascher, and R. I. Jaffee, eds.), pp. 221–229. McGraw-Hill, New York.

Baker, G. A., Jr. (1972). Converging bounds for the free energy in certain statistical mechanical problems. *J. Math. Phys.* **13**, 1862–1864.

Baker, G. A., Jr. (1973). The existence and convergence of subsequences of Padé approximants. *J. Math. Anal. Appl.* **43**, 498–528.

Baker, G. A., Jr. (1975). A theorem on the convergence of Padé approximants. To be published.

Baker, G. A., Jr., and Chisholm, J. S. R. (1966). The validity of perturbation series with zero radius of convergence. *J. Math. Phys.* **7**, 1900–1902.

Baker, G. A., Jr., and Gammel, J. L. (eds.) (1970). "The Padé Approximant in Theoretical Physics." Academic Press, New York.

Baker, G. A., Jr., and Gammel, J. L. (1971). Application of the principle of the minimum maximum modulus to generalized moment problems and some remarks on quantum field theory. *J. Math. Anal. Appl.* **33**, 197–211.

Baker, G. A., Jr., Gammel, J. L., and Wills, J. G. (1961). An investigation of the applicability of the Padé approximant method. *J. Math. Anal. Appl.* **2**, 405–418.

Baker, G. A., Jr., Gilbert, H. E., Eve, J., and Rushbrooke, G. S. (1967). High temperature expansions for the spin-$\frac{1}{2}$ Heisenberg model. *Phys. Rev.* **164**, 800–817.

Basdevant, J. L. (1968). "Padé Approximants, Ecole Internationale de la Phsique des Particules Elementaires: Herceq Novi (Yougoslavie)." Cent. Rech. Nucl., Strasbourg, France.

Basdevant, J. L. (1973). Strong interaction physics and the Padé approximation in quantum field theory. *In* "Padé Approximants" (P. R. Graves-Morris, ed.), pp. 77–100. Inst. of Phys., London.

Basdevant, J. L., and Lee, B. W. (1969). Padé approximation and bound states: Exponential potential. *Nucl. Phys. B* **13**, 182–188.

Basdevant, J. L., Bessis, D., and Zinn-Justin, J. (1969). Padé approximation in strong interactions, two-body Pion and Kaon systems. *Nuovo Cimento A* [10], **60**, 185–238.

Beardon, A. F. (1968a). The convergence of Padé approximants. *J. Math. Anal. Appl.* **21**, 344–346.

Beardon, A. F. (1968b). On the location of poles of Padé approximants. *J. Math. Anal. Appl.* **21**, 469–474.

Beckenbach, E. F., and Bellman, R. (1965). "Inequalities." Springer-Verlag, Berlin and New York.

Bernstein, S. (1928). Sur les fonctions absolument monotones. *Acta Math.* **52**, 1–66.

Bessis, D. (1973). Topics in the theory of Padé approximants. *In* "Padé Approximants" (P. R. Graves-Morris, ed.), pp. 19–44. Inst. of Phys., London.

Bessis, D., and Pusterla, M. (1967). Unitary Padé approximants for the S matrix in strong coupling field theory and application to the calculation ofthe ρ and f_0 Regge trajectories. *Phys. Lett. B* **25**, 279–281.

Bessis, D., and Pusterla, M. (1968). Unitary Padé approximants in strong coupling field theory and application to the calculation of the ρ- and f_0-Meson Regge trajectories. *Nuovo Cimento A* [10] **54**, 243–294.

Bessis, D., Turchetti, G., and Wortman, W. R. (1972). N-N interaction from Lagrangian field theory. *Phys. Lett. B* **39**, 601–604.

Brush, S. G. (1967). History of the Lenz-Ising model. *Rev. Mod. Phys.* **39**, 883–893.

Bulirsch, R., and Stoer, J. (1964). Fehlerabschätzungen und Extrapolation mit rationalen Funktionen bei Verfahren Vom Richardson-Typus. *Numer. Math.* **6**, 413–427.

Carleman, T. (1926). "Les Fonctions Quasi Analytiques." Gauthier-Villars, Paris. [English transl.: J. L. Gammel, Los Alamos Sci. Lab. Rep., LA-4702-TR (1971).]

Cartan, H. (1928). Sur les systèmes de fonctions holomorphes à variétés linéaires et leurs applications. *Ann. Sci. École Norm. Sup.* [3], **45**, 255–346.

Caser, S., Piquet, C., and Vermeulen, J. L. (1969). Padé approximants for a Yukawa potential. *Nucl. Phys. B* **14**, 119–132.

Cauchy, M. A.-L. (1821). "Cours d'Analyse de L'École Royale Polytechnique; Ire Partie. Analyse Algébrique." L'imprimerie Royal, Paris.

Chisholm, J. S. R. (1963). Solution of linear integral equations using Padé approximants. *J. Math. Phys.* **4**, 1506–1510.

Chisholm, J. S. R. (1966). Approximation by sequences of Padé approximants in regions of meromorphy. *J. Math. Phys.* **7**, 39–44.

Chisholm, J. S. R. (1973). Rational approximants defined from double power series. *Math. Comput.* **27**, 841–848.

Chisholm, J. S. R., and McEwan, J. (1974). Rational approximants defined from power series in N variables. *Proc. Roy. Soc. A* **336**, 421–452.

Common, A. K. (1968). Padé approximants and bounds to series of Stieltjes. *J. Math. Phys.* **9**, 32–38.

Common, A. K. (1969). Properties of Legendre expansions related to series of Stieltjes and applications to π–π scattering. *Nuovo Cimento A* [10], **63**, 863–891.

Common, A. K. (1970). The solution of the N/D equations using the Padé approximant method. *In* "The Padé Approximant in Theoretical Physics" (G. A. Baker, Jr. and J. L. Gammel, eds.), pp. 241–256. Academic Press, New York.

Common, A. K., and Graves-Morris, P. R. (1974). Some properties of Chisholm approximants. *J. Inst. Math. Appl.*, to be published.

Copley, L. A., and Masson, D. (1967). Padé-approximant calculation of π–π scattering. *Phys. Rev.* **164**, 2059–2062.

Copson, E. T. (1948), "An Introduction to the Theory of Functions of a Complex Variable." Oxford Univ. Press, London and New York.

de Montessus de Ballore, R. (1902). Sur les fractions continues algébriques. *Bull. Soc. Math. France* **30**, 28–36.

de Montessus de Ballore, R. (1905). Sur les fractions continues algébriques. *Rend. Circ. Mat. Palermo* **19**, 1–73.

Dienes, P. (1957). "The Taylor Series, an Introduction to the Theory of Functions of a Complex Variable." Dover, New York.

Domb, C., and Sykes, M. F. (1961). Use of series expansion for the Ising model susceptibility and excluded volume problem. *J. Math. Phys.* **2**, 63–67.

Dwight, H. B. (1947). "Tables of Integrals." Macmillan, New York.

Edrei, A. (1939). Sur les déterminants récurrents et les singularités d'une fonction donnée par son developpement de Taylor. *Compositio Math.* **7**, 20–88.

Edrei, A. (1953). Proof of a conjecture of Schoenberg on the generating function of a totally positive sequence. *Can. J. Math.* **5**, 86–94.

Euler, L. (1737). De fractionibus continuis. *Comm. Acad. Sci. Imper. Petropol.* **9**.

Ferer, M., Moore, M. A., and Wortis, M. (1971). Some critical properties of the nearest-neighbor classical Heisenberg model for the fcc lattice in finite field for temperatures greater than T_c. *Phys. Rev. B* **4**, 3954–3963.

Fisher, M. E. (1962). On the theory of critical point density fluctuations. *Physica* **28**, 172–180.

Fisher, M. E. (1967). The theory of equilibrium critical phenomena. *Rep. Prog. Phys.* **30**, P. II, 615–731.

Fisher, M. E., and Burford, R. J. (1967). Theory of critical-point scattering and correlations. I. The Ising model. *Phys. Rev.* **156**, 583–622.

Fisher, M. E., and Hiley, B. J. (1961). Configuration and free energy of a polymer molecule with solvent interaction. *J. Chem. Phys.* **34**, 1253–1267.

Fleischer, J., (1972). Analytic continuation of scattering amplitudes and Padé approximants. *Nucl. Phys. B* **37**, 59–76.

Fleischer, J. (1973). Nonlinear Padé approximants for Legendre series. *J. Math. Phys.* **14**, 246–248.

Fogli, G., Pellicoro, M. F., and Villani, M. (1971). A summation method for a class of series with divergent terms. *Nuovo Cimento A* [11], **6**, 79–97.

Fogli, G. L., Pellicoro, M. F., and Villani, M. (1972). An approach to the radiative corrections in Q.E.D. in the framework of the Padé method. *Nuovo Cimento A* [11], **11**, 153–177.

Franklin, P. (1940). "A Treatise on Advanced Calculus." Wiley, New York.

Frobenius, G. (1881). Ueber Relationen zwischen den Näherungsbrüchen von Potenzreihen. *J. für Math. (Crelle)* **90**, 1–17.

Gammel, J. L. (1970). Private communication.

Gammel, J. L. (1974). Continuation of functions beyond natural boundaries. *Rocky Mt. J. Math.* **4**, 203–206.

Gammel, J. L., and McDonald, F. A. (1966). Application of the Padé approximant to scattering theory. *Phys. Rev.* **142**, 1245–1254.

Gammel, J. L., Rousseau, C. C., and Saylor, D. P. (1967). A generalization of the Padé approximant. *J. Math. Anal. Appl.* **20**, 416–420.

Gargantini, J., and Henrici, P. (1967). A continued fraction algorithm for the computation of higher transcendental functions in the complex plane. *Math. Comput.* **21**, 18–29.

Garibotti, C. R. (1972). Schwinger variational principle and Padé approximants. *Ann. Phys. (N.Y.)* **71**, 486–496.

Garibotti, C. R., and Villani, M. (1969a). Continuation in the coupling constant for the total K and T matrices. *Nuovo Cimento A* [10], **59**, 107–123.

Garibotti, C. R., and Villani, M. (1969b). Padé approximant and the Jost function. *Nuovo Cimento A* [10], **61**, 747–754.

Garibotti, C. R., Pellicoro, M. F., and Villani, M. (1970). Padé method in singular potentials. *Nuovo Cimento A* [10] **66**, 749–766.

Gaunt, D. S., and Guttmann, A. J. (1973). Series expansions: Analysis of coefficients. *In* "Phase Transitions and Critical Phenomena" (C. Domb and M. S. Green, eds.), Vol. 3. Academic Press, New York, to be published.

Gauss, C. F. (1813). Disquisitiones generales circa seriem infinitam

$$1 + \frac{\alpha\beta}{1\cdot\gamma} x + \frac{\alpha(\alpha + 1)\beta(\beta + 1)}{1\cdot 2\cdot\gamma\cdot(\gamma + 1)} xx + \text{etc.}$$

Comment. Soc. Regiae Sci. Goettingensis Recentiores **2**.

Genz, A. C. (1973). Applications of the ϵ-algorithm to quadrature problems. *In* "Padé Approximants and Their Applications" (P. R. Graves-Morris, ed.), pp. 105–116. Academic Press, New York.

Gillemin, E. A. (1957). "Synthesis of Passive Networks." Wiley, New York.

Goldberger, M. L., and Watson, K. M. (1964). "Collision Theory." p. 918. Wiley, New York.

Gordon, R. G. (1968). Error bounds in equilibrium statistical mechanics. *J. Math. Phys.* **9**, 655–663.

Graffi, S., Grecchi, V., and Simon, B. (1970). Borel summability: Application to the anharmonic oscillator. *Phys. Lett. B* **32**, 631–634.

Gragg, W. B. (1968). Truncation error bounds for g-fractions. *Numer. Math.* **11**, 370–379.

Gragg, W. B. (1970). Truncation error bounds for π-fractions. *Bull. Amer. Math. Soc.* **76**, 1091–1094.

Gragg, W. B. (1972). The Padé table and its relation to certain algorithms of numerical analysis. *SIAM (Soc. Ind. Appl. Math.) Rev.* **14**, 1–62.

Graves-Morris, P. R. (1973). Padé approximants and potential scattering. *In* " Padé Approximants" (P. R. Graves-Morris, ed.), pp. 64–76. Inst. of Phys., London.

Graves-Morris, P. R., Jones, R. H., and Makinson, G. J. (1974). The calculation of some rational approximants in two variables. *J. Inst. Math. Appl.*, to be published.

Grenander, U., and Szegö, G. (1958). "Toeplitz Forms and Their Applications." Univ. of California Press, Berkeley.

Griffiths, R. B. (1972). Rigorous results and theorems phase. *In* "Transitions and Critical Phenomena" (C. Domb and M. S. Green, eds.), Vol. 1, pp. 7–109. Academic Press, New York.

Hadamard, J. (1892). Essai sur l'étude des fonctions données par leur développement de Taylor (deuxieme partie). *J. de Math.* [4], **8**, 101–186.

Hamburger, H. (1920). Ueber eine Erweiterung des Stieltjes'schen Momentenproblems. I. *Math. Ann.* **81**, 235–319.

Hamburger, H. (1921). Ueber eine Erweiterung des Stieltjes'schen Momentenproblems. II, III. *Math. Ann.* **82**, 120–164, 168–187.

Haymaker, R. W., and Schlessinger, L. (1970). Padé approximants as a computational tool for solving the Schrodinger and Bethe-Salpeter equations. *In* "The Padé Approximant in Theoretical Physics" (G. A. Baker, Jr., and J. L. Gammel, eds.). Academic Press, New York.

Hille, E. (1962). "Analytic Function Theory," Vol. 2. Ginn, Boston.

Hunter, D. L., and Baker, G. A., Jr. (1973). Methods of series analysis I. Comparison of current methods used in the theory of critical phenomena. *Phys. Rev. B* **7**, 3346.

Hurewicz, W., and Wallman, H. (1941). "Dimension Theory." Princeton Univ. Press, Princeton, New Jersey.

Jacobi, C. G. J. (1846). Uber die Darstellung einer Reihe Gegebner Werthe durch eine Gebrochne Rationale Function. *J. Reine Angew. Math.* (*Crelle*) **30**, 127–156.

Jahnke, E., and Emde, F. (1945). "Tables of Functions with Formulae and Curves." Dover, New York.

Jones, R. H. (1974). General rational approximants in N variables. Submitted to *J. Approx. Theory*.

Jones, W. B., and Thron, W. J. (1971). A posteriori bounds for the truncation error of continued fractions. *SIAM* (*Soc. Ind. Appl. Math.*) *J. Numer. Anal.* **8**, 693–705.

Karlin, S. (1968). "Total Positivity," Vol. 1. Stanford Univ. Press, Stanford, California.

Langhoff, P. W., and Karplus, M. (1970). Application of Padé approximants to dispersion force and optical polarizability computations. *In* "The Padé Approximant in Theoretical Physics" (G. A. Baker, Jr., and J. L. Gammel, eds.). Academic Press, New York.

Legras, J. (1966). Résolution numerique des grandes systèmes differentiels linéaires. *Numer. Math.* **8**, 14–28.

Loeffel, J. J., Martin, A., Simon, B., and Wightman, A. S. (1969). Padé approximants and the anharmonic oscillator. *Phys. Lett. B* **30**, 656–658.

Longman, I. M. (1973). Use of Padé table for approximate Laplace transform inversion. *In* "Padé Approximants and their Applications" (P. R. Graves-Morris, ed.), pp. 131–133. Academic Press, New York.

Luke, Y. L. (1969). "The Special Functions and Their Approximations." Vols. 1 and 2. Academic Press, New York.

Mandelstam, S. (1958). Determination of the pion-nucleon scattering amplitude from dispersion relations and unitarity. General theory. *Phys. Rev.* **112**, 1344–1360.

Markoff, A. (1884). Demonstration de certaines inéqalités de Tchebychef. *Math. Ann.* **24**, 172–180.

Marty, F. (1931). Recherches sur la repartition des valeurs d'une fonction meromorphe. *Ann. Fac. Sci. Univ. Toulouse* [3] **23**, 183–261.

Masson, D. (1967). Analyticity in the potential strength. *J. Math. Phys.* **8**, 2308–2314.

Masson, D. (1970a). Hilbert space and the Padé approximant. *In* "The Padé Approximant in Theoretical Physics" (G. A. Baker, Jr., and J. L. Gammel, eds.), pp. 197–217. Academic Press, New York.

Masson, D. (1970b). Approximate N/D solutions using Padé approximants. *In* "The Padé Approximant in Theoretical Physics" (G. A. Baker, Jr., and J. L. Gammel, eds.), pp. 231–240. Academic Press, New York.

Milne-Thomson, L. M. (1951). "The Calculus of Finite Differences." Macmillan, New York.

Montel, P. (1927). "Leçons Sur les Familles Normales de Fonctions Analytiques et Leurs Applications." Gauthier-Villars, Paris.

Moore, M. A., Jasnow, D., and Wortis, M. (1969). Spin-spin correlation function of the three-dimensional Ising ferromagnet above the Curie temperature. *Phys. Rev. Lett.* **22**, 940–943.

Morse, P. M., and Feshbach, H. (1953). "Methods of Theoretical Physics," Pt. I. McGraw-Hill, New York.

Muir, T. (1960). "A Treatise on the Theory of Determinants," revised and enlarged by W. H. Metzler. Dover, New York.

Newcomb, R. W. (1966). "Linear Multipost Synthesis." McGraw-Hill, New York.

Newton, R. (1960). Analytic properties of radial wave functions. *J. Math. Phys.* **1**, 319–347 (1960).

Nuttall, J. (1967). Convergence of Padé approximants for the Bethe-Salpeter amplitude. *Phys. Rev.* **157**, 1312–1316.

Nuttall, J. (1970). The convergence of Padé approximants of meromorphic functions. *J. Math. Anal. Appl.* **31**, 147–153.

Nuttall, J. (1972). Orthogonal polynomials for complex weight functions and the convergence of related Padé approximants. Private communication.

Nuttall, J. (1973). The convergence of Padé approximants for a class of functions with branch points. Submitted to *J. Approx. Theory*.

Onsager, L. (1944). Crystal statistics. I. A two-dimensional model with an order–disorder transition. *Phys. Rev.* **65**, 117–149.

Ostrowski, A. (1925). Über folgen analytischer Funktionen und einige Verschärfungen des Picardschen Satzes. *Math. Z.* **24**, 215–258.

Padé, H. (1892). (Thesis) Sur la représentation approchée d'une fonction pour des fractions rationnelles. *Ann. Sci. École Norm. Sup. Suppl.* [3], **9**, 1–93.

Padé, H. (1899). Mémoire sur les développements en fractions continues de la fonction exponentielle pouvant servir d'introduction à la théorie des fractions continues algébriques. *Ann. Sci. École Norm. Sup.* **16**, 395–426.

Padé, H. (1900). Sur la distribution des réduites anormales d'une fonction. *Comp. Rend.* **130**, 102–104.

Parlett, B. (1968). Global convergence of the basic QR algorithm on Hessenberg matrices. *Math. Comput.* **22**, 803–817.

Peirce, B. O. (1910). "A Short Table of Integrals." Ginn, Boston, Massachusetts.

Perron, O. (1954). "Die Lehre von den Kettenbrüchen," 3rd ed., Vols. 1 and 2. Teubner, Stuttgart.

Pommerenke, C. (1973). Padé approximants and convergence in capacity. *J. Math. Anal. Appl.* **41**, 775–780.

Pringsheim, A. (1910). Über Konvergenz und Funktionen-Theoretischen Charakter gewisser Limitärperiodischer Kettenbrüche. *S.-B. Kgl. Bayer. Akad. Wiss. München Math.-Phys. Kl.* **6**, 1–52.

Rutishauser, H (1954). Der quotient-differenzen-algorithmus. *Z. Angew. Math. Phys.* **5**, 233–251.

Saff, E. B. (1969). On the row convergence of the Walsh array for meromorphic functions. *Trans. Amer. Math. Soc.* **146**, 241–257.

Saff, E. B. (1972). An extension of Montessus de Ballore's theorem on the convergence of interpolating rational functions. *J. Approximation Theory* **6**, 63–67.

Schoenberg, I. J. (1948). Some analytic aspects of the problem of smoothing. *In* "Studies and Essays presented to R. Courant on his 60th Birthday, Jan. 8, 1948," pp. 351–370. Wiley (Interscience), New York.

Schoenberg, I. J. (1951). On Pólya frequency functions, I: The totally positive functions and their Laplace transforms. *J. Anal. Math.* **1**, 331–74.

Schwarz, H. A. (1869). "Gesammelte Mathematische Abhandlungen," Vol. 2, pp. 109–110.

Scott, W. T., and Wall, H. S. (1940). A convergence theorem for continued fractions. *Trans. Amer. Math. Soc.* **47**, 155–172.

Shanks, D. (1955). Nonlinear transformations of divergent and slowly convergent sequences. *J. Math. and Phys.(Cambridge, Mass.)* **34**, 1–42.

Shohat, J. A., and Tamarkin, J. D. (1963). "The Problem of Moments." Ameri. Math. Soc., Providence, Rhode Island.

Simon, B. (1970). Coupling constant analyticity for the anharmonic oscillator (appendix by A. Dicke). *Ann. Phys. (N. Y.)* **58**, 76–136.

Stieltjes, T. J. (1884). Quelques recherches sur les quadratures dites mécaniques. *Ann. Sci. École Norm. Sup.* [3], **1**, 409–426.

Stieltjes, T. J. (1889). Sur la réduction en fraction continue d'une série précédent suivant les Pouissances descendants d'une variable. *Ann. Fac. Sci. Univ. Toulouse* **3**, H, 1–17.

Stieltjes, T. J. (1894). Recherches sur les fractions continues. *Ann. Fac. Sci. Univ. Toulouse* **8**, J, 1–122; **9**, A, 1–47.

Stoer, J. (1961). Über Zwei Algorithmen zur Interpolation mit Rationalen Funktionen. *Numer. Math.* **3**, 285–304. [English transl.: W. B. Gragg, On two algorithms for interpolation with rational functions. (Math. Dept., Univ. Calif., La Jolla, Calif.)]

Tani, S. (1965). Padé approximant in potential scattering. *Phys. Rev. B* **139**, 1011–1020.

Tani, S. (1966a). Complete continuity of kernel in generalized potential scattering I. Short range interaction without strong singularity. *Ann. Phys. (N. Y.)* **37**, 411–450.

Tani, S. (1966b). Complete continuity of kernel in generalized potential scattering II. Generalized Fourier series expension. *Ann. Phys. (N. Y.)* **37**, 451–486.

Tjon, J. A. (1973). Application of Padé approximants in the three-body problem. *In* "Padé Approximants and Their Applications" (P. R. Graves-Morris, ed.), pp. 241–252. Academic Press, New York.

Trudi, N. (1862). "Teoria de'Determinanti e loro Applicazioni." Libreria Sci. e Ind. de B. Pellerano, Napoli.

Tschebycheff, P. (1858). Sur les fractions continues. *J. de Math.* **8**, 289–323.

Tschebycheff, P. (1874). Sur les valeurs limites des intégrales. *J. Math. Pures Appl.* [2], **19**, 157–160.

Van Vleck, E. B. (1904). On the convergence of algebraic continued fractions, whose coefficients have limiting values. *Trans. Amer. Math. Soc.* **5**, 253–262.

Varga, R. S. (1961). On higher order stable implicit methods for solving parabolic partial differential equations. *J. Math. and Phys.* **40**, 220–231.

Villani, M. (1972). A summation method for perturbative series with divergent terms. *In* "Cargèse Lectures in Physics" (D. Bessis, ed.), Vol. 5, pp. 461–474. Gordon & Breach, New York.

Vitali, G. (1903). Sulle serie di funzioni analitiche. *Rend. R. Ist. Lombard.* [2], **36**, 771–774.

Wall, H. S. (1948). "Analytic Theory of Continued Fractions." Van Nostrand-Reinhold, Princeton, New Jersey.

Wallin, H. (1972). The convergence of Padé approximants and the size of the power series coefficients. *Applicable Anal.* to be published.

Wallis, J. (1655), "Arithmetica Infinitorium." *In* "Opera Mathematica" Vol. I, pp. 355–478. Oxoniae e Theatro Shedoniano (1695). Reprint by Georg Olms Verlag, Hildeshein, New York (1972).

Walsh, J. L. (1967). On the convergence of sequences of rational functions. *SIAM (Soc. Ind. Appl. Math.) J. Numer. Anal.* **4**, 211–221.

Watson, P. J. S. (1973). Algorithms for differentiation and integration. *In* "Padé Approximants and Their Applications" (P. R. Graves-Morris, ed.), pp. 93–98. Academic Press, New York.

Weinberg, L. (1962). "Network Analysis and Synthesis." McGraw-Hill, New York.

Wheeler, J. C., and Gordon R. G. (1970). Bounds for averages using moment constraints. *In* "The Padé Approximant in Theoretical Physics" (G. A. Baker, Jr., and J. L. Gammel, eds.), pp. 99–128. Academic Press, New York.

Wilson, R. (1927). Divergent continued fractions and polar singularities. *Proc. London Math. Soc.* **26**, 159–168.

Wilson, R. (1928a). Divergent continued fractions and polar singularities. II. Boundary pole multiple. *Proc. London Math. Soc.* **27**, 497–512.

Wilson, R. (1928b). Divergent continued fractions and polar singularities. III. Several boundary poles. *Proc. London Math. Soc.* **28**, 128–144.

Wilson, R. (1930). Divergent continued fractions and non-polar singularities. *Proc. London Math. Soc.* **30**, 38–57.

Wuytack, L. (1974). Extrapolation to the limit by using continued fraction interpolation. *Rocky Mt. J. Math.* **4**, 395–397.

Wynn, P. (1956). On a device for computing the $e_m(S_n)$ transformation. *Math. Tables and Other Aids to Comput.* **10**, 91–96.

Wynn, P. (1966). Upon systems of recursions which obtain among the quotients of the Padé table. *Numer. Math.* **8**, 264–269.

Ynduráin, F. J. (1973). The moment problem and applications. *In* "Padé Approximants" (P. R. Graves-Morris, ed.), pp. 45–63. Inst. of Phys., London.

Zinn-Justin, J. (1971a). Strong interaction dynamics with Padé approximants. *Phys. Rept. C* **1**, No. 3, 55–102.

Zinn-Justin, J. (1971b). Convergence of the Padé approximant in the general case. Proceedings of the Colloquium on Advanced Computing Methods, Marseille, **2**, 88–102.

INDEX